# Lecture Notes in Physics

Edited by J. Ehlers, München, K. Hepp, Zürich,
H. A. Weidenmüller, Heidelberg, and J. Zittartz, Köln
Managing Editor: W. Beiglböck, Heidelberg

## 42

# H II Regions and Related Topics

Proceedings of a Symposium Held at
Mittelberg, Kleinwalsertal, Austria
January 13–17, 1975

Edited by T. L. Wilson and D. Downes

# Springer-Verlag
# Berlin Heidelberg GmbH 1975

**Editors**

Dr. T. L. Wilson
Dr. D. Downes
Max-Planck-Institut für Radioastronomie
Auf dem Hügel 69
53 Bonn 1/BRD

Library of Congress Cataloging in Publication Data

Symposium on H II Regions and Related Topics, Mittelberg, Austria (Bregenz), 1975.
   Proceedings of a symposium on H II regions and related topics.

   (Lecture notes in physics ; 42)
   Sponsored by the Deutsche Forschungsgemeinscahft,
the Max-Planck-Gesellschaft zur Förderung der Wis-
senschaften e.V., and the European Physical Society.
   Bibliography:  p.
   Includes index.
   1.  Interstellar hydrogen--Congresses.  2.  Stars
--Evolution--Congresses.  3.  Radio sources (Astron-
omy)--Congresses.  4.  Interstellar matter--Con-
gresses.  I.  Wilson, Thomas L    II.  Downes, D
III.  Deutsche Forschungsgemeinschaft (Founded 1949)
IV.  Max-Planck-Gesellschaft zur Förderung der
Wissenschaften.  V.  European Physical Society.
VI.  Title.  VII.  Series.
QB791.5.S9 1975        523.1'12        75-30666

ISBN 978-3-540-07409-0    ISBN 978-3-540-37927-0 (eBook)
DOI 10.1007/978-3-540-37927-0

TABLE OF CONTENTS

Part IV:             EVOLVED H II REGIONS AND OB-STARS

Chairman: R. Kippenhahn

Part V:         LARGE-SCALE DISTRIBUTION OF H II REGIONS

Chairman: B.F. Burke

Part VI:                SELECTED GALACTIC H II REGIONS

Chairman: M.J. Seaton

Part VII:        H II REGIONS ASSOCIATED WITH NUCLEI OF GALAXIES

Chairman: P.G. Mezger

---

\*
These manuscripts were not received.

# INTRODUCTION

The Symposium on H II Regions and Related Topics was held at Mittelberg, Kleinwalsertal, Austria, during January 13-17, 1975. This meeting
dealt with recent radio, infrared and optical observations of H II
regions and associated molecular clouds and maser sources, as well as
the interpretation of these observations.

Separate sessions were devoted to current ideas on star formation
and radio and optical observations of external galaxies. Our understanding of these topics has greatly advanced during the past decade
and the current generation of large radio telescopes in Europe has
provided a new impetus to the research field. This meeting brought
together an international group of scientists involved in nearly all
areas related to H II regions.

The Symposium was sponsored by the Deutsche Forschungsgemeinschaft,
the Max-Planck-Gesellschaft zur Förderung der Wissenschaften e.V., and
the European Physical Society. Financial support was given by the first
two organizations. The idea of holding this meeting and the initial
planning came from P.G. Mezger. The Scientific Program Committee
consisted of  F.D. Kahn, R. Kippenhahn, H.J. Habing, J. Lequeux,
P.G. Mezger, E.E. Salpeter and M.J. Seaton. All of the work required
to make this a smoothly running symposium  was very capably handled
by the Local Organizing Committee, R. Schwartz, H. Steppe and U. Wilson,
with the help of E.Dietz and H. Höch.

The papers in this volume were prepared by the authors themselves,
and we have included none of the discussion which followed the presen-
tations. Textual changes were required for only a few manuscripts and
R. Booth, R. Genzel and C.M. Walmsley helped with the editing.

## LIST OF PARTICIPANTS

Aitken, D.K., London

Altenhoff, W.J., Bonn

Appenzeller, I., Heidelberg

Baars, J., Dwingeloo

Barlow, M.J., Sussex

Becklin, E.E., Pasadena

Bedijn, P.J., Leiden

Bieging, J., Bonn

Biermann, P., Bonn

Biermann, L., Munich

Booth, R.S., Jodrell Bank

Borgmann, J., Roden

Burke, B.F., Cambridge, Mass.

Burbidge, E.M., La Jolla

Burbidge, G.R., La Jolla

Cesarsky, D.A., Meudon

Churchwell, E., Bonn

Conti, P.S., Boulder

Courtès, G., Marseilles

Davies, R.D., Jodrell Bank

Deharveng, L., Marseilles

Downes, D., Bonn

Duinen, R.J. van, Groningen

D'Odorico, S., Asiago

Dyson, J.E., Manchester

Elsässer, H., Heidelberg

Emerson, J.P. London

Encrenaz, P., Meudon

Felli, M., Florence

Fricke, K.J., Bonn

Goldsworthy, F.A., Leeds

Grasdalen, G.L., Tucson

Grewing, M., Bonn

Gull, S.F., Cambridge, UK

Habing, H.J., Leiden

Harper, D.A., Chicago

Harris, S., Cambridge, UK

Harten, R.H., Leiden

Higgs, L.A., Ottawa

Hills, R.E., Cambridge, UK

Hoerner, S. von, Green Bank

Israel, F.P., Leiden

Jennings, R.E., London

Jong, T. de, Leiden

Jones, B., London

Kahn, F.D., Manchester

Kardashev, N.S., Moscow

Kegel, W.H., Heidelberg

Kesteven, M., Jodrell Bank

Kippenhahn, R., Munich

Krügel, E., Munich

Lee, T.J., Edinburgh

Lequeux, J., Meudon

Lemke, D., Heidelberg

Lortet, M.C., Meudon

Martin, A.H.M., Cambridge, UK

Matthews, H.E., Groningen

Meaburn, J., Manchester

Mezger, P.G., Bonn

Michel, K.W., Lindau

Olthof, H., Groningen

Panagia, N., Bologna

Pankonin, V., Bonn

Perinotto, M., Florence

Petrosian, V., Stanford

Pottasch, S.R., Groningen

Pauls, T.A., Bonn

Rowan-Robinson, M., London

Salem, M., London

Salpeter, E.E., Ithaca

Schmidt-Kaler, Th., Bochum

Scoville, N.Z., Amherst

Seaton, M.J., London

Setti, G., Bologna

Shaver, P.A., Groningen

Soifer, B.T., La Jolla

Sorochenko, R.L., Moscow

Smith, L.F., Bonn

Sume, A., Onsala

Ulrich, M.H., Austin

Vidal, J.L., Toulouse

Walmsley, C.M., Bonn

Weliachew, L., Meudon

Wehrse, R., Heidelberg

Weinberger, R., Heidelberg

Wendker, H.J., Hamburg

Werner, M., Pasadena

Wilson, T.L., Bonn

Wink, J.E., Bonn

Winnewisser, G., Bonn

Wynn-Williams, C.G., Cambridge, UK

Yorke, H.W., Munich

Zuckerman, B., Maryland

# FORMATION OF MOLECULES AND DUST

E. E. SALPETER

Cornell University[*], Ithaca, N.Y., and California Institute of Technology[+], Pasadena, California

## ABSTRACT

The theory of formation of various molecules, both on surfaces of interstellar dust-grains and in gas-phase reactions, is reviewed. Destruction of molecules, the special problems of very dark clouds and the life-history of dust-grains are also reviewed briefly.

[*] Work supported in part by National Science Foundation Grant MPS72-05056 with Cornell University.

[+] Temporarily at California Institute of Technology as a Fairchild Scholar.

# I. INTRODUCTION

My main topic is the formation and destruction of interstellar molecules. Fortunately, recent reviews have been published on this subject: Short summaries will be found in two general reviews on interstellar molecules (Winnewisser and Mezger 1974; Zuckerman and Palmer 1974) and a very detailed review is given by Watson (1975). For this reason I shall only summarize the highlights in Sect. II for the general interstellar medium and in Sect. III for the very dark clouds which present special problems.

The reviews cited above give a detailed literature survey on molecule formation, so I give here only a few references to individual papers: First of all, the classic paper by Bates and Spitzer (1951); Hollenbach, Werner and Salpeter (1971) and Lee (1972) for $H_2$-production; Herbst and Klemperer (1974) for gas-phase recombination and Watson and Salpeter (1972) for grain-surface recombination of other molecules. Accretion onto grains is discussed by Carrasco et al (1973) and Field (1974), photo-ejection from them by Greenberg (1973). A few more recent references are given below.

General reviews on grains are given by Aannestad and Purcell (1973) and Greenberg (1975). Some classic papers on grain formation are by Hoyle and Wickramsinghe (1962) and by Donn and Stecher (1965), numerical calculations are described by Fix (1969); recent papers and current problems are discussed by Salpeter (1974). Grain destruction is discussed by Aannestad (1973) and observations pertaining to grains by Woolf (1973).

## II. MOLECULES IN CLOUDS WITHOUT EXCESSIVE SHIELDING

We consider only HI-regions, which contain few photons beyond 13.6 eV, the Lyman edge of hydrogen. In these regions dust-grains are certainly undepleted, i.e. their density is proportional to the total number density n of hydrogen-nuclei, and their temperatures should be below 40°K. The situation is particularly simple in regions where the (neutral) hydrogen is mainly atomic: It is likely that a large fraction of all atoms which hit a grain-surface stick to it long enough to form a molecule with another incident atom. Because of the abundance and mobility of atomic hydrogen, the surface-recombination process most likely results in simple hydrides such as $CH^+$, CH, NH, OH and especially in molecular hydrogen itself. The rate $r_{sr}$ per X-atom for surface-recombination to form the diatomic molecule XH is of the order of $r_{sr} \sim n/10^9$ years.

In an unshielded HI-region most molecules are photo-dissociated by the general diluted starlight quite rapidly (in 30 to 1000 years, say). In a cloud-region partially shielded from starlight by an optical depth $\tau_{vis}$ in the visible, the relevant ultraviolet photons ($\sim 7$ eV to 13.6 eV) are attenuated by a factor of order $\sim e^{2.5\tau_{vis}} \sim 10^{\tau_{vis}}$. Typical dissociation rates are then of order $r_{phd} \sim (10^{\tau_{vis}} \times 100 \text{ years})^{-1}$ and molecules in general are expected to be less abundant than atoms when the parameter

$$\eta \equiv n \times 10^{\tau_{vis}}/10^7 \qquad (1)$$

is small compared with unity. The stellar photons in the near UV not only dissociate molecules but can also eject molecules intact from a grain surface to which they have been adsorbed. The photoejection rate is uncertain and depends on the nature of the grain and of the molecule, but typical rates (Greenberg 1973) are larger than dissociation rates, possibly by factors of $\sim 10^3$. If $\eta \lesssim 10^3$, an appreciable fraction of molecules which stuck to a grain surface are then ejected by starlight and remain in the gas-phase.

The dependence of the photodissociation rate on the optical depth $\tau_{vis}$ for molecular hydrogen differs radically from the simple factor $10^{\tau_{vis}}$ in eqn. (1): In HI-regions molecular hydrogen cannot be dissociated by continuum radiation, but only by the radiation in a few intense absorption lines. In a sufficiently large column of hydrogen the $H_2$ produced absorbs enough of the line radiation, at a slightly greater depth the $H_2$ abundance is greater because there are fewer photons and the $H_2$ in turn depletes the photons even more. The end result is an "all or nothing" phenomenon, somewhat analogous to the occurrence of Strömgren spheres, so that the fractional abundance of $H_2$ is very low for $\tau_{vis} \lesssim 0.5$ and quite high for $\tau_{vis} > 1$, say. Copernicus observations of $H_2$-abundances (Spitzer and Jenkins 1975) bear out this "all or nothing" phenomenon at least qualitatively. In unshielded HI regions carbon (and also silicon and sulphur) is singly ionized. The recombining carbon also presents self shielding to the radiation between 11.3 eV and 13.5 eV and for $\tau_{vis} > \sim 3$ the carbon would be mainly neutral. In fact, we shall see that CO is probably even more abundant than C in such dense clouds.

The situation for clouds of increasing optical depth $\tau_{vis}$ (which is usually correlated with increasing hydrogen number density n) is summarized schematically in Table 1.

Table 1. Typical conditions in clouds of various optical depths

| | | | |
|---|---|---|---|
| $\eta$ | $10^{-4}$ | $\gtrsim 0.1$ | $>10^{3}$ |
| $n$ | $10^{2}$ | $\gtrsim 10^{3}$ | $10^{4}$ |
| $\tau_{vis}$ | 1 | 3 | 6 |
| | H | $H_2$ | $H_2$ |
| | $C^+(Si^+,S^+)$ | $C^+(Si^+,S^+)$ | C or CO | most condensed; |
| | O, N | O, N | $H_2$ | cosmic rays |

For a cloud with density n (in H cm$^{-3}$) and radius R (in pc) the optical depth is $\tau_{vis} \sim nR/500$. A cloud with $n \sim 10^2$, $R \sim 5$ pc, $\tau_{vis} \sim 1$ and mass $M \sim 1500\ M_\odot$ represents the threshold for an appreciable abundance of $H_2$. For darker clouds <u>most</u> of the hydrogen is molecular and in the gas phase, but other molecules (especially CO) become abundant only for $\tau_{vis} > \sim 3$. For very dark clouds ($\tau_{vis} > 6$, say) the attenuated starlight should be negligible, cosmic rays play a role and the special features of such regions are discussed in Sect. III.

We turn next to reactions that take place in the gas-phase, which could dominate since the simple gas-kinetic collision rate between atoms ($\sim n/100$ years) is about $10^7$ times faster than the recombination rate on grain surfaces. We have to distinguish "exchange reactions", which reshuffle atoms and radicals between preexisting molecules, from genuine recombination which builds molecules from simple atoms or ions. Genuine recombination of two neutral atoms to form a diatomic molecule is too slow to be of interest in all cases. The radiative recombination $C^+ + H \rightarrow CH^+ + h\nu$ is of considerable interest (Herbst and Klemperer 1973), but at best its rate is comparable with the overall recombination rate on grain surfaces (and might be a few times smaller). For cloud conditions in the second column of Table 1 (hydrogen largely molecular, but carbon still mainly $C^+$) another radiative recombination is important, $C^+ + H_2 \rightarrow CH_2^+ + h\nu$. This reaction is at least as fast as surface recombination and <u>might be</u> as much as a hundred times faster (Black et al 1975; Stecher and Williams 1974), $r_{rad} \sim n_{H_2}/10^7$ years. Such a fast rate, if substantiated, would have a profound effect on the abundances of most other

molecules as well, since many molecules are connected by a
complex network of exchange reactions:

Most exchange reactions involving positive ions are
particularly fast, if they are exothermic. Of particular
importance are ion-exchange associations of the form

$$X^+ + H_2 \begin{cases} \nearrow & XH + H^+ \\ \searrow & XH^+ + H \end{cases}; \quad r \sim \frac{n_{H_2}}{(10 \text{ to } 100) \text{ years}} \qquad (2)$$

for X = D, N, O, $H_2$, CH, etc. However, these reactions do
not apply for X = C, Si and S because the low ionization
potentials which make $C^+$, $Si^+$ and $S^+$ abundant also make the
exchange reactions endothermic. Positive molecular ions
are usually destroyed rapidly by dissociative recombination
if the electron concentration $n_e$ is at all appreciable;
reactions of form $XH^+ + e \rightarrow X + H$ often have rates of order
$n_e/1$ year.

Hydrides of the more common elements C, N, O, Si, S,
etc. are easily made as discussed above, but molecules involving two atoms of C or heavier can also be produced by
exchange reactions fairly rapidly. For instance, a favorable
rate r (per C-nucleus) of producing CO comes from the
reactions

$$\begin{matrix} C^+ + OH \searrow \\ CH^+ + O \nearrow \end{matrix} \quad CO + H^+; \quad r \sim n_0/30 \text{ years}, \qquad (3)$$

where $n_0$ is the total abundance of oxygen (nuclei per $cm^3$).
Carbon-monoxide is an exceptionally abundant interstellar
molecule because its production rate is particularly fast

and its photodissociation rate particularly slow (CO also having an exceptionally high binding energy). Various production mechanisms have been suggested for formaldehyde; one example is a reaction-chain ending with

$$CH_3 + O \rightarrow H_2CO + H; \quad r \sim n_0/10^3 \text{ years} \, . \qquad (4)$$

Calculating abundances of the more complex molecules is somewhat beyond the "state of the theoretical art" at the moment: There are various competing formation chains of gas-phase reactions involving smaller neutral molecules and/or ions, as well as different destruction mechanisms. Furthermore, there is the possibility of catalytic reactions (Anders et al 1974) taking place on surfaces of special dust grains which would make some special polyatomic molecules efficiently, but not others. The compounding of many numerical uncertainties makes purely theoretical calculations unprofitable at the moment.

## III. MOLECULES IN VERY DARK CLOUDS

The last column in Table 1 refers to "very dark" clouds with visual extinction $\tau_{vis} \gtrsim 6$. Such clouds are of great interest observationally, since the most prolific "molecular clouds" are of this type, but they also present special problems and opportunities to the theorist: If such a cloud has a simple geometry (almost spherical shape without any stars inside) the large extinction implies that the amount of starlight which penetrates into the cloud is negligible. If, in addition, this state of affairs has persisted for a long enough time then it is a puzzle why

many molecules are so abundant in the gas-phase--most molecules (except $H_2$) should be condensed out on grain-mantles:
A certain fraction of molecules made on grain surfaces
remain there after the formation. This difficulty alone
could be overcome if most molecules were made (a) on very
small grains where the sticking-fraction could be very
small (Duley 1973, Purcell 1975, Allen and Robinson 1975) or
(b) in gas-phase reactions. However, the problem remains
even in this since molecules in the gas-phase hit grain
surfaces fairly often in a dense cloud and a certain fraction of them will stick. For "normal" grain-temperatures
and favorable surface conditions on the grain some abundant
molecular species may re-evaporate from the surface, but if
the cloud were really "dark and inert" the grain temperatures would be so low that few molecules besides $H_2$ could
reevaporate.

Cosmic rays of moderately high energies
($\gtrsim 100$ MeV/nucleon) are likely to penetrate even into dense
clouds. Their ionization power is estimated to be
$\zeta \sim (10^{10}$ years$)^{-1}$ ionizations per $H_2$ molecule. These
cosmic-rays initiate the reaction chains discussed below; it
is possible (Erents 1973) that they also eject molecules
adsorbed onto dust-grains very efficiently, in which case
the high molecular abundance in the gas-phase would have a
simple explanation. Nevertheless, one may have to consider
complications which are not included in present-day, steady-
state calculations: Dark clouds lie in close proximity to
bright stars and H II-regions, the interfaces may be highly
convoluted and turbulence may interchange material from a
cloud-interior and from an interface more rapidly than the
slowest molecule-grain equilibration time. In other words,
stellar UV may yet turn out to be important even for

seemingly dark clouds, but at present we ignore this possibility.

Although the ionization-rate of cosmic rays is not extremely great, it is appreciable and about as large for atoms or molecules with a <u>high</u> ionization-potential as it is for "easily-ionized" material. In particular, $He^+$-ions are produced in this manner and the high electron-affinity (~25 eV) of the helium-ions enables them to dissociate and ionize neutral molecules. The ionized radicals produced in this manner can undergo further reactions; for instance, with XY standing for an abundant inert molecule such as CO, $N_2$ or $O_2$, we have

$$\left. \begin{array}{l} He^+ + XY \rightarrow X^+ + Y + He \ , \\ X^+ + H_2 \rightarrow XH^+ + H \end{array} \right\} \qquad . \qquad (5)$$

The ensuing reaction-chain is rather complex, since each positive molecular ion may undergo dissociative recombination or react further with neutral molecules, such as

$$XH^+ + H_2 \rightarrow XH_2^+ + H \quad . \qquad (6)$$

For instance, with XY standing for $N_2$ in eqn. (5), hydride-ions up to $NH_4^+$ can be built up which then recombines dissociatively with an electron to form $NH_3$.

Many other molecules can be built up, starting with cosmic rays and simple molecules. For instance, $He^+$ produces $C^+$ out of CO and $NH_3$ out of $N_2$, so that HCN can be produced by

$$C^+ + NH_3 \rightarrow H_2CN^+ + H \ , \qquad \Bigg\}$$

$$H_2CN^+ + e \rightarrow HCN + H \ . \qquad (7)$$

Besides $He^+$, cosmic rays also produce $H_2^+$ at a rate (per $cm^3$) about 5 times faster. $H_2^+$ reacts very rapidly with $H_2$ to form $H_3^+$, which should be one of the more common positive ions in dark clouds. However, the reaction

$$H_3^+ + CO \rightarrow HCO^+ + H \qquad (8)$$

is fast enough so that $HCO^+$ is expected to be more abundant than $H_3^+$, at least in very dense clouds.

There is some controversy whether lower energy cosmic rays (or X-rays) could be produced abundantly somewhere and penetrate into clouds to raise the ionization rate $\zeta$ far above the value $\sim(10^{10}$ years$)^{-1}$ for the more energetic cosmic rays. Some information on $\zeta$ can be obtained from observed molecular abundances, most elegantly from the observed abundance ratio of HD to $H_2$ (Watson 1973, Dalgarno et al 1973, Watson 1975) in clouds: This ratio is expected to be appreciably smaller than the overall D/H isotope ratio, because self-shielding prevents the photodissociation of $H_2$ but not of the less abundant HD (at least in clouds which are not extremely dark). The observed HD-abundance gives the effective production-rate of HD according to the production scheme

$$H^+ + D \overset{\longleftarrow}{\longrightarrow} D^+ + H \ , \qquad \Bigg\}$$

$$D^+ + H_2 \rightarrow HD + H^+ \ . \qquad (9)$$

The inverse reactions need not be considered (because the photodestruction of HD depresses HD and $D^+$) and the HD production rate is proportional to the abundance of $H^+$. This abundance in turn depends on the cosmic ray ionization rate $\zeta$ and the observations are compatible with modest values of order $(10^{10}$ years$)^{-1}$ for many clouds.

## IV.  FORMATION AND DESTRUCTION OF DUST GRAINS

In discussing the life-history of solid particles in interstellar space one has to consider separately refractory materials and volatile substances. The situation is simplest for refractory grains (or grain-cores) with condensation-temperatures $T_{cond}$ above 1000°K, say, since they are made or replenished easily but destroyed only rarely.

One place where refractory grains are made "ab initio" (rather than solids condensing out on preexisting grains) are the atmospheres of cool supergiants: Observationally, supergiants are known with surface temperatures $T_{eff}$ down to 2,500°K and lower, infrared emission indicates the presence of silicate grains in the cooler atmospheres and outflow of mass is indicated for the most luminous giants and for supergiants. Theoretically, the situation is simple because material flows upwards from the hot interior where all substances are in gaseous form and solids condense out of the vapor phase (by homogeneous nucleation) when the temperature falls a little below $T_{cond}$. The situation would particularly simple if $T_{eff}$ were as low as $T_{cond}$; however, grains still form above the photosphere as long as

$T_{eff} \lesssim 2T_{cond}$, say. For supergiants, at least, the luminosity is great enough that radiation pressure acting on the
grains leads to an outflow of both grains and gas.

The thermochemistry of grain production in cool
stellar atmospheres is affected by the great stability of
CO-molecules in the gas-phase: Whichever of C and O is
less abundant (by number of atoms) is almost completely
bottled up in CO and not available for grain formation.  In
the majority of stars oxygen is somewhat more abundant than
C so that no carbon-rich minerals are formed; the most
abundant materials which form refractory grains under these
circumstances are magnesium silicates and iron.  In a rarer
class of stars C is slightly more abundant than O (oxygen
having been depleted by the C,N,O bi-cycle) and graphite-
like carbon-particles (possibly containing some hydrogen)
can be formed.  Of all the elements heavier than He which
flow into interstellar space from cool stars, only 10 to
20% by mass is in the form of refractory grains and the
rest is gaseous, mainly in CO, $NH_3$ and $H_2O$.

Refractory grains, i.e. minerals containing Mg and Si
(and, especially, the less abundant Ca and Al), most probably survive in normal H II-regions and in "normal" cloud-
cloud collisions (collision speeds $\lesssim$ 30 km/sec).  These
grains are destroyed, and their constituents fed into the
interstellar gas, only in (a) the blast-wave of young super-
novae remnants (younger than $10^4$ years with shock-speeds
> 100 km/sec) and (b) in the formation of new stars and
subsequent mass-loss from hot stars (e.g. stellar winds
from massive stars).  These destruction processes are rare
enough so that the lifetime of a refractory grain (or grain-

core) in interstellar space is not much shorter than the
lifetime of the Galaxy, $\sim 10^{10}$ years.

The situation is radically different for volatile
solids, containg the "ices" $CH_4$, $NH_3$ and $H_2O$ or CO and more
complex molecules: They are <u>not</u> formed in the atmospheres
of even the very coolest stars (although they may or may not
be formed in the outflow from such stars a few stellar radii
further out). They are likely to condense out in cool,
dense, dark clouds; refractory, smaller grains already exist
in these clouds, so the volatiles will form a mantle around
these refractory cores. As discussed in the previous
section, the quantitative details of molecule condensation
are controversial, but it is likely that the galactic-
spiral wave compressions are enough to make condensations
possible every $10^8$ years or so. In <u>very</u> old supernova
remnants ($> 10^5$ years) high densities may also be encountered
where condensation may take place. Solids as volatile as
the "ices" are likely to be destroyed by sputtering
(Aannestad 1973) in the more energetic cloud-cloud
collisions (collision speeds in excess of 30 or 40 km/sec).

The so-called "dirty-ice mantles" are thus made and
destroyed many times in the lifetime of the Galaxy, the
formation rate being at least $\sim(10^8$ years$)^{-1}$ and the
destruction rate probably even greater. The life-history of
actual interstellar grains is complicated further by the
possibility of solid materials with binding-energies and
volatility intermediate between silicates (or graphite or
iron) and ices. Polymers built up of organic molecules (C,
O and H being abundant) are particularly likely: There are
some suggestions of such polymers even forming directly on

grain-surfaces (Wickramsinghe 1975), but more generally the effect of stellar UV on a mixture of ices should produce various complex hydrocarbon compounds (Donn and Jackson 1970). A grain may then have three components to it: 1) An innermost inert refractory core; 2) an outermost fragile envelope of "dirty ice" which comes and goes rapidly (compared with $10^{10}$ years); and 3) a slowly-growing, intermediate mantle of "dirty oil or tar".

## REFERENCES

Aannestad, P. 1973, Ap. J. (Suppl.) 25, 205.
Aannestad, P. and Purcell, E. M. 1973, Ann. Rev. Astr. Ap. 11, 309.
Allen, M. and Robinson, G. W. 1975 (to be published).
Anders, E., Hayatsu, R. and Studier, M. 1974, Ap. J. 192, L101.
Bates, D. and Spitzer, L. 1951, Ap. J. 113, 441.
Black, J. H., Dalgarno, A. and Oppenheimer, M. 1975 (Center for Ap. Preprint No. 217).
Carrasco, L., Strom, S. E. and Strom, K. M. 1973, Ap. J. 182, 95.
Dalgarno, A. and Black, J. H. 1973, Ap. J. 184, L101.
Donn, B. and Stecher, T. P. 1965, Ap. J. 142, 1681.
Donn, B. and Jackson, W. M. 1970, Bull. A.A.S. (June).
Duley, W. W. 1973, Ap. and Sp. Sci. 23, 43.
Erents, S. K. and McCracken, G. M. 1973, J. Appl. Phys. 44, No. 7 (July).
Field, G. B. 1974, Ap. J. 187, 453.
Fix, J. D. 1969, MNRAS 146, 37 and 51.
Greenberg, J. M. 1975, Les Houches Lecture Notes.
Greenberg, L. 1973, IAU Symp. No. 52 (ed. Greenberg and van de Hulst), Dordrecht-D. Reidel.
Herbst, E. and Klemperer, W. 1973, Ap. J. 185, 505.
Hollenbach, D., Werner, M. and Salpeter E. E. 1971, Ap. J. 163, 165.
Hoyle, F. and Wickramasinghe, N. 1967, Nature 214, 969.
Lee, T. J. 1972, Nature (Phys. Sci.) 237, 99.
Purcell, E. M. 1975 (unpublished).
Salpeter, E. E. 1974, Ap. J. 193, 579 and 585.
Spitzer, L. and Jenkins, E. B. 1975, Ann. Rev. Astr. Ap. 13 (in press).

Stecher, T. P. and Williams, D. A. 1974, MNRAS 168, 51P.

Watson, W. D. and Salpeter, E. E. 1972, Ap. J. 175, 659.

Watson, W. D. 1973, Ap. J. 181, L129.

Watson, W. D. 1975, Les Houches Lecture Notes.

Winnewisser, G., Mezger, P. G. and Breuer, H. D. 1974, Topics in Current Chemistry 44, 1.

Woolf, N. J. 1973, IAU Symp. No. 52 (ed. Greenberg and van de Hulst), Dordrecht-D. Reidel.

Zuckerman, B. and Palmer, P. 1974, Ann. Rev. Astr. Ap. 12, 279.

# INTERPRETATION OF THE CARBON RECOMBINATION LINE

C. M. WALMSLEY

Max-Planck-Institut für Radioastronomie

Bonn, FRG

## ABSTRACT

The ionization equilibrium of carbon and heavier elements in a neutral gas cloud close to an early type star is discussed using a simple model for the transfer of the ultra-violet radiation. The balance between atomic and molecular hydrogen is also considered. It is concluded that the H II region exciting star may be, but should not be assumed to be, the source of the photons which ionize carbon in the region giving rise to the observed recombination lines. The origins of the narrow hydrogen and heavier element lines are also discussed.

## INTRODUCTION

Some convergence of opinion appears to have been reached in theoretical discussions of the carbon line regions. Three recent articles (Dupree 1974, Zuckerman and Ball 1974, Hoang-Binh and Walmsley 1974) all come to the conclusion that the carbon lines are formed in dense ($n_H > 10^3$ cm$^{-3}$) clouds, which are intimately connected with H II regions and their associated molecular clouds. On the observational side, an important step forward has been the direct determination of the angular extent of the C II regions (Balick et al., 1974; Wilson et al., 1975). This, and the assumption of spherical symmetry, allow a determination of the r.m.s. electron density. A typical value for $N_e$ is 10 cm$^{-3}$ and, since carbon provides most of the electrons, the neutral gas density must be at least $3 \cdot 10^4$ cm$^{-3}$. The temperature is a less certain quantity but fits to the frequency dependence of the C line suggest that $T \sim 100$ K (Hoang-Binh and Walmsley, 1974). Measurements of the 109 $\alpha$/ 137 $\beta$ ratio suggest even larger values in particular for W 3. (Walmsley and Hoang-Binh, 1975). The problem is complicated by the fact that in Orion A, and probably also in W3, the line velocity becomes more negative at lower frequencies, the stimulated term in the transfer equation becomes dominant and one sees mainly gas in front of the continuum source. It follows that one is seeing different regions at the lower frequencies and hence that the models used to fit the frequency dependence should not be taken too literally.

In this contribution, I assume that the results summarized above are roughly correct and examine some of the consequences for the physical conditions in the carbon line regions. In section 2, I discuss the carbon ionization equilibrium and in section 3 the associated problem of the balance between atomic and molecular hydrogen within the ionized carbon ( C II ) region. In section 4, I discuss the theoretical interpretation of the narrow hydrogen and heavy element (Mg, Si, S, Fe ) lines.

Section 2    Carbon ionization equilibrium in neutral gas close to an early

type star

It is not clear whether the exciting star of the H II region is the source of
the photons which ionize carbon.  Figure 1 shows for zero-age main sequence
(ZAMS) stars the photon luminosity in the range 912-1101 Å as a function of
spectral type.  For comparison, the luminosity of Lyman continuum photons as
well as the luminosity in the wavelength range 1101-1526 Å are also shown.  The
latter is roughly the wavelength range relevant for the ionization of Mg, Si, and
Fe.  The curves are based upon the non LTE models of Auer and Mihalas(1972)
and the stellar radii given by Mezger et al. (1974) .  Evidently, early O stars
have a stronger 912-1101 luminosity than later types but the weighting is not so
strong as for the Lyman continuum.  The slope of the curves partly reflects
the increase of stellar radius with earlier spectral type on the ZAMS.  For more evolved stars, which do not show such an increase, later types may be more
important as far as the carbon ionizing flux is concerned.  In any case, a photon luminosity of $10^{49}$ sec$^{-1}$ seems a reasonable estimate for O stars in the
range 912-1101 Å .  The flux actually incident upon the carbon cloud may be diminished due to absorption by dust within the H II region (Mezger et al. , 1974)
so that this may be an overestimate.

Also the dust will compete with the carbon and with molecular hydrogen for photons within the neutral gas.  The details of this competition are dependent upon
dust to gas ratio, albedo, and carbon abundance.  The incident UV flux is so
high that, in contrast to the dark cloud case, the fraction of carbon which is
neutral at the exposed surface is very low ($\sim 10^{-4}$ ).  As a result, the dust
absorbs most of the photons even in the wavelength range 912-1101 Å.  Zuckerman and Ball (1974) have therefore simply assumed that carbon is ionized in a
slab whose dust absorption optical depth is 10 for $\lambda \sim 1000$ Å.  One can test the
validity of this by calculating the carbon ionization equilibrium as a function of
optical depth.  I have done this using essentially the one dimensional transfer
approach suggested by Werner (1970) .  However, I have included the effect of
the "heavy elements" (Mg, Si, S, Fe ) , which causes carbon to be slightly more

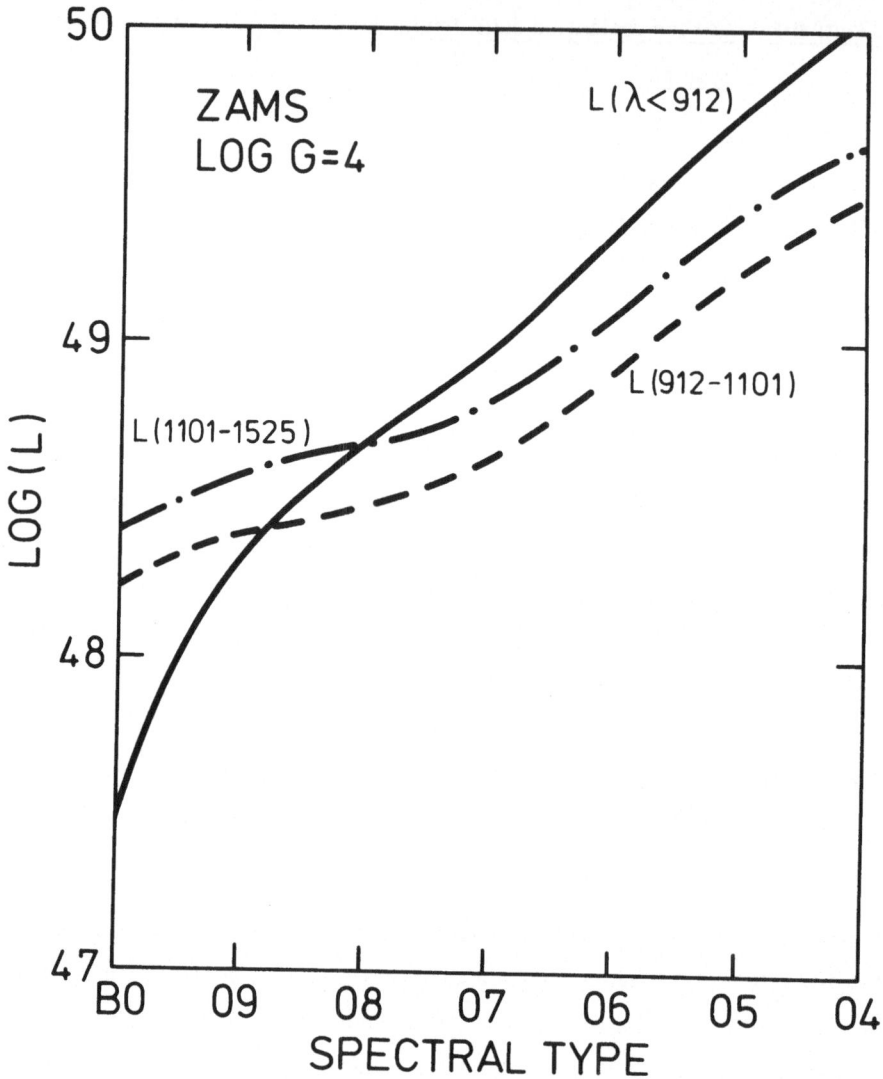

Fig. 1  Ultra-violet photon luminosity in the wavelength ranges λ<912Å
(full line), 912 Å <λ <1101Å (dashed line), and 1101 Å <λ < 1526 Å
(dash dot line) plotted as a function of spectral type for zero-age main
sequence stars with log g = 4 · 0.

neutral than one finds by a simple application of equation (10) from Werner's
article. Molecular hydrogen opacity is neglected in this calculation and the
dust is treated as a pure absorber. The dust albedo may be high close to 1000 Å

(Mezger et al. , 1974) and I have tried roughly to simulate the effect of the scattering by using an effective dust optical depth $\tau_d$ ( treating the dust as an absorber) equal to 0.3 times the extinction optical depth at 1000 Å. The factor $0 \cdot 3$ is based on the result (Code, 1973) that in a two stream approximation to the radiative transfer, the UV flux falls off as $\exp -\sqrt{1 - \gamma} \ \tau$ where $\gamma$ is the albedo. Hence, the value chosen corresponds to an albedo of $0 \cdot 9$. The effective dust absorption coefficient is then $n\, \bar{\sigma}$ where $n$ is the hydrogen number density and $\bar{\sigma} = 0.3\, n_g\, \sigma_g / n \approx 5\ 10^{-22}\ \mathrm{cm}^2$ if one assumes a normal dust to gas ratio and applies the results of Jenkins and Savage (1974) and of Bless and Savage (1972). I have assumed solar abundances, a temperature of 100 K, and a neutral cloud of density $10^4\ \mathrm{cm}^{-3}$ situated 0.5 pc from a star of 912-1101 luminosity $L = 10^{48}$ photons $\mathrm{sec}^{-1}$. This allows me to calculate the carbon ionization degree as a function of the dust visual extinction optical depth $(A_v /1.086)$. The carbon ionization rate $\zeta_c$ at the exposed surface is $3.7\ 10^{-7}\ \mathrm{sec}^{-1}$ and one should note that the results are only a function of $\zeta_c / n$ where $n$ is the density. Thus I would obtain the same result for $\zeta_c = 3.7\ 10^{-6}\ \mathrm{s}^{-1}$ and $n = 10^5 \mathrm{cm}^{-3}$. A typical situation is shown in figure 2 and parameters for four models are given in Table 1.

## Table 1

Depth of carbon ionized (C II) layer and emission measurements for a cloud of density n at a distance 0.5 pc from an early type star of luminosity L photons $\mathrm{s}^{-1}$ in the range 912-1101 Å.

| $\dfrac{L}{s^{-1}}$ | $\dfrac{n}{cm^{-3}}$ | $\tau_d$(CII) | DEPTH of CII (pc) | EMISSION MEASURE (pc cm$^{-6}$) | | | | |
|---|---|---|---|---|---|---|---|---|
| | | | | CII | SiII | MgII | SII | Fe II |
| $10^{48}$ | $10^4$ | 7.7 | 0.6 | 7.0 | 1.0 | 0.9 | 0.6 | 1.2 |
| $10^{48}$ | $10^5$ | 5.4 | 0.04 | 48 | 7.3 | 6.4 | 4.4 | 9.1 |
| $10^{49}$ | $10^4$ | 9.9 | 0.8 | 9.2 | 1.2 | 1.1 | 0.7 | 1.6 |
| $10^{49}$ | $10^5$ | 7.7 | 0.06 | 70 | 9.7 | 9.0 | 5.9 | 12 |

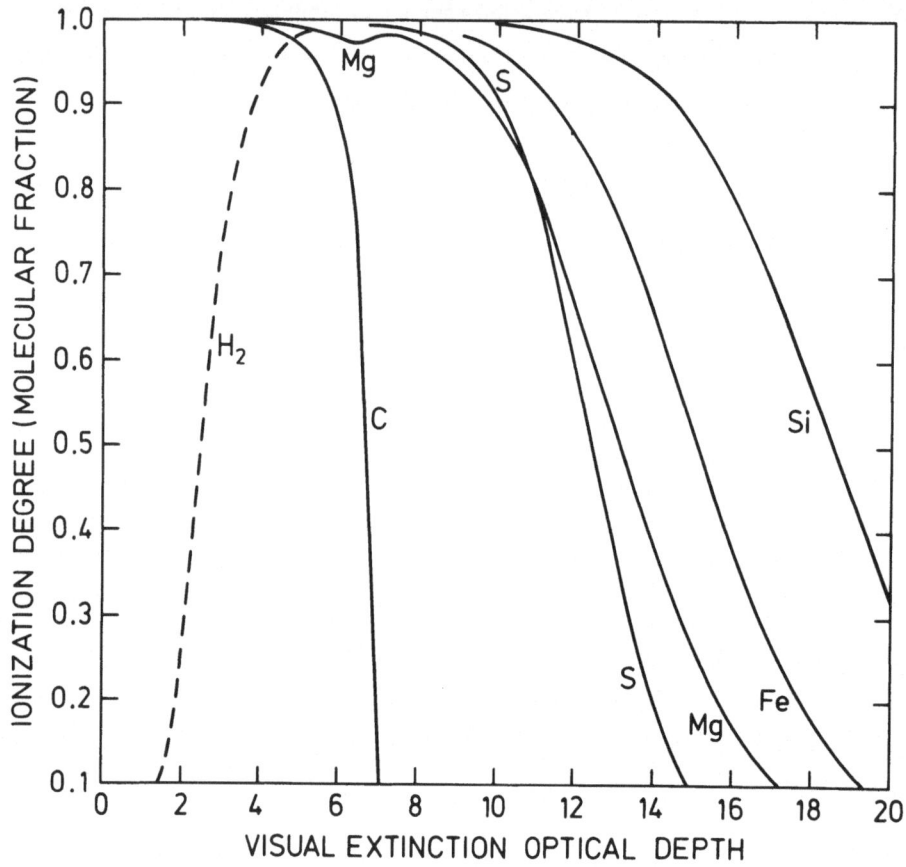

<u>Fig. 2</u>  Ionization degree for C, S, Mg, Fe, and Si plotted as a function of depth in a neutral cloud of density $10^4$ cm$^{-3}$ situated 0. 5 pc from a star emitting $10^{48}$ photons sec$^{-1}$ in the wavelength range 912–1101 Å. The fraction of hydrogen in molecular form is shown by the dashed line.

The heavy element ionization equilibrium has been computed considering only the effect of dust upon the radiation field and assuming an effective optical depth for each element proportional to the extinction optical depth at the photoionization threshold. The unshielded photoionization rates were presumed proportional to the values given by de Boer et al. (1973) for the Habing field case. The sharp ionization front for carbon is a consequence of self shielding. One notes

from table 1 that the optical depth $\tau_d$ (CII) at which carbon is 50 percent ionized is in the range 5 - 10 . In agreement with the result of Zuckerman and
Ball (1974), the depth of the carbon ionized layer is $\sim 0 \cdot 05$ pc if the neutral
density n is $10^5$ cm$^{-3}$ . It is interesting that the carbon emission measure is
rather insensitive to L. Since an emission measure of order 50 pc cm$^{-6}$ is
needed to explain the high frequency carbon results (for T = 100 K ) , it would
seem that the neutral density should be $\sim 10^5$ cm$^{-3}$ independent of the spectral
type of the exciting star. Also, one might expect that not only the exciting star
of the H II region but also stars of somewhat later spectral type would contribute to the carbon ionization. However, the study made by Penston (1973) of stars
in the Orion nebula region does not show up any obvious candidates apart from
the exciting stars of M 42 and M 43 .

The above results suggest that the ionized carbon is confined to a layer just
beyond the H I/ H II ionization front. One would therefore expect a fairly good
correlation between carbon and radio continuum contours. However, the C85 $\alpha$
map made by Balick et al. (1974) of Orion shows carbon considerably more extended on the north-east (dark bay) side than the continuum. On the basis of
table 1, one would therefore expect a density of $\sim 10^4$ cm$^{-3}$ allowing the C II
to extend over $\sim 0 \cdot 5$ pc . This could be compatible with observations if clumps
of density $\gtrsim 10^5$ cm$^{-3}$ exist within this more extended region and supply the
observed carbon line. On the other hand, one might conceivably be getting extra carbon ionizing photons from M 43 which is also on this side of the nebula.

### Section 3   Molecular hydrogen abundance in the C II region

Another question of some importance is whether hydrogen is mainly molecular within the ionized carbon (C II) region. One can attempt to answer this by
extending the results of Jura (1974) to cover the case where dust absorption
cannot be neglected. Thus if I include a term exp $( - \tau_d )$ in Jura's equation
(A3) , I derive a differential equation for the fraction $f = 2 n ( H_2 ) / n$ of hydrogen which is in molecular form.

$$\frac{df}{d\tau_d} = f\,(1-f) + e^{2\tau_d}\,A\,(1-f)^3 \tag{1}$$

Here, $\tau_d$ is the effective optical depth at 1000 A as defined previously and A is a dimensionless constant equal to $R^2 n^2/(\bar{\sigma}\,\beta^2 I^2)$. In this expression, R is the recombination rate of atomic hydrogen on grain surfaces, I is the un-shielded $H_2$ dissociation rate and $\beta = 4.2\,10^5$ cm$^{-1}$ is a constant defined by Jura. For the kind of situation which we have been discussing, $I \sim 10^{-7}$ sec$^{-1}$ and $A \sim 10^{-2}$. A fraction $\sim 10^{-5}$ of the hydrogen is molecular on the unshielded side of the C II region. Equation 1 does not appear to have an analytic solution although for $A > 1$, the second term on the right hand side will dominate. In this case one finds

$$(1-f)^{-2} - 1 = A\,(e^{2\tau_d} - 1) \tag{2}$$

This reduces to Jura's result for the case when $\tau_d$ is small. I have used a Runge-Kutta procedure to integrate equation (1) and a typical result ($A \sim 4.2\,10^{-3}$) is shown in figure 2. Equation (2) is a reasonable approximation for most purposes. One sees that hydrogen is molecular in rather more than half of the C II region. An interesting point is that a large fraction of the molecular hydrogen may be in an excited vibrational state and therefore directly observable. The number density of such excited $H_2$ will be of the order $10^{-10} n^2$ cm$^{-3}$ where n is the total density. Also, if the temperature is really $\geq 100$ K, the rotational transitions should be strong.

Section 4   The narrow hydrogen and heavy element lines

One might hope to gain independent information about physical conditions in the carbon line regions from the, apparently associated, narrow hydrogen and heavy element lines (Ball et al. 1970, Chaisson et al. 1972). The first question to be answered is whether these lines come from the same neutral gas which produces the carbon. There are three observational tests which one might apply. In the first place, the velocity and width of the lines should be compatible. Approximately this seems to be the case but there is mounting evidence that the narrow hydrogen line in NGC 2024 is slightly offset in velocity ( $0 \cdot 5 - 1 \text{ km s}^{-1}$ ) relative to carbon (Zuckerman and Ball, 1974; Wilson et al. , 1975 ). Secondly, one might hope at high frequencies to have sufficient angular resolution to compare the spatial variation of the line strengths. Such measurements are extremely difficult due to the extreme weakness of the lines. However, Wilson et al. (1975), find that at 11 cm narrow hydrogen, like carbon, peaks to the south of the continuum peak in NGC 2024. MacLeod et al. (1975), find the hydrogen peaks further south than carbon at 3 cm. Moreover, they also claim that the heavy element line has a different spatial distribution from carbon. The third check upon whether the heavy element, narrow hydrogen, and carbon lines come from the same region is to compare the frequency dependence of their line strengths. For lines produced in the same volume of space, the same excited level populations or $b_n$ factors should be applicable and hence under optically thin conditions, one should obtain the same frequency dependence for the line intensity. At low frequencies, the carbon lines may become slightly optically thick and this will enhance the strength of the line from the more abundant ion (Hoang-Binh and Walmsley, 1974) . A compilation of results has been made by Pedlar and Hart (1974), and more recent observations are available made by Chaisson (1974) and by Chaisson and Lada (1974). Though the scatter is large, there is no indication of a trend and it seems probable that the lines come from roughly similar regions. On the basis of the velocity shift in NGC 2024, it seems certain that for the narrow hydrogen the overlap is not complete.

This poses a problem concerning the origin of the ionization which gives rise to the narrow hydrogen line. It has been proposed that this line is formed in the H I/H II ionization front (Zuckerman and Ball, 1974). However the thickness of the ionization front should be in the range 1 - 10 astronomical units and to produce an observable line over this path length would require an electron density of $\sim 10^3 \text{ cm}^{-3}$. This makes it very hard to understand the similar behaviour of the narrow hydrogen and carbon lines as a function of frequency. One might consider then the possibility that the hydrogen is ionized by cosmic rays.

Supposing that the carbon and narrow hydrogen lines come from the same region and have the same strength, I estimate that the cosmic ray ionization rate required to explain the observed line is of order $10^{-14}$ - $10^{-13}$ $sec^{-1}$ . This is for an atomic hydrogen density of $10^3$ - $10^4$ $cm^{-3}$ and the requirement is increased if, as seems probable from the previous discussion, hydrogen is molecular in part of the C II zone. This cosmic ray flux is much greater than that measured at earth, which corresponds to an ionization rate of $\sim 10^{-17}$ $sec^{-1}$ (Spitzer, 1968 ). One might imagine that a large cosmic ray flux could be produced if, due to irregularities in the magnetic field, the cosmic rays remain attached to the gas during the condensation of the cloud out of the general interstellar medium. However the lifetime of cosmic rays in a dense cloud is very short ( 4000 $E^{1.5}/n$ years where E is in MeV)and hence cosmic rays, if present, have to be continuously supplied. Another possibility is X-rays. The cross-section per hydrogen atom for 1 keV X-rays is $\sim 2 \ 10^{-22}$ $cm^2$ (Brown and Gould, 1970 ) and if one believes in the indications that the narrow hydrogen line comes from a region similar to that responsible for carbon, one can exclude X- rays softer than about $0 \cdot 5$ keV . For larger energies, there is a considerable amount of observational material available. In fact, an Uhuru source 3U 0527-05 is close to the position of the Orion nebula (Giacconi et al. , 1974) . At a distance of 500 pc, its luminosity would be $2 \ 10^{33}$ $erg \ sec^{-1}$ and this corresponds, in a neutral medium, to about $3 \ 10^{43}$ ionizations per second (Silk and Werner, 1969 ) . The narrow hydrogen line in Orion has only been observed at low frequencies (Pedlar and Hart 1974, Chaisson and Lada 1974 ) . Its spatial distribution is unknown and, as a consequence,it is difficult to make a reasonable estimate of the ionization implied by the recombination line observations. A rough estimate suggests that 3U 0527-05 could supply the observed hydrogen ionization if it is close by. However, there is no known X-ray source in the vicinity of NGC 2024 or W 3 and I regard the possibility that the narrow hydrogen line is produced by X-ray sources close to H II regions as unlikely. As far as the heavy element line is concerned, the results in figure 2 show

that Magnesium, Silicon, Sulfur, and Iron can be ionized within the C I zone.
The assumption has been made here that they do not turn into molecules. Oppenheimer and Dalgarno (1974) claim that Magnesium and Iron are chemically inert. From table 1, one finds that the ratio of the sum of the heavy element
emission measures to the carbon emission measure varies between $0 \cdot 5$ and
$0 \cdot 57$ as compared with an assumed abundance ratio of $0 \cdot 4$. The heavy element line thus obtains a considerable contribution from the C I zone. Note that
the present discussion does not take into account temperature gradients. Because of this and because of the large ultra-violet flux, the heavy element line
is not so strong relative to carbon as in previous models (Walmsley, 1973).
There is direct evidence now (Morton, 1974) for depletion of heavy elements in
neutral regions. I have taken the abundances given by Morton for the $\zeta$ Oph
cloud and have used them in the models of table 1. The heavy element emission measure is in this case mostly due to Sulfur and is approximately $0 \cdot 15$
of the carbon emission measure. The carbon emission measure is $\sim 3$-4 pc
$cm^{-6}$ for $n = 10^5$ $cm^{-3}$ with the depleted abundances. High frequency observations of the heavy element to carbon intensity ratio suggest a value of $\sim 0 \cdot 1$ in
NGC 2024 and $0 \cdot 35$ in W3 (Chaisson, 1974). However MacLeod et al. (1975)
find that the ratio at the peak is low relative to offset positions in NGC 2024.
This may be comprehensible on the basis of figure 2.


Section 5    Conclusion

The discussion above has centred upon high density neutral gas close to
an early type star. The characteristics of this gas are indicated by the high
frequency carbon line observations. The low frequency lines (18,21 cm) may
well be mainly due to lower density regions in front of the continuum source.
The main conclusion is that the carbon line emission measure is not very sensitive to the spectral type of the star which provides the carbon ionizing photons. This is partly due to the intrinsic stellar properties but also because
the extent of the C II region is governed by the dust. Hydrogen is atomic over

rather less than half the C II region. The heavier elements are ionized to a depth twice as great as carbon.

Whether this results in an observable contribution to the heavy element line from the C I region depends upon the temperature but it is certainly a possibility. The origin of the narrow hydrogen line is an open question and none of the explanations proposed so far seem satisfactory.


ACKNOWLEDGEMENT

I would like to thank Drs. Wilson, Pankonin, Churchwell, and Krügel for their comments upon this manuscript. This work was supported by the Sonderforschungsbereich Radioastronomie at the University of Bonn.

## REFERENCES

Auer, L. H. , Mihalas, D. , 1972, Ap. J. Suppl. 24, 193

Ball, J. A. , Cesarsky, D. , Dupree, A. K. , Goldberg, L. , Lilley, A. E. ,
    1970, Ap. J. (Letters) 162, L 25

Balick, B. , Gammon, R. H. , Doherty, L. H. , 1974, Ap. J. , 188, 45

Bless, R. C. , Savage, B. D. , 1972, Ap. J. 171, 293

de Boer, K. S. , Koppenaal, K. , Pottasch, S. R. , 1973, Astr. & Ap. 28, 145

Brown, R. L. , Gould, R. J. , 1970, Phys. Rev. D 1, 2232

Chaisson, E. J. , Black, J. H. , Dupree, A. K. , Cesarsky, D. A. , 1972
    Ap. J. (Letters) 173, L 131

Chaisson, E. J. , Lada, C. J. , 1974, Ap. J. , 189, 227

Chaisson, E. J. , 1971, Ap. J. , 170, 81

Chaisson, E. J. , 1974, Ap. J. , 191, 411

Code, A. D. , 1973, I. A. U. Symposium No. 52 . Interstellar Dust and
    related topics, ed. J. M. Greenberg, H. C. Van de Hulst (D. Reidel) P 505

Dupree, A. K. , 1974, Ap. J. , 187, 25

Giacconi, R. , Murray, S. , Gursky, H. , Kellog, E. , Schreier, E. ,
    Matilsky, T. , Koch, D. , Tananbaum, H. , 1974, Ap. J. Suppl. 27, 37

Hoang-Binh, D. , Walmsley, C. M. , 1974, Astr. & Ap. 35, 49

Jenkins, E. B. , Savage, B. D. , 1974, Ap. J. , 187, 243

Jura, M. , 1974, Ap. J. , 191, 375

MacLeod, J. M. , Doherty, L. H. , Higgs, L. A. , 1975, Astr. & Ap. , in press

Mezger, P. G. , Smith, L. F. , Churchwell, E. , 1974, Astr. & Ap. 32, 269

Morton, D. C. , 1974, Ap. J. (Letters), 193, L 35

Oppenheimer, M. , Dalgarno, A. , 1974, Ap. J. , 192, 29

Pedlar, A. , Hart, L. , M. N. R. A. S. , 168, 577

Penston, M. V. , 1973, Ap. J. , 183, 503

Silk, J. , Werner, M. W. , 1969, Ap. J. , 158, 185

Spitzer, L. , 1968, Diffuse Matter in Space, J. Wiley, P 111

Walmsley, C. M. , Hoang-Binh, D. , 1975, Paper given at I. A. U. regional
    meeting in Trieste, Sept. 1974

Walmsley, C. M. , 1973, Astr. & Ap. 25, 129

Werner, M. W. , 1970, Ap. Letters, 6, 81

Wilson, T. L. , Thomasson, P. , Gardner, F. , 1975, Astr. & Ap. , in press

Zuckerman , B. , Ball, J. A. , 1974, Ap. J. , 190, 35 .

# THE STELLAR POPULATION OF DARK CLOUDS

G. L. GRASDALEN

Kitt Peak National Observatory*
Tucson, Arizona

## ABSTRACT

Our present knowledge of stars associated with dark
clouds is reviewed with emphasis placed on the observational
properties and large-scale behavior of young stars.

*Operated by the Association of Universities for Research in
Astronomy, Inc., under contract with the National Science
Foundation.

The objects associated with dark clouds can be divided
into three main categories: the visible, the hidden, and the
placental populations. These divisions are obviously rather
arbitrary and refer mainly to observational rather than
evolutionary characteristics of the objects. While this
ordering may apparently obscure our understanding of the
pre-main-sequence evolution of individual objects, I feel
that this ordering may be useful in illuminating our understanding or lack of understanding of the star-formation
process on a relatively large scale. Further, it may help
in providing a larger context for new, especially radio
molecular-line, observations. The processes of pre-main-
sequence evolution are reviewed in great detail by Strom,
Strom and Grasdalen (1975).

The extremely young clusters discussed by Walker (1956,
1957, 1959, 1961, 1969) represent the largest aggregates of
visible stars associated with dark clouds. Optically, each
of these clusters appears superposed on a dark cloud. For
at least one cluster, NGC 2264, the small and uniform amount
of reddenings indicates that the cluster is in front of the
dark cloud and can be only slightly embedded within the dark
cloud. In general, these clusters give the impression that
they have largely freed themselves of interstellar material,
presumably either by using it very efficiently to form stars
or by physically removing it from the immediate cluster
vicinity.

The color-magnitude diagram for these clusters can be
divided into three main categories: 1) the main sequence,
2) the pre-main-sequence stars on their Henyey tracks, 3)
the domain of the T Tauri stars. The upper main sequences
of these clusters are well populated from the O stars to the
late B or early A stars. Walker has demonstrated that the
luminosity function for the cluster main-sequence stars
agrees well with the initial birth-rate function derived by
Salpeter (1955) from the field luminosity function.

At approximately the same luminosity as the coolest stars on the main sequence there is a reasonably tight sequence of stars stretching from the main sequence to the late F or early G stars.  This sequence can be readily identified with stars on their Henyey (equilibrium radiative) tracks toward the main sequence.  The presence of these stars allows us to assign an approximate age to each cluster equal to the contraction time of the least massive star which has arrived on the main sequence.  Typical ages are several million years.  Observationally these contracting stars are recognized by their position above the main sequence in the color-magnitude diagram.  Upon close examination, they are seen to possess vestiges of shell phenomena, small infrared excesses (Strom, Strom and Yost 1971) and weak Balmer emission lines (Strom, Strom and Yost 1971; Smith 1972).  This lack of striking peculiarity makes these stars very difficult to recognize in the less populous groups.

Finally, the coolest and least luminous stars do not exhibit any obvious sequence or order.  For stars apparently cooler than the early G stars, there is only an upper bound to the luminosity; otherwise the entire range in luminosity from well below the main sequence to substantially above the main sequence is occupied.  It is within this domain that the T Tauri stars fall.  These strong emission-line and generally strongly variable stars are the most readily identified pre-main-sequence objects.  However, they are also among the most difficult pre-main-sequence stars to decipher.  Herbig (1962) has given an excellent review of their optical properties.  One of their most vexing properties is the presence of a strong ultraviolet and blue continuum which often masks the true photospheric properties (Walker 1972; Grasdalen et al. 1975; Rydgren, Strom and Strom 1975).  The physical origin of the T Tauri stars' blue continua and their generally strong infrared excesses are still in question, although Rydgren et al. (1975) have

presented strong arguments for believing they are both due
to gaseous shells.  Because of the spectacular observational
characteristics of the T Tauri stars, the presence in the
same region of the color-magnitude diagram of apparently
non-variable, non-emission-line stars is often overlooked.
As for the Henyey track stars, these quiescent companions of
the T Tauri stars have only been recognized in the richest
groups, the extremely young clusters.

The only class of pre-main-sequence star not present in
the extremely young clusters is Herbig's group of Ae and Be
stars associated with nebulosity (Herbig 1960).  By definition these stars are associated with dark clouds.  Recently
they have been shown spectroscopically to lie above the main
sequence in locations appropriate for high-mass stars on
their Henyey tracks toward the main sequence (Strom et al.
1972).  Since the theoretical contraction times for these
high-mass stars are quite short, typically 3-5 x $10^5$ years,
their absence from the extremely young clusters reinforces
the idea that these stars are very young.  Unfortunately,
these stars have so far only been identified in rather
sparse groups.  They are accompanied occasionally by main-
sequence stars of similar luminosity and apparently always
by much fainter variable or emission-line stars analogous to
the T Tauri stars.  As yet we have only fragmentary evidence
(Aveni and Hunter 1972) on the total visible population in
the small groups associated with the Ae and Be stars.

The T associations are the classical cases of stars
associated with dark clouds (Joy 1945).  These groups are
characterized by the paucity of early-type main-sequence or
pre-main-sequence stars.  There are very strong selection
effects operating against the discovery of Henyey-track
stars or the quiescent companions of the T Tauri stars.
They would not have been discovered from surveys for emission-line stars, and unless they happen to illuminate
reflection nebulae, they would be totally missed.  One is

suspicious therefore that a large fraction of the visible
population of the T associations has yet to be discovered.
This general lack of reasonably well-understood stars makes
the age dating of T associations an almost entirely subjec-
tive judgment.

A perhaps crucial exception to these general statements
may be the dark cloud in the Chamaeleon (Henize and Mendoza
1973). This dark cloud contains virtually every recognized
variety of pre-main-sequence star: a normal AO main-sequence
star, an emission-line A star, a large number of variables
and emission-line stars, as well as two stars apparently on
their Henyey tracks recognizable as members in this case
because they illuminate reflection nebulae. Because of the
cloud's nearby location and its rich population, it will
provide an ideal case for comparison with the extremely
young clusters.

The extremely young clusters appear sufficiently transparent so that an accurate census can be obtained for each
cluster. However, this is by no means obvious in the case
of the T associations. From visual inspection, one gains
the impression that the visible stars are only those that
happen to lie relatively close to the cloud's surface. To
verify this impression surveys at wavelengths where the
cloud is reasonably transparent must be made.

To date, the Ophiuchus dark cloud is the most
thoroughly studied case. Surveys at an infrared wavelength
of 2.2 $\mu$ (Grasdalen, Strom and Strom 1973; Vrba et al. 1975)
have revealed a large number of stars embedded within the
dark cloud. Based on the visible population alone, the
Ophiuchus cloud would be classed as a T association; there
are only a few main-sequence stars visible. These infrared
surveys have demonstrated that present within the dark cloud
is an entire cluster closely analogous to the extremely
young clusters of Walker. The similarity extends to the

space density and luminosity function of the upper main-
sequence stars.

In the case of Ophiuchus our attention was first drawn
to the possibility of embedded stars by the faint patch of
nebulosity near source 1.  The presence of faint reflection
nebulae has in general been a fruitful indicator of embedded
stars.  Similar cases are Allen's source near B10 in the
Taurus clouds (Allen 1972), the reflection nebula in Serpens
(Worden and Grasdalen 1974), the peculiar object, Haro 13a,
in Orion (Allen et al. 1975).

The increasing realization that a substantial number of
stars associated with dark clouds are so deeply embedded
within the cloud that they are extremely faint or invisible
has three unsettling consequences:

1)  The conventional interpretation that the T associations and the young clusters are representatives of two
different initial luminosity functions is seriously under-
mined.  It is certainly too soon to assert that a universal
luminosity function prevails, but we can no longer be confi-
dent that a variety do exist.

2)  The physical mechanisms that produce rich, visible
galactic clusters can no longer be assumed to be universal.
The Ophiuchus cluster is roughly the same age as the
extremely young clusters, yet there is little prospect that
it will shortly bloom into a visible cluster.

3)  We cannot as yet draw a sharp line between dark
clouds in which star formation has begun and those which
have not yet formed stars.  It is becoming increasingly
difficult to point to opaque clouds in which we have not yet
found evidence of star formation.  This result may have
severe implications for our ability to disentangle the
physical conditions conducive to star formation from those
produced by the process of star formation itself.

By placental population I mean those pre-main-sequence objects which are sufficiently young that they are still surrounded by the material from which they have directly formed. From the observed density of dark clouds, and an assumed efficiency of star formation of less than 25%, we deduce expected sizes of these placental formations in the range from $10^{-2}$ to $10^{-1}$ pc. This implies, unless the peculiar velocity of the formed star is extremely small, that only objects less than $5 \times 10^{5}$ years old will still be associated with their placental cloud. The only objects we have discussed so far that can be presumed to be that young are the Herbig Ae and Be stars. However, these hot, luminous stars show no obvious evidence for association with placental material. Their reflection nebula only demonstrate their close connection with the dark-cloud material, they do not exhibit any obvious natal characteristics.

The discovery (Strom, Strom and Grasdalen 1974) of an infrared source and a Herbig-Haro object within an obvious substructure of the Corona Austrina dark cloud provides the first optical evidence of a placental structure. The optical object is seen as an obscuring, nearly-circular object superposed on a background of reflection nebulosity. Within the "white spot" there is a small nebulosity whose spectrum identifies it as an Herbig-Haro object (Herbig 1951; Haro 1952). These facts can be understood with only a mild extension of the picture of embedded star with associated reflection nebula developed previously. The Herbig-Haro nebula (object) is hypothesized to be the reflection nebula produced by the embedded, hence invisible at optical wavelengths, infrared source.

This hypothesis predicts that all Herbig-Haro objects should be accompanied by infrared sources. For a number of objects, such associated infrared sources have been discovered (Strom, Grasdalen and Strom 1974). Unfortunately, in these other cases the presumed underlying placental structure cannot be determined by optical means; the Herbig-Haro nebulosities appear on an undifferentiated background. However, radio molecular-line observations by Lada et al. (1974) have served to delineate these structures quite clearly. In the cases studied, small regions of high molecular-line intensity enclose the infrared source and the Herbig-Haro nebulosity.

Beyond the small physical sizes of these structures, their youth is also demonstrated by their close association with Herbig Ae and Be stars.

The physical correctness of the reflection nebula hypothesis has been demonstrated by detailed studies of the linear polarization of two objects (Strom, Strom and Kinman 1974; Vrba, Strom and Strom 1975).

The main importance of the Herbig-Haro objects for this review is that they are the youngest objects with any optical manifestation, or the possibility of more less-direct comparison with visible objects that can be studied directly. It is apparent that a variety of as yet exotic objects lie deeply embedded in dark clouds (e.g., Wynn-Williams, Becklin and Neugebauer 1975). The interpretation of these objects will likely rest heavily on our ability to deal quantitatively with the overlying inter-stellar extinction, an ability that may be best cultivated by studies of these less-obscured, more-nearly-normal objects.

REFERENCES

Allen, D. A. 1972, Ap. J. (Letters), 172, L55.
Allen, D. A., Strom, K. M., Grasdalen, G. L., Strom, S. E., and Merrill, K. M. 1975, M.N.R.A.S., in press.
Aveni, A. F., and Hunter, J. H. 1972, A. J., 77, 17.
Grasdalen, G., Joyce, R., Knacke, R. F., Strom, S. E., and Strom, K. M. 1975, A. J., 80, 117.

Grasdalen, G. L., Strom, K. M., and Strom, S. E. 1973,
　　Ap. J. (Letters), 184, L53.
Haro, G. 1952, Ap. J., 115, 572.
Henize, K. G., and Mendoza V, E. E. 1973, Ap. J., 180, 115.
Herbig, G. H. 1951, Ap. J., 113, 697.
───────────. 1960, Ap. J. Suppl., 4, 337.
───────────. 1962, Adv. Astr. and Ap., 1, 47.
Joy, A. H. 1945, Ap. J., 102, 168.
Lada, C. J., Gottlieb, C. A., Litvak, M. M., and
　　Lilley, A. E. 1974, Ap. J., 194, 609.
Rydgren, A. E., Strom, S. E., and Strom, K. M. 1975, submit-
　　ted to Ap. J.
Salpeter, E. G. 1955, Ap. J., 121, 161.
Smith, M. 1972, Ap. J., 176, 617.
Strom, K. M., Strom, S. E., and Grasdalen, G. L. 1974,
　　Ap. J. (Letters), 187, L83.
Strom, K. M., Strom, S. E., and Kinman, T. D. 1974, Ap. J.
　　(Letters), 191, L93.
Strom, K. M., Strom, S. E., and Yost, J. 1971, Ap. J., 165,
　　479.
Strom, S. E., Grasdalen, G. L., and Strom, K. M. 1974,
　　Ap. J., 191, 111.
Strom, S. E., Strom, K. M., and Grasdalen, G. L. 1975, Ann.
　　Rev. Astr. and Ap., 13, in press.
Strom, S. E., Strom, K. M., Yost, J., Carrasco, L., and
　　Grasdalen, G. 1972, Ap. J., 173, 353.
Vrba, F. J., Strom, K. M., Strom, S. E., and
　　Grasdalen, G. L. 1975, Ap. J., 197, 77.
Vrba, F. J., Strom, S. E., and Strom, K. M. 1975, Pub.
　　A.S.P., 87, 337.
Walker, M. F. 1956, Ap. J. Suppl., 2, 365.
───────────. 1957, Ap. J., 125, 636.
───────────. 1959, ibid., 130, 57.
───────────. 1961, ibid., 133, 438.
───────────. 1969, ibid., 155, 447.
───────────. 1972, ibid., 175, 89.
Worden, S. P., and Grasdalen, G. L. 1974, Astr. and Ap., 34,
　　37.
Wynn-Williams, C. G., Becklin, E. E., and Neugebauer, G.
　　1975, Ap. J., 187, 473.

# OBSERVATIONS OF THE H AND C RECOMBINATION LINES TOWARD NGC 2024

T.L. Wilson

Max-Planck-Institut für Radioastronomie
Bonn, FRG

This report deals with radio observations of the carbon and narrow hydrogen recombination lines. The theoretical discussion of these data are contained in the paper by Walmsley, in this volume.

Maps of the carbon line region toward NGC 2024 have been made at 3 frequencies, and the results are given in Table 1. All of the refer-

A Summary of Carbon Line Map Results for NGC 2024

| Frequency | Center position of C-line source | | Full width to half power (corrected for beam) | Line width[*] | Radial velocity[*] | Reference |
|---|---|---|---|---|---|---|
| | $\alpha$ | $\delta$ | | | | |
| (GHz) | (1950.0) | | | (km s$^{-1}$) | (km s$^{-1}$) | |
| 2.7 | 05$^h$ 39$^m$ 12.4$^s$ | -01$^o$ 56.5' | 3.6' x 3.6' | 4.4 ± 1.0 | 10.0 ± 0.8 | Wilson et al. (1975) |
| 10.6 | 05 39 10.3 | -01 56.6 | 4.2' x 2.6' (NE to SW) | 5.1 ± 1.0 | 10.4 ± 0.3 | MacLeod et al. (1975) |
| 14.7 | 05 39 11 | -01 57.5 | 0.8' x 0.8' | 2.8 ± 1.0 | 11.0 ± 0.2 | Rickard et al. (1975) |

[*] averaged over map

ences are at this time in the press or in preparation. MacLeod et al.
give no errors on the carbon line cloud size and position. Wilson et al.
give maximum probable errors of ± 2' for the size, and ± 1' for the
position, and Rickard et al. report a one sigma size error of 1'. The
MacLeod et al. and Wilson et al. results are in fair agreement. Rickard
et al. have interpreted their results in terms of a second carbon line
region. However, the uncertainties in their results would permit
agreement between their carbon line cloud parameters and the data measured by MacLeod et al. and Wilson et al.

The narrow hydrogen line has a full width to half power of
$\lesssim 10$ km s$^{-1}$, whereas the hydrogen recombination line from the H II
region has a line width of $\sim 26$ km s$^{-1}$. A number of authors have
assumed that this narrow hydrogen recombination line is formed in exactly the same volume of space as the carbon line, and have used the
difference in the widths of the two lines to determine the kinetic
temperature and turbulence of the line emitting region. However,
Zuckerman and Ball (1974) have pointed out that the radial velocity of
the narrow hydrogen recombination line appears to be slightly lower than
the radial velocity of the carbon recombination line. High quality
measurements of the velocities of these lines since that paper give the
following velocity differences, i.e. radial velocity of hydrogen minus
radial velocity of carbon:

| OBSERVING FREQUENCY $\nu_o$ | VELOCITY DIFFERENCE $\Delta v(H-C)$ | REFERENCE |
|---|---|---|
| 1.4 GHz | -0.5 ± 0.5 | Pedlar & Hart, 1974 |
| 2.7 GHz | -0.7 ± 0.8 | Wilson et al., 1975 |
| 1.4 GHz | -0.5 ± 0.1 | Wilson & Thomasson (unpubl.) |

All errors are one sigma. MacLeod et al. (1975) do not give the
velocity difference between the narrow hydrogen and carbon lines, but
these authors do comment that the position where the narrow hydrogen
line and carbon line have their maxima are spatially separated, and that
the peak of the narrow hydrogen recombination line emission lies south

of the peak of the carbon line emission.

MacLeod, J.M., Doherty, L.H., Higgs, L.A., 1975, Astron. Astrophys.,
        in press.

Pedlar, A., Hart, L., 1974, Mon. Not. R. astron. Soc., 168, 577.

Rickard, L.J., Zuckerman, B., Palmer, P., Turner, B.E., 1975, in prep.

Wilson, T.L., Thomasson, P., Gardner, F.F., 1975, Astron. Astrophys.,
        in press.

Zuckerman, B., Ball, J., 1974, Astrophys. J., 190, 35.

# THE IONIZATION OF CARBON IN DUSTY H I REGIONS

D. A. Cesarsky and J. Lepine

DERAD, Observatoire de Meudon, FRANCE

I. INTRODUCTION. The investigation of the ionization of
carbon is relevant for the understanding of several
astrophysical processes that occur in H I regions, where
C II is the main source of electrons. The electron
density is a parameter of interest for the consideration
of electronic excitation of certain molecules. The
column density of C II is observed by means of UV-
absorption lines; the emission measure is needed for the
interpretation of radio recombination lines.

II. CALCULATIONS. We have considered an isothermal region
of constant density n. The density of carbon is given by
$n(C) = n A_C$ and the degree of ionization is defined as
$x = n(C\ II)/n(C)$. We consider a plane-parallel geometry, the
radiation field outside the region is characterizd by the
ionization rate $\zeta_0 (sec^{-1})$. At any point inside the region,
the ionization rate is given by the following expression:

$$\zeta = \zeta_0 \; e^{-\tau} \qquad\qquad (1)$$

where $\tau$ is the optical depth measured from the surface inwards. The ionizing radiation is absorbed by the dust and by the neutral carbon; therefore, $\tau$ is given by the following equation:

$$d\tau = \{ \; n(d) \; \sigma(d) + n(C \; I) \; \sigma(C \; I) \; \} \; ds =$$

$$= (A - x) \; B \; n \; ds \qquad\qquad (2)$$

where $A = n(d) \; \sigma(d)/n(C) \; \sigma(C \; I) + 1$, $B = 3 \; x \; 10^{18} \sigma(C \; I) \; A_C$, $n(d)$ and $\sigma(d)$ are the number density and absorption cross-section of the dust, $\sigma(C \; I)$ is the photo-ionization cross-section of C and s is the path length. Finally, the ionization equilibrium equation at each elementary volume in the cloud is given by the following expression:

$$\frac{\Gamma}{n \; A_C} = \frac{x^2}{1 - x} \; , \; \Gamma = \zeta/\alpha \qquad\qquad (3)$$

where $\alpha$ is the recombination coefficient, the only temperature dependent parameter. Combining equations 1 through 3 we obtain the variation of x with depth in the cloud:

$$dx = - \; B \; n \; \frac{(1 - x) \; (A - x) \; x}{2 - x} \; ds \qquad (4)$$

This equation was already obtained by Werner (1970); upon integration it gives the degree of ionization as a function of s and $x_0$, the degree of ionization at the surface, s = 0. Here, we are interested in the column density of carbon, $N(C \; II) = n \; A_C \int x \; ds$ and in the emission measure, EM = $n^2 \; A_C^2 \int x^2 \; ds$. Both can be obtained analytically from eq.4; they are given by

$$N(C\ II) = \frac{A_C}{B} \{ \frac{2 - A}{A - 1} \ln \frac{A - x_o}{A} - \frac{1}{A - 1} \ln (1 - x_o) \} \quad (5)$$

$$EM = \frac{n\ A_C^2}{B} \{ \frac{A(2 - A)}{A - 1} \ln \frac{A - x_o}{A} - \frac{1}{A - 1} \ln (1 - x_o) -$$
$$- x_o \} \quad (6)$$

Numerical calculations of $x(s, x_o)$, $N(C\ II)$ and EM were performed for several values of n, $A_C$ and $\Gamma$. For illustrative purposes, the calculations of $x(s, x_o)$ are shown as $L_{CII}$, the depth in the cloud at which x falls to 0.05. The parameter $L_{CII}$ is useful in assesing the validity of the plane parallel geometry: if the actual radius of the H I region exceeds $L_{CII}$ the approximation is valid and the computed $N(C\ II)$ and EM have to be doubled to allow for the ionization of the near and far sides of the region. A typical result is shown in figure 1, where we have plotted $L_{CII}$, $N(C\ II)$ and EM as a function of n, for $\Gamma = 10$ cm$^{-3}$, $A_C = 4 \times 10^{-4}$ and A = 1.25 ("normal dust", Werner, 1970). Inspecting figure 1 we first note that the plane-parallel geometry is accurate for most astronomical situations where $n > 10^3$ cm$^{-3}$, since $L_{CII} = 1$ pc. Secondly, we note that $N(C\ II)$ is a decreasing function of n, whereas EM increases with the density; this is due to the fact that $L_{CII}$ decreases faster than $n^{-1}$ but not as fast as $n^{-2}$. Thirdly we see that EM tends to a limiting value, EM = $0.03\Gamma/A$. Since A cannot be smaller than unity, the only way to obtain a high EM is increasing $\Gamma$.

III. APPLICATION TO ONE ASTRONOMICAL PROBLEM. In this brief summary, we shall concentrate only on the problem of the carbon recombination line emission in the direction of H II regions. The estimate of the EM based on the observations is strongly dependent on the line-emission-model. A lower limit, however, can be estimated to be EM = 3 pc cm$^{-6}$. Even the most favorable case to obtain large values of EM, namely no dust and no depletion of carbon, implies that $\Gamma > 50$. This value largely exceeds the available interstellar C-ionizing radiation. We therefore suspect that the association between C-recombination emission and H II regions is probabily due to the fact that in the vicinity of H II regions there is an important flux of C-ionizing radiation. The radiation either escapes from the H II region itself or it is produced by stars in the same region that are not hot enough to produce an observable H II region but that still radiate the less energetic C-ionizing photons. If the C II region surrounds the H II region, of radius R, the variation of $\zeta$ due to the spherical symmetry is given by $d\zeta/\zeta = 2\ dR/R$. Adopting R = 1 pc and $dR = L_{CII} = 0.05$ pc, we note that the variation of $\zeta$ due to the spherical geometry is of the order of 10%; consequently, the plane-parallel geometry adopted here should give a good estimate of the properties of the ionized carbon.

A similar analysis has been made of other astronomical data, as outlined in the introduction. The results will be discussed elsewhere (Cesarsky and Lepine, 1975, in preparation.)

REFERENCES

Werner,M. 1970, Ap. Letters, 6, 81.

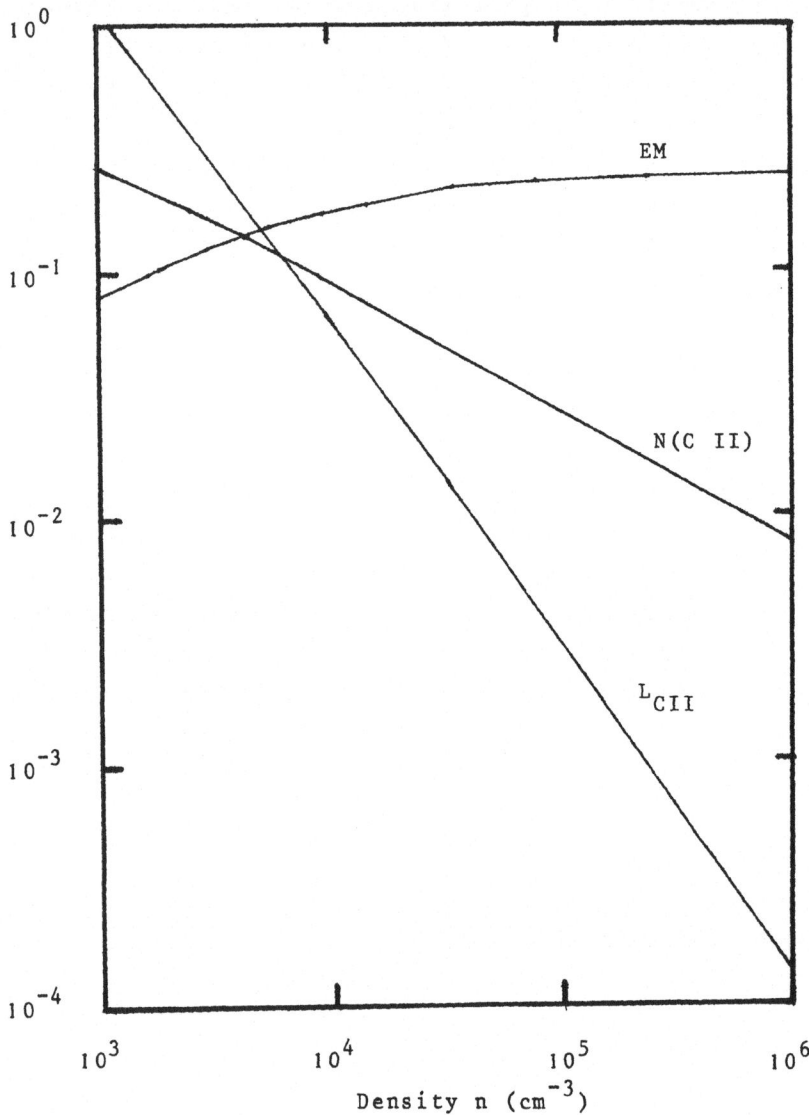

Figure 1. The variation of $L_{CII}$ (pc), N(C II) (pc cm$^{-3}$), and EM (pc cm$^{-6}$) as a function of the total density n.

# RECENT INTERSTELLAR MOLECULAR LINE WORK

G. WINNEWISSER

Max-Planck-Institut für Radioastronomie,
Bonn, Germany

ABSTRACT


A summary of recent interstellar molecular line work
will be presented. Transitions of the follwing molecules have
been detected in Sgr B2: Vinylcyanide, $H_2C_2HCN$, formic acid,
HCOOH, dimethyl ether $(CH_3)_2O$ and isotopically labelled
cyanoacetylene – $^{13}C$, $HC^{13}CCN$ and $HCC^{13}CN$. The data on cyanoacetylene give an <u>upper</u> limit to the abundance ratio $^{12}C/^{13}C$
of 36 $\pm$ 5. A short discussion of the interstellar chemistry
leads to the conclusion that hydrocarbons such as acetylene,
HCCH, ethylen, $H_2CCH_2$ and ethane $H_3CCH_3$ should be present in
interstellar clouds.

Precise frequency predictions derived from laboratory data have lead to a number of new interstellar molecular detections. The new interstellar molecules and their transitions are seen in emission only in the direction of the galactic center radio source Sgr B2. The observed position for all detections discussed in this summary was Sgr B2 (OH). (R.A. (1950): $17^h44^m11^s$ and Dec (1950): $-28^o22'30"$.) A summary of the molecules, their transitions and rest frequencies are presented in Table I.

Larger molecules whose interstellar transitions of the parent molecule are strong enough provide good possibilities for the determination of isotopic abundance ratios, since their lines are generally not affected significantly by saturation. The J = 1 - 0 interstellar rotational line of cyanoacetylene $H^{12}C^{12}C^{12}CN$ in Sgr B2 is sufficiently strong for an interstellar observation of the various cyanoacetylene - $^{13}C$ isotopes to be detected. The J = 1 - 0 transitions of $H^{12}C^{13}C^{12}CN$ and $H^{12}C^{12}C^{13}CN$ were consequently detected in Sgr B2 (Gardner and Winnewisser 1975). The peak antenna temperature $T_A$ of each line is 0.060 $\pm$ 0.005 K to within 5 - 10 per cent. The ratio of $T_A$ for the $^{12}C$ line to $T_A$ for the $^{13}C$ line is 38 $\pm$ 5. An upper limit to the abundance ratio $^{12}C/^{13}C$ of 36 $\pm$ 5 is obtained for the gas in front of the continuum source, i.e. the $^{13}C$ isotope in the Sgr B2 cloud is overabundant by a factor of about 2 1/2 relative to terrestrial conditions. Similar values have been obtained by Wannier et al., 1975, for 14 different clouds.

The detection of interstellar vinyl cyanide by Gardner and Winnewisser, 1975, marks the discovery of the first molecule containing a carbon-carbon double bond. This reactive molecule has been discovered in emission by its $2_{11} - 2_{12}$ transition with a peak antenna temperature of 0.36 K. It probably results from maser amplification of the continuum. Preliminary detection of the $4_{13} - 4_{14}$ transition (Whiteoak et al., 1975) supports this assignment.

TABLE I

Recent New Molecular Detections

| Molecule | Transition | | Rest Frequency MHz | Telescope |
|---|---|---|---|---|
| | | F' F" | | |
| $H_2C_2HCN$ vinyl cyanide | $2_{11}-2_{12}$; | 1-1 | 1371.709 | Parkes |
| | | 3-3 | 1371.794 | |
| | | 2-2 | 1371.947 | |
| | $4_{13}-4_{14}$ | 3-3 | 4572.305 | Parkes |
| | | 5-5 | 4572.347 | |
| | | 4-4 | 4572.5o9 | |
| $(CH_3)_2O$ dimethyl ether | $2_{02}-1_{11}$ | AE+EA | 9118.838 | Parkes |
| | | EE | 9119.670 | |
| | | AA | 9120.517 | |
| $H^{12}C^{13}C^{12}CN$ $H^{12}C^{12}C^{13}CN$ cyano acetylene | 1 - 0 1 - 0 | | 9059.718[+] 9060.501[+] | Parkes |
| HCOOH formic acid | $2_{11}-2_{12}$ | | 4916.312 | Effelsberg |

[+]Only relative spacing known to that accuracy.

Interstellar formic acid, HCOOH, has been detected by its $2_{11} - 2_{12}$ rotational K-doubling transition (Winnewisser and Churchwell, 1975). This emission line results probably from an inversion of the $K_a = 1$ doublets, similar to vinyl cyanide and has a peak antenna temperature of 0.04 K and a full width to half power of 24 km s$^{-1}$. The line is velocity shifted by +64$\pm$2 km s$^{-1}$. Until now the only evidence for the existence of interstellar formic acid was a possible detection of the $1_{11} - 1_{10}$ transition at 1.6 GHz by Zuckerman et al., (1971). The column density of the $2_{12}$ level is uncertain but under certain various assumptions the level column density is estimated to range between $10^{12} < N(2_{12}) < 10^{15} cm^{-2}$.

Interstellar dimethyl ether $(CH_3)_2O$ has been detected by Snyder et al., (1974) through various millimeter wave transitions in the direction of the molecular cloud in the Orion nebula. We would like to report here the detection of the $2_{11} - 2_{02}$ transition in Sgr B2 (Winnewisser and Gardner, 1975). The central EE line has an antenna temperature of 0.05 K and may show time dependent intensity variation. Preliminary determination of the central velocity from different observations leads to values between +60 and +70 km s$^{-1}$.

Chemically, cyanoacetylene and vinylcyanide are related in so far as they are derivatives of the hydrocarbon chain, acetylene, ethylene and ethane. Although these molecules can not be detected in the microwave or millimeterwave region, due to their lack of an electric dipole moment, derivatives of them can be with one or more hydrogen atoms replaced by different functional groups. Since the nitriles such as the -CN radical, hydrogen cyanide, HCN, cyanoacetylene HCCCN and methyl cyanide $CH_3CN$, were already known to exist in interstellar clouds and are known to be relatively abundant in Sgr B2, the search for vinyl cyanide seemed natural.

However, this particular class of nitrile molecules lends itself to forming the corresponding isonitriles, which are stable organic compounds containing divalent carbon. Although it has been suggested that the unidentified line U (90.665) could arise from the HNC molecule (Snyder and Buhl, 1971) whose laboratory spectrum is unknown, there is little definite knowledge whether the isonitriles are synthesized under interstellar conditions. With the definite existence of vinylcyanide in interstellar space, it becomes possible to determine whether the corresponding isonitrile, vinyl iso-cyanide, $H_2C_2HNC$ exists under interstellar conditions. Since the molecule is chemically stable and spectroscopically al-most identical to vinyl cyanide (Yamada and Winnewisser, 1975, Gerry and Winnewisser, 1973), its presence in interstellar clouds seems to depend solely on the prevailing chemistry in interstellar clouds.

Vinylcyanide provides the first interstellar evidence for the reactive vinyl radical, $H_2C = CH$, which suggests that ethylene, the simplest olefine, is present in the inter-stellar medium. Analogously, one may conclude that the existence of the ethynyl radical, -CCH, (Tucker et al., 1974) indicates that acetylene will also be a widely distributed interstellar molecule. Similar arguments favour ethane to be an inter-stellar molecule, already known by medium resolution infra-red measurements to be a constituent of the atmospheres of Jupiter and Saturn (Ridgway 1974 and Tokunaga et al., 1975).

ACKNOWLEDGEMENTS

It is my pleasure to acknowledge the fine collaboration I had with Drs. E. Churchwell and F.F. Gardner.

## REFERENCES

Gardner, F.F., and Winnewisser, G. 1975,Ap.Letters, 197,73.
Gardner, F.F., and Winnewisser, G. 1975,Ap.Letters, 195,127.
Gerry, M.C.L., and Winnewisser, G. 1973,J.Mol.Spectrosc.,
    48, 1.
Ridgway, S.T. 1974, Ap.Letters, 187, 41.
Snyder, L.E.and Buhl, D., Schwartz, P.R., Clark, F.O.,
    Johnson, D.R., Lovas, F.J.,and Giguere, P.T. 1974,
    Ap.Letters 191,79.
Tokunaga, A., Knacke, R.F.,and Owen, T. 1975, Ap.Letters 197,
    77.
Tucker, K.D., Kutner, M.L.,and Thaddeus, P. 1974, Ap.Letters
    193, 115.
Wannier, P.G., Penzias, A.A., Linke, R.A.,and Wilson, R.W.
    1975, preprint.
Whiteoak, J.B., Gardner, F.F.,and Winnewisser, G. 1975 (to
    be published).
Winnewisser, G.,and Churchwell, E. 1975, Ap.Letters, (in
    press).
Winnewisser, G.,and Gardner, F.F. 1975, (to be published).
Yamada, K.,and Winnewisser, M. 1975, Z. Naturforsch., (in
    press).
Zuckerman, B., Ball, J.A.,and Gottlieb, C.A. 1971, Ap.Letters
    163, 41.

# A REVIEW OF STAR FORMATION

S. VON HOERNER

National Radio Astronomy Observatory
Green Bank, West Virginia, USA

I.  ## Some Empirical Aspects

1. Galactic history
2. Various places of star formation
3. Single or in clusters?
4. Stellar rotation

II.  ## Raw Material and Observations

1. Overall properties
2. Equipartition of energies
3. Direct observations

III.  ## Instabilities

1. Thermal instability
2. Layer instability
3. Gravitational instability
4. Repeated fragmentation

IV.  ## Obstacles for Star Formation

1. Magnetic energy
2. Angular momentum problem
3. Prevented star formation

V.  ## Lines of Development

VI.  ## Remaining Problems

## I.   SOME EMPIRICAL ASPECTS

Star formation is still going on at present, as we see from the
existence of many young stars and some extremely young ones, at a rate
of about three solar masses per billion years and per $pc^2$ of the galactic plane in the solar neighbourhood:

$$R_* = (3 \pm 1) \times 10^{-9} \; \frac{M_\odot}{year \; pc^2} \qquad (1)$$

From the mass (and age) distribution of stars, we know that this rate
has been about constant during the last five billion years, but had been
about 10 - 30 times higher at the very beginning. About half of all
stars existing today had been formed during the first billion years
(Schmidt 1959, Salpeter 1959, von Hoerner 1960). Comparing the distribution of gas and of young stars perpendicular to the galactic plane,
Schmidt finds that the rate of star formation is in proportion to some
power of the available gas density:

$$R_* \sim \rho^n, \; with \; n \stackrel{\sim}{\sim} 2 \qquad (2)$$

During their development, the more massive stars must shed most of their
mass back to the interstellar matter, partly continuously and partly in
explosion, before they can retire as white dwarfs with not more than one
solar mass. During the development and the explosion, some part of the
stellar mass is converted from hydrogen to helium and heavier elements.
The chemical composition of the interstellar matter, and of the stars
formed from it, thus becomes a function of time. The very oldest stars
should contain no heavy elements, the youngest stars should be the
richest.

Although a general correlation of this type is observed, it is not
as strong as expected. It seems there are not enough stars with low
metal content. We call $N(Z)dZ$ the present number of stars with metal
content in $Z...Z+dZ$. For a very crude estimate, we assume that the mass
fraction q of all stars formed is "frozen-in" in small stars which do
not develop, while the fraction 1 - q goes into massive stars which
develop fast ($<< 10^{10}$ years) shedding practically all their mass back.

With Eq. (2) and n = 2, these approximations yield $-d\rho/dt = qR(t) \sim \rho^2(t)$ or, omitting all constants,

$$\rho(t) = (1 + t)^{-1} \text{ and } R_*(t) = (1 + t)^{-2} \tag{3}$$

For $Z \ll 1$ we have in general $dZ/dt = p(1 - q) R_*/\rho$, where p is the fraction (of mass shed back) which is converted into metals. With our approximations, then $dZ/dt = R_*/\rho = (1 + t)^{-1}$, or $Z(t) = \ln(1 + t)$. For stars formed at time t with metal content Z, we have in general $N(Z)dZ = qR_*(t)dt$; and with the equations just derived, neglecting constants, the presently observed distribution should decrease as

$$N(Z) = e^{-Z} \tag{4}$$

This disagrees with van den Bergh (1962), who finds N(Z) increasing with Z, as derived from the observed distribution of UV-excesses. Since the chemical history of our Galaxy is an important question, one should try to resolve this disagreement with a more careful investigation.

## 2. Various Places of Star Formation

Wherever we see much gas, we mostly see young stars, too: in all galaxies of type S, SB and Irr (but not in E and S0). And within the galaxies, we see again gas and young stars concentrated together in spiral arms and denser clouds (but not in cores and halos). In general, we may assume a development in these galaxies similar to our own case, with a high formation rate at the beginning and a more constant one at present.

As to equation (2), the Small Magellanic Cloud has been investigated by Sanduleak (1969) who compares bright stars with 21-cm observations, finding

$$n = 1.84 \pm 0.14, \text{ for SMC} \tag{5}$$

and in a similar way, van Genderen (1969) finds $n = 1.66 \pm 0.31$ from Cepheids and 21-cm observations. For the Andromeda Nebula, however, Hartwick (1971) compares H II regions with 21-cm observations and obtains

$$n = 3.05 \pm 0.12, \text{ for M31} \qquad (6)$$

A more detailed investigation of M33 by Madore, van den Bergh and Rogstad (1974) compares bright stars and H II regions with 21-cm observations. They find about the same exponent n for stars and for H II
regions, but a different value for the inner and the outer parts of the
nebulae:

$$n = \begin{cases} 0.74 \pm 0.20 \text{ for } R < 12' \\ \\ 2.48 \pm 0.08 \text{ for } 12' < R < 24' \end{cases} \text{ for M33} \qquad (7)$$

Very different results are obtained for some special cases, with more
recent outbursts of star formation. The spectral light distribution of
the peculiar double galaxy NGC 4676 indicates a rapid, brief period of
star formation in some of its components (including the long tail),
during a violent tidal interaction about $10^8$ years ago, as found by
Stockton (1974). Two compact dwarf galaxies, IZw18 and IIZw40, investigated by Searle and Sargent (1972), are metal-poor although the normal
helium content and their mass is mainly in the form of gas. The spectral light distribution shows that their present rate of star formation
is much higher than in the past, indicating either very young galaxies
or occasional and rare outbursts of star formation.

In summary, we find that $R \sim \rho^2$ may be regarded as a useful but rough
approximation for most cases. Actually, n seems to depend on the location, and sometimes there are rapid outbursts in special cases, Physically, one should expect that R depends not only on the density, but on
several other factors as well (turbulence, chemical composition, dust
abundance, magnetic fields, stellar radiation) which leaves a wide field
for further studies.

## 3. Single or in clusters?

Empirically, we see most of the very young stars in associations;
some young and medium old stars occur in open clusters; and the vast
majority of medium and old stars are the more evenly distributed field

stars, except for a small fraction of the oldest stars which make up the globular clusters. Associations are gravitationally unstable and dissolve with lifetimes of about $10^7$ years. If they had been stable originally, they must have contained 300 times more gas than the stars we see today.

Open clusters are stable but dissolve, too; either by internal exchange of energy in case of high density, or by external exchange with field stars in case of low density, both with lifetimes of $10^7$ - $10^9$ years. Question: have all the field stars been formed in open clusters and associations, or do we need an additional formation of single stars, too? For an answer, we need the formation rate of clusters, as well as their content of stars of various mass.

The problem is that we observe only the brightest stars of open clusters, and only the very bright ones of associations, all fainter ones being completely outnumbered by the general background of the field stars; for details, see Figure 20 of von Hoerner (1960). What, then, is the distribution of stellar masses, $f(M)dM$, in clusters? The simplest assumption is that $f(M)$ of associations and open clusters is the same, as observed in the latter for bright stars; and that $f(M)$ of both, for faint stars, is the same as that observed for field stars. This gives about (von Hoerner 1967)

$$f(M) = \begin{cases} M^{-0.7}, & \text{for } M \leq 0.5 \ M_\odot, \\ M^{-2.8}, & \text{for } M \geq 1.0 \ M_\odot. \end{cases} \qquad (8)$$

Under this assumption, and with the cluster formation rate as derived from their observed age distribution, von Hoerner (1960) finds that, of all stars, about

$$\begin{array}{l} 20 \text{ \% are formed in open clusters} \\ 80 \text{ \% are formed in associations} \end{array} \qquad (9)$$

An additional formation of single field stars is not necessary. But because of the uncertainly involved, it cannot be excluded, either.

The assumption of the same mass-distribution of faint stars, in clusters and in field stars, was criticized by van den Bergh and Sher (1960) and van den Bergh (1961) who observe a deficiency of faint stars in open clusters. But it may well be that many low-mass stars are dynamically "evaporated" from a cluster, before they have time to complete their construction and to show up as stars. On the other side, it could also be that f(M) actually is a function of time, depending on metal content for example, such that relatively more light stars were formed in the past than now. Unfortunately, we do not have a theory for f(M), either.

Within two young clusters, Iben and Talbot (1966) find that the smaller stars have formed first, massive stars later, and that all star formation is stopped shortly after the most massive stars occurred; which is a bit of a puzzle, since the formation of heavier stars seems easier theoretically and thus should start earlier in a contracting cloud. This question deserves a new investigation.

## 4. Stellar Rotation

It is well known that early-type stars rotate faster than late-type ones; but is rotation then a function of stellar mass or of age? Kraft (1967) found that stars of equal type (F5 - 6) rotate faster in young clusters than in old ones, and still faster as field stars, which makes it clearly a function of age, with a deceleration time scale of $4 \times 10^8$ years. Bernacca and Perinotto (1974) investigate 1000 main sequence stars in clusters and field. They confirm Kraft's result of an age effect, with

$$\text{rotation deceleration} = 1 \times 10^9 \text{ years, for} \\ \text{type A9 and later} \tag{10}$$

But their Table 7 also shows a strong mass-effect: within the same cluster, less massive stars rotate slower, even in a cluster as young as the Pleiades ($1.3 \times 10^7$ years).

For the angular momentum problem in the theory of star formation, it is of help to see that even a faint star on the main sequence still has

some means to transport its angular momentum away. But a complete
theory should also be able to explain the mass effect: why do the less
massive young stars rotate more slowly?

## II. RAW MATERIAL AND OBSERVATIONS

### 1. Overall Properties

The interstellar matter of the solar neighbourhood has in the average
about

$$\rho = 1.2 \times 10^{-24} \text{ g/cm}^3 = 0.7 \text{ atoms cm}^{-3}, \qquad (11)$$
$$T = 3000 \text{ K},$$
$$H = 3 \times 10^{-6} \text{ Gauss.}$$

It consists mainly of hydrogen, with about 30 % (mass) of helium, and
2 % of heavier elements, about half of which are condensed in dust
grains. Hot and cold clouds will be discussed later.

As to the dynamics, we see an irregular turbulence of about

$$v = 7 \text{ km/sec}$$

and two regular features: a differential galactic rotation gradient of
5 - 10 (km/sec)/kpc locally and 25 on the average, and density waves
and shock fronts (spiral theory).

### 2. Equipartition of Energies

Calculating the energy density of thermal, turbulent, magnetic and
radiative energies, we find

$$E_{th} \approx E_{turb} \approx E_{mag} \approx E_{rad} \approx 4 \times 10^{-13} \text{ erg/cm}^3 \qquad (13)$$

Theoretically, one expects an equipartition of the energy exchanges or that coupling mechanisms between the various energy reservoirs (thermal, turbulent, magnetic) are stronger or faster than the mechanisms of energy input (differential galactic rotation, stellar radiation) and/ or energy output (infrared radiation from dust, optical and radio from gas). Equation (13) thus indicates that this condition is fulfilled.

Regarding star formation, we then should always assume equipartition of energies for the original gas clouds.

## 3. Direct Observations

Which are the places where we think that star formation will start soon, and what do they look like? As to optical observations, we may be suspicious of all especially dense clouds, as the first state, and of the so-called elephant trunks (Rosette Nebula) and later on, the globules (Bok, Cordwell and Cromwell 1971) as a later state, shortly before star formation. The large globules have sizes between 0.1 and 1 pc and masses of 2 - 70 $M_\odot$, while the small ones of 0.01 - 0.1 pc have 0.1 - 1 $M_\odot$, both thus covering the full range of stellar masses. And after star formation, we see very young stars during their contraction toward the main sequence, many of them with emission lines or as T-Tauri stars.

More recently, a new important impact came from radio observations. Mezger and Höglund (1967) found a close spatial correlation between OH emission clouds and H II regions, and a closer investigation led to the discovery of the compact H II regions (Mezger et al. 1967, Ryle and Downes 1967) of $\leq$ 0.5 pc radius and $\geq 10^4$ cm$^{-3}$ density.

It was immediately assumed that these compact H II regions are ionized by extremely young massive stars hidden within, and that the formation of further stars may be triggered in or near these regions (Mezger and Robinson 1968, Hjellming 1970). Many surveys discovered more of these interesting objects, for example Churchwell et al. (1969) and Turner (1969). Several molecules were observed in these regions (see the review of Mezger (1971) who discusses the connection with the formation of protostars).

The dense H II regions are mostly rich in <u>structure</u> Each of the 0-stars of the Orion complex seems to have its own small radio component (Webster and Altenhoff 1970), superbright radio "knots" of an H II region are described by Miley et al. (1970), and the general picture is very similar to the optical appearance of young star associations with their patchy structure and many subgroups as described by Blaauw (1964). Several compact H II regions are embedded in a large massive H I cloud of about 200 000 $M_\odot$ which seems to contract (Bridle and Kesteven 1970). Knapp (1974) investigates 88 interstellar dust clouds in the 21-cm line, looking for very narrow absorption lines as indicators of small cold H I clouds, which she finds in about half of all cases.

Even our galactic center seems to be a place of rapid star formation according to Martin and Downes (1972) who find seven small bright components of $\leq 0.5$ pc diameter in an extended source of size 10 pc. Of other galaxies, most have quiet cores of old stars, and some have active cores with violent explosions. Our Galaxy then seems to be of a medium active type.

Another important impact to the field of star formation came from <u>infrared</u> observations, from about 1963 on (see the review by Becklin and Neugebauer 1967). Several objects, emitting 80 % or more of their total energy in the infrared, are found in H II regions and OH sources (Wynn-Williams et al. 1974, Becklin et al. 1974) and with Herbig-Haro objects (Strom et al. 1974a, 1974b). The connection with star formation is discussed by several of these authors and by Cohen (1973).

Infrared and radio observations can be understood if one assumes that a young star or group of stars is embedded in these objects, where the Lyman continuum ionizes hydrogen and helium (radio observations) whereas the total energy output of the stars is finally absorbed by the dust and reemitted (infrared observations), see Harper and Low (1971), Wynn-Williams and Becklin (1974), and Mezger, Smith and Churchwell (1974). Since UV and total luminosities of young stars are known, the radio and optical data even indicate which spectral type is embedded. Compact H II regions and infrared objects thus would show places of almost completed star formation, where we observe the left-overs from it, before they are blown away by the increasing stellar radiation.

The cocoon stars (Reddish 1968) seem to be an intermediate case, where matter still is falling into the central star, decelerated by the radiation pressure to form a dense shell of 10-100 AU radius before the dust is heated enough to evaporate. Observational and theoretical aspects are discussed by Davidson (1970), Strom et al. (1971) and Baglin et al. (1973); the theory of cocoons will be described by Kahn (1975).

III. INSTABILITIES

1. Thermal Instability

If interstellar matter is gradually condensed or expanded, such that the temperature regulates itself by an equilibrium between gains (UV, X-rays, cosmic rays) and losses (emission from gas and dust), one obtains stable situations for hot and cold matter, but a wide region of instability in between (Field 1965, Goldsmith et al. 1969) which leads to the two-phase model of the interstellar medium. Many more detailed investigations have been carried out; for example by Bergeron and Souffrin (1971) regarding various heating processes by Penston (1970) and Grewing and Walmsley (1974) about a depletion of heavy elements, and Biermann et al. (1972) regarding the transition times.

The general picture is sketched in Figure 1a, showing the pressure-density diagram with its unstable region (broken line) and fast transitions (dash-dotted). If this transition time is very short as compared to the duration of a full compression-expansion cycle of external distortions, we expect an observed distribution as shown in Figure 1b, with two sharp peaks and with a central region of zero probability. But if the external distortions occur more frequently, Figure 1b would become more smeared out. The actual observations indicate some smoothing but not **too much**, with well-defined maxima for the cold and hot phases:

$$\text{2-phase model:} \quad \begin{cases} \text{hot } T = 7000 \text{ K}, \ \rho = 0.2 \text{ cm}^{-3}; \\ \\ \text{cold } T = \quad 40 \text{ K}, \ \rho = \quad 30 \text{ cm}^{-3} \end{cases} \tag{14}$$

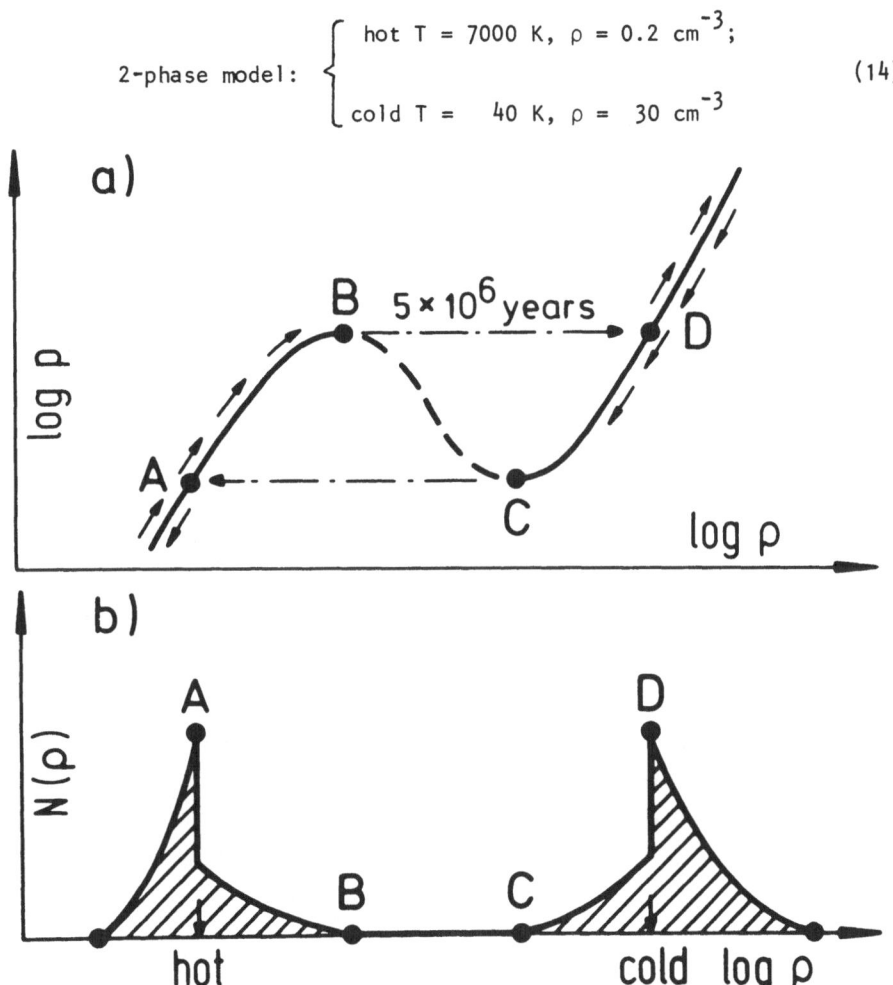

Fig. 1   The two-phase model of interstellar matter.

   a) Pressure-density diagram, with unstable region (dashed) and fast transitions (dash-dotted).

   b) The expected distribution of density.

The most important external distortions are the density waves of Lin's theory (see Roberts 1969) and their shock fronts (Shu et al. 1972, Roberts and Yuan 1970, Biermann 1974). An excellent confirmation is the Westerbork map of M51 (Mathewson et al. 1972) and its comparison with the optical picture.

## 2. Layer Instability

Near a massive young star, we have the ionized H II region of lower density, surrounded by a compressed denser H I region being pushed outward by the radiation pressure. Acceleration of a dense layer above a lighter one is an unstable situation (Spitzer 1954) where each small perturbation grows into the so-called elephant trunks as seen in the Rosette Nebula. These trunks are unstable themselves and must disintegrate into several round blobs or globules of dense H I, which then contract further under their own gravity plus the pressure of the surrounding H II (Ebert 1955, Bonnor 1956).

Actually, there are several other types of instability which may be of importance, but will be neglected here, as for example the Rayleigh-Taylor instability from magnetic fields and gravity (Parker 1966, Lerche and Parker 1967).

## 3. Gravitational Instability (Jeans)

If the potential energy of a dense cloud gets larger than its thermal energy (negative total energy), the cloud is no longer stable and starts to collapse, first slowly and then in free fall. Since, for given $\rho$ and $T$, the potential energy goes with a higher power $(R^5)$ of the radius than the thermal energy $(R^3)$, there is a limiting Jeans mass: smaller ones are stable, larger ones collapse:

$$\text{Jeans mass } M_J = \frac{30\ 000\ M_\odot}{\sqrt{\rho}\ cm^3} \left(\frac{T}{100\ K}\right)^{3/2} ; \tag{15}$$

$$\text{Jeans radius } R_J = 80\ pc\ \left(\frac{cm^{-3}}{\rho}\right)^{1/2} \left(\frac{T}{100\ K}\right)^{1/2} \tag{16}$$

During contraction, M stays constant; R decreases, but probably not a great deal because of the angular momentum problem and the repeated fragmentation, to be discussed later, maybe by a factor 5 or so. We apply Jeans criterion to the two-phase model and obtain:

| 2-phase | $\rho(cm^{-3})$ | $T(K)$ | $M_J(M_\odot)$ | $\frac{1}{5} R_J (pc)$ | (17) |
|---------|-----------------|--------|----------------|------------------------|------|
| hot  | .2 | 7000 | $4 \times 10^7$ | 300 |
| cold | 30 | 40 | $1.4 \times 10^3$ | 1.8 |

The first line resembles nothing known, thus indicating that stars do not form out of hot gas. But the second line agrees with open clusters, in both mass and radius.

Associations are not gravitationally bound (at least not in their present state) which makes the application of Jeans' criterion impossible (or difficult). Globular clusters, however, are well-defined bound objects. If we apply equations (15) and (16) backwards, adopting M = $2 \times 10^5$ $M_\odot$ and R = 30 pc and assuming again contraction by a factor 5, we obtain for the original state of the gas, at the moment of Jeans' instability:

$$\begin{array}{ll} T = 360 \text{ K} & \\ \rho = 1.0 \text{ cm}^{-3} & \end{array} \quad \text{for globular clusters} \quad (18)$$

where $\rho$ looks reasonable for regions of positive density fluctuations in the original round gas cloud of the protogalaxy, whereas the value of T then would need some extra explanation.

Regarding the formation of single stars, we assume a favourably low temperature of T = 10 K, and obtain the needed density from equation (15) as:

| sp. | $\rho(cm^{-3})$ | (19) |
|-----|-----------------|------|
| O5 | $6 \times 10^2$ |
| A0 | $8 \times 10^4$ |
| G0 | $8 \times 10^5$ |
| M8 | $2 \times 10^8$ |

This shows that single O-stars may well form at some special dense and cold places, that it gets more and more difficult for less massive stars, and that the small stars certainly need some extra mechanism of formation.

## 4. Repeated Fragmentation

Jeans' critical mass is $M \sim T^{3/2} \rho^{-1/2}$. If a large optically thin cloud then gets unstable and contracts, the temperature will stay about constant because of its external regulation by gain and loss. The density, however, increases, which makes the critical mass decrease: after the original cloud has somewhat contracted, smaller masses within it become unstable and contract, which means the cloud breaks up into a number n of fragments. Within each fragment, the process then is repeated, and so on (Hoyle 1953).

The limit of fragmentation is reached when the energy released by the free fall (potential to thermal) cannot be transported fast enough out of the opaque fragment. Estimates of Hoyle (1953) and Mestel and Spitzer (1956) gave about 0.5...1.5 $M_\odot$ for these last fragments. Adopting 1 $M_\odot$ and T = 10 000 K, one then finds a radius of a few astronomical units, in nice agreement with mass and size of our solar system:

$$
\begin{aligned}
&M = 1\ M_\odot \\
\text{last fragment = protostar:}\quad &R = 3\ \text{AU} \\
&\rho = 1.4 \times 10^{-7}\ \text{g/cm}^3
\end{aligned}
$$

A difficulty was pointed out by Layzer (1963). Fragmentation needs density fluctuations which imply velocity fluctuations; and forming fragments may hit and destroy each other before they have contracted enough. This objection should be taken seriously. It seems to indicate that one needs rather large density fluctuations to start with, for increasing the mean path and the collision time as compared to the internal free fall time of the fragments. A very rough estimate showed that an original turbulence with Mach numbers of 3 - 10 would yield enough density contrast for allowing repeated fragmentation; but this question would deserve a more careful treatment.

There is another unsolved question. Although repeated fragmentation yields the right value for the average stellar mass (and for the size of our planetary system), we still do not know how to explain the observed mass distribution $f(M)$ within clusters. Some attempts have been made, but none of them satisfactory. The importance was pointed out in connection with the formation of young clusters and old field stars.

## IV. OBSTACLES TO STAR FORMATION

### 1. Magnetic Energy

We split up each step of the repeated fragmentation into its two-half steps: contraction and then fragmentation. During contraction with frozen fields, the field strength $H \sim R^{-2}$ and the magnetic energy $E_m = (1/6)H^2 R^3 \sim R^{-1}$, while the potential energy $E_p = GM^2 R^{-1} \sim R^{-1}$, thus

$$E_m/E_p = \text{const, during contraction.} \qquad (21)$$

After fragmentation, each of the n fragments has $E_m \sim n^{-1}$, while $E_p = GM^2 R^{-1} \sim R^5 \sim n^{-5/3}$, thus

$$E_m/E_p \sim n^{2/3} \text{ during one fragmentation} \qquad (22)$$

After s fragmentation steps, we have a total of $N = n^s$ fragments, and the energy ratio has increased unfavourably as

$$E_m/E_p \sim N^{2/3} \text{ in total} \qquad (23)$$

Fragmentation thus will be <u>stopped</u> whenever $E_m < E_p$ would get violated. If we start with the equipartition of equation (13), and with virial equilibrium for the original cloud before concentration, we then have originally $(E_m/E_p)_o = 1/6$ which means fragmentation can only produce a maximum number $N_{max}$ of fragments

$$N_{max} = (E_p/E_m)_o^{3/2} = 6^{3/2} = 15, \text{ for equipartition.} \qquad (24)$$

We thus must demand, at least at some time during the whole process, a very low temperature with

$$\text{gliding fields, for open clusters} \tag{25}$$

in order to surpass equation (24). Gliding times, as a function of frag-mental mass, radius and temperature, should be calculated and compared to contraction times, for finding out whether condition (25) actually is fulfilled.

For globular clusters, the temperature was most probably much too hot for gliding fields. But the original magnetic fields may have been very small, fulfilling the condition $(E_p/E_m)_o^{3/2} \geq N$. With $N = 10^6$, $M = 3 \times 10^5$ $M_\odot$ and $R = 50$ pc, the conditions means

$$H \leq 1 \times 10^{-7} \text{ Gauss, originally,} \tag{26}$$

which may well have been the case, before equipartition had been established.

## 2. The Angular Momentum Problem

A cloud looses energy by radiation and wants to contract. But if the angular momentum stays constant, the rotational velocity increases as $v_r \sim R^{-1}$ but must stay below the Keplerian velocity $v_K = (GM/R)^{1/2}$ which increases only as $v_K \sim R^{-1/2}$. Three-dimensional contraction thus is stopped whenever $v_r = v_K$ is reached, and thereafter the cloud can only contract to a thin disk of constant radius.

If a cloud of 1 $M_\odot$ and $\rho = 30$ cm$^{-3}$ should contract to a stellar density of 1 g/cm$^3$, its radius must contract by a factor $3 \times 10^7$. But if the original cloud took part in the galactic rotation and keeps its angular momentum, it can only contract by a factor of 50 before the limit $v_r = v_K$ is reached. Likewise, if the original cloud consisted of n turbulence elements with random motions $v_t$, its net angular momentum is $J = RMv_t/\sqrt{n}$, with $n \approx 8$, say, and one can show that the limit again is reached when the radius contracted by a factor of 50, if the internal turbulence was in equipartition with the thermal energy.

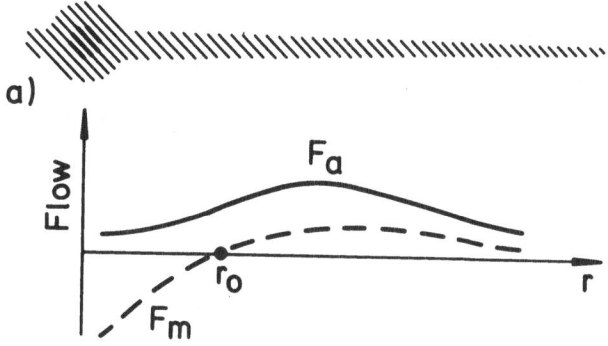

a)

b)

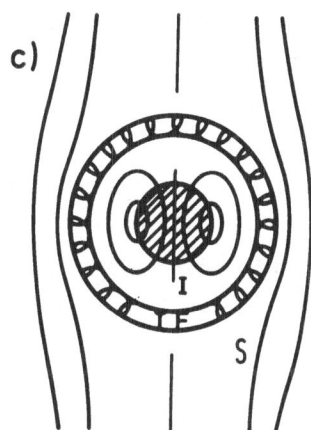

c)

Fig. 2: Transport of angular momentum.

   a) Gaseous rotating disk with turbulent friction. Matter flows
      inward for $r < r_0$ and outward for $r > r_0$ ($F_m$); but angular
      momentum ($F_a$) flows outward throughout.

   b) Connected magnetic fields get twisted and carry angular
      momentum into the surroundings.

   c) Disconnected fields of a later state. The inner part (I) ro-
      tates rigidly, the outer surrounding (S) not at all. Both
      are connected by a layer with turbulent friction (TF).

Fragmentation helps the problem since the original angular momentum of the parent cloud is split up into orbital and internal angular momentum of the fragments, where only the internal part hurts. But an estimate shows that one cannot get more than a few hundred fragments, rotating at the Keplerian limit and thus unable to contract any further to final stellar dimensions.

All this shows clearly that a considerable transport of angular momentum is needed, and three methods have been suggested. First, von Weizsäcker (1948) investigated the flat gaseous disk rotating in its own potential and found that turbulent friction causes an outward flow of angular momentum at any distance r, while matter flows inward for the inner parts and outward for the outer parts, thus creating a massive central core and a gradually thinning-out disk which carries almost all angular momentum away, see Fig. 2a. Detailed quantitative investigations confirmed this general result, under various assumptions regarding the potential, for example a large central mass (Lüst 1952), without central mass in only two dimensions (Trefftz 1952) and all three dimensions (von Hoerner 1952), and a growing central mass with an ellipsoidal disk (von Hagenow 1955). All these estimates agreed in yielding a massive central core and enough transport of angular momentum, with time scales of $10^8 - 10^9$ years for which only very rough estimates could be given.

Second, magnetic fields may yield an effective transport, especially during a first phase where the field lines of the rotating cloud are still connected with the surroundings (Ebert 1960), into which a rotation moves with Alfven speed, see Fig. 2. The time scale of this transport was estimated to be $10^7 - 10^8$ years, and various delay factors are discussed by von Hoerner (1967).

Third, in a later phase the field lines of the rotating protostar or star will be disconnected from the surroundings (Lüst and Schlüter 1955). Turbulent friction between the rigidly rotating inner part and the non-rotating surrounding will slow down the rotation, Fig. 2c, which may take $10^5 - 10^6$ years for a protostar and $10^6 - 10^7$ years for a final star like the Sun.

We thus have three means of angular momentum transport, which all sound very reasonable and should certainly work the right way. The very weak point is the crude estimate for the time scales. It would be worthwhile to repeat all these estimates, with realistic models and on modern computers.

## 3. Prevented Star Formation

If the magnetic energy is too large, which is the case if we start with equipartition and have no gliding fields, a cloud can contract unlimited according to equation (21), but cannot fragment any more according to equations (22) and (24). Angular momentum may not be a problem in this case since the strong fields will yield enough transport. The only thing left for the cloud then is to contract until either nuclear reactions start an explosion, or until it all disappears behind its Schwarzschild radius, whatever comes first. The first case holds for smaller masses, the second one for larger ones. In an intermediate case, the nuclear reactions start first but the radius is already so small that the nuclear energy cannot compete with the potential one; The cloud makes only a hiccup instead of an explosion and then disappears. The critical mass, between explosion and black hole, can be estimated with about $3 \times 10^6$ $M_\odot$ (to be published elsewhere).

This case of too much magnetic energy should occur occasionally, and the most interesting case just below of the critical mass indicates objects like globular clusters or the dense centers of galaxies. Maybe violent core explosions and strong radio galaxies could be explained this way. And since globular clusters were formed early, this would explain the larger number and/or higher radio luminosity far back in time as indicated by the number counts of radio sources.

Second, if the transport of angular momentum is not effective enough (weak or fast-gliding fields), clouds or fragments can finally only contract to thin disks which then have a long lifetime, as compared to the free fall time before, and which thus will hit and destroy each other, leading to a disruption. One may expect this case to occur frequently, maybe explaining the low efficiency of star formation. In each cycle

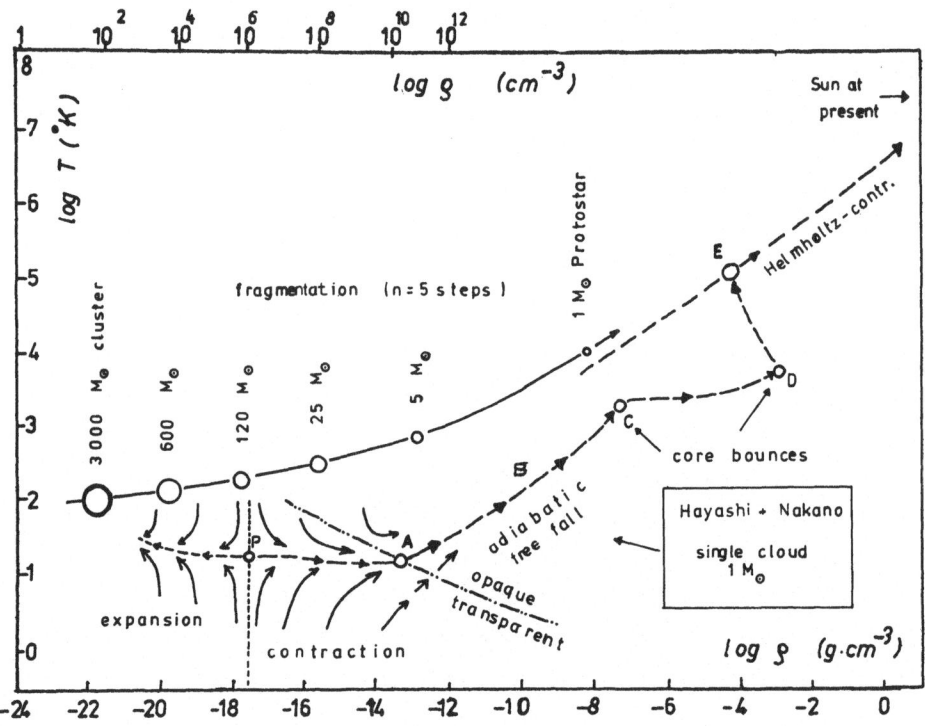

Fig. 3: Lines of development.

Upper curve: repeated fragmentation, with n = 5 steps;

Lower curve: single star, 1 M$_\odot$ (Hayashi and Nakano 1965).

of a spiral density wave, only a small fraction of the available gas is actually turned into stars.

## V. LINES OF DEVELOPMENT

We would like to describe the development, from cloud to protostar to star, in the temperature-density diagram of Fig. 3, for a star of average mass. There are two cases: repeated fragmentation leading to an open cluster, and fragmentation of single field stars.

For the first case, we may start with the cold phase of the two-phase model as given in (17), $\rho = 30$ cm$^{-3}$ and $T = 40$ K, and with its critical Jeans mass of 1400 $M_\odot$. The endpoint of fragmentation is the protostar of (20), with $\rho = 1.4 \times 10^{-7}$ g/cm$^3$ and roughly $T = 10\ 000$ K. But the way in between is uncertain. For convenience, we use a number $n = 5$ of fragments in each step. The temperature must rise, first slowly and then steeper with the increasing opacity; details have not been calculated as far as I know, and the points entered in Fig. 3 are only a guess. The further development is determined by the radiative transport of energy and virial equilibrium between thermal and potential energy, which means a slow Helmholtz contraction. The star settles on the main sequence when nuclear burning takes over.

The second case was treated in a good and detailed investigation by Hayashi and Nakano (1965), for an original cloud of 1 $M_o$ in a wide range of $\rho$ and T; their Figures 1 and 4 are combined in our Fig. 3. Regarding the original state, there is a critical point P at about $T = 20$ K and $\rho = 10^6$ cm$^{-3}$. Clouds of a different T will first undergo a fast cooling or heating towards the critical temperature. Then comes a slow contraction or expansion away from the critical density, which means that a minimum original density of $10^6$ cm$^{-3}$ is needed for future contraction. Since this is $3 \times 10^4$ times higher than the average cold-phase density, one definitely needs some special mechanism to provide it (globules, fragmentation,...?) which is not discussed by Hayashi and Nakano.

The slow contraction then goes over into a free fall with about constant temperature. At point A the cloud becomes opaque, and a while

later we have free fall with adiabatic heating, until point C. The
center then is hot enough for its own equilibrium and stops its fall.
The core bounces and produces strong shock waves at C and D, which
reorganizes the density distribution on the way to E. From then on,
we have slow Helmholtz contraction.

The critical density at point O was investigated by Miki and Nakano
(1974) for masses larger than 1 $M_\odot$ and under various heating and cooling
mechanisms. Under normal assumptions they obtain $\rho = 10^5$ cm$^{-3}$ for M =
10 $M_\odot$, $10^3$ cm$^{-3}$ for 100 $M_\odot$, and 100 cm$^{-3}$ for 100 $M_\odot$, and 100 cm$^{-3}$ for
1000 $M_\odot$, not too different from our values (19) of Jeans' criterion.

We have agreed to draw the line here, and all the more recent theoretical work on the contraction of clouds and protostars and their fine
tracks in an HR-diagram will be reviewed by Dr. Kippenhahn.

VI. <u>REMAINING PROBLEMS</u>

Finally, I would like to list a number of problems or tasks to be
considered for future work, in the order of their appearance in the
present review.

1. If all metals are made by stellar evolution, we should see a
larger number of old metal-poor stars than we do. The rough estimate of
equation (4) and the observations mentioned deserve a more careful new
investigation.

2. Equations (2), (5), (6) and (7) indicate that the rate of star
formation $R_* \sim \rho^n$, with n = 1...3 depending on location. A detailed
theory should try to explain the exponent, and to include also parameters other than just the density (chemical composition, dust, radiation, magnetic fields...).

3. Is there an independent formation of single field stars, or do
all stars form in clusters and associations? This hinges on several sub-
questions:

a) What is the mass distribution f(M) for faint stars in open
clusters and associations?

b) A theoretical derivation of f(M), including external parameters (composition, dust, radiation, fields), and leading
maybe to a time-dependence.

c) Although most O-stars appear in associations, some do not
(Aveni and Hunter 1967). What about other young stars, such as
T-Tauri stars?

4. Iben and Talbot (1966) claim that small stars form first and
massive ones later, contrary to expectations from Jeans' criterion in
contracting clouds. This needs repeating with more recent data about
stellar evolution and atmospheres.

5. Why do young stars rotate faster than small stars of the same
age (Bernacca and Perinotto 1974)?

6. If elephant trunks and globules are the normal precursors of
star formation, one would expect to see more of them than we actually
know of. Theoretical estimates of their lifetimes are required.

7. The rich, complex but irregular structure of associations and
large H II regions deserves explanation. In general: under which conditions do we get globular clusters, open clusters, associations? What
makes the latter look different? Have associations originally been more
massive and stable?

8. Why do we have gas and star formation in some cores of spiral
nebulae and not in others?

9. Where exactly in a spiral density wave do we observe dark clouds,
large and compact H II regions, infrared objects and very young stars?

10. For the two-phase model, and a more detailed distribution of density and temperature, we would like to see more systematic observational data.

11. Starting conditions for globular clusters, equation (18).

12. Can Layzer's (1963) criticism of repeated fragmentation be solved, for example by a high density contrast from originally large Mach numbers?

13. Gliding times for magnetic fields are needed, during the single steps of repeated fragmentation, to be compared with contraction times.

14. Did galaxies form with original magnetic fields? See equation (26).

15. Although we have three reasonable methods for the needed transport of angular momentum, we do not have good modern estimates for their time scales.

16. There may be a number of massive black holes in our Galaxy from prevented star formation. How to find them observationally?

17. Detailed model calculations are missing for repeated fragmentation.

18. Regarding recent model calculations of contracting clouds and protostars, one would hope that it becomes possible to include also the two obstacles of star formation: magnetic energy and transport of angular momentum.

## REFERENCES

Aveni, A.F. and Hunter, J.H.  1967, Astron. J. 72, 1019.
Baglin, A., Berruyer, N. and Morel, P.J. 1973, Astrophys. Lett. 15, 9.
Becklin, E.E. and Neugebauer, G. 1967, Charlottesville Symposium on
    H II regions.
Becklin, E.E., Frogel, J.A., Kleinmann, D.E., Neugebauer, G., Persson,

S.E. and Wynn-Williams, C.G. 1974, Ap. J. 187, 487.

Bergeron, J. and Souffrin, S. 1971, Astron. & Astrophys. 11, 40.

Bergh, S., van den and Sher, D. 1960, Pub. Dunlap Obs. 7, 203.

Bergh, S., van den 1961, Ap. J. 134, 554.

Bergh, S., van den 1962, A. J. 67, 486.

Bernacca, P.L. and Perinotto, M. 1974, Astron. & Astrophys. 33, 443.

Biermann, P., Kippenhahn, R., Tscharnuter, W., and Yorke, H. 1972,
    Astr. & Astrophys. 19, 113.

Biermann, P. 1973, Astron. & Astrophys. 22, 407.

Blaauw, A. 1964, Ann. Rev. Astr. & Astrophys. 2, 213.

Bok, B.J., Cordwell, C.S. and Cromwell, R.H. 1971, in Symposium on
    Dark Nebulae, Globules and Protostars.

Bonnor, W.B. 1956, Mon. Not. 116, 351.

Bridle, A.H. and Kesteven, M.J. 1970, Astron. J. 75, 902.

Churchwell, E., Felli, M. and Mezger, P.G. 1969, Astr. Letters 4, 33.

Cohen, M. 1973, Mon. Not. 164, 395.

Davidson, K. 1970, Astrophys. Space Sci. 6, 422.

Ebert, R, 1955, Z. Astrophys. 37, 217.

Ebert, R. 1960, in "Die Entstehung von Sternen", Springer Verlag
    (Ed. Heckmann).

Field, G.B. 1965, Ap. J. 142, 531.

Genderen, A.M., van 1969, BAN Suppl. 3, 221.

Goldsmith, D.W., Habing, H.J. and Field, G.B. 1969, Ap. J. 158, 173.

Grewing, M. and Walmsley, C.M. 1974, Astron. & Astrophys. 30, 281.

Hagenow, K.U., von 1955, Z. Naturf. 10a, 631.

Harper, D.A. and Low, F.J. 1971, Ap. J. 165, L9.

Hartwick, F.D. 1971, Ap. J. 163, 431.

Hayashi, C. and Nakano, T. 1965, Progr. Theor. Phys. 34, 754.

Hjellming, R.M. 1970, Mem. Soc. Roy. Sci. Liège 19, 105.

Hoerner, S., von 1952, Z. Astrophys. 31, 165.

Hoerner, S., von 1960, in "Die Entstehung von Sternen", Springer Verlag
    (Ed. Heckmann).

Hoerner, S.. von 1967, Charlottesville Symposium on H II regions.

Hoyle, F. 1953, Ap. J. 118, 513.

Iben, I., and Talbot, R.J. 1966, Ap. J. 144, 968.

Kahn, F. 1975, This symposium.

Knapp, G.R. 1974, Astron. J. 79, 527.

Kraft, R.P. 1967, Ap. J. 150, 551.

Layzer, D. 1963, Ap. J. 137, 351.

Lerche, I. and Parker, E.N. 1967, Ap. J. 149, 559.

Lüst, R. 1952, Z. Naturf. 7a, 87.

Lüst, R. and Schlüter, A. 1955, Z. Astrophys. 38, 190.

Madore, B.F., van den Bergh, S. and Rogstad, D.H. 1974, Ap. J. 191, 317.

Martin, A.H.M. and Downes, D. 1972, Astrophys. Lett. 11, 219.

Matheson, D.S., van der Kruit, P.C. and Brouw, W.N. 1972, Astron. &
    Astrophys. 17, 468.

Mestel, L. and Spitzer, L. 1956, Mon. Not. 116, 503.

Mezger, P.G., Altenhoff, W., Schraml, J., Burke, B.F., Reifenstein,
    E.C. and Wilson, T.L. 1967, Ap. J. 150, L157.

Mezger, P.G. and Höglund, B. 1967, Ap. J. 147, 490.

Mezger, P.G. and Robinson, B.J. 1968, Nature 220, 1107.

Mezger, P.G. 1971, in "Highlights of Astronomy", de Jager ed.

Mezger, P.G., Smith, L.F. and Churchwell, E. 1974, Astron. & Astrophys.
    32, 269.

Miki, S. and Nakano, T. 1974, preprint.

Miley, G.K., Turner, B.E., Balick, B. and Heiles, C. 1970, Ap. J. <u>160</u>, L119.

Parker, E.N. 1966, Ap. J. <u>145</u>, 811.

Penston, M.V. 1970, Ap. J. <u>162</u>, 771.

Reddish, V.C. 1968, in Interstellar Ionized Hydrogen (ed. Y. Terzian) Benjamin, N.Y.

Roberts, W.W. 1969, Ap. J. <u>158</u>, 123.

Roberts, W.W. and Juan, C. 1970, Ap. J. <u>161</u>, 877.

Ryle, M. and Downes, D. 1967, Ap. J. <u>148</u>, L17.

Salpeter, E.E. 1958, Ap. J. <u>129</u>, 608.

Sanduleak, N. 1969, A. J. <u>74</u>, 47.

Schmidt, M. 1959, Ap. J. <u>129</u>, 243.

Searle, L. and Sargent, W.L.W. 1972, Ap. J. <u>173</u>, 25.

Shu, F.H., Milione, V., Gebel, W., Yuan, C., Goldsmith, D.W. and Roberts, W.W. 1972, Ap. J. <u>173</u>, 557.

Strom, S.E., Grasdalen, G.L. and Strom., K.M. 1974b, Ap. J. <u>191</u>, 111.

Spitzer, L. 1954, Ap. J. <u>120</u>, 1.

Stockton, A. 1974, Ap. J. <u>187</u>, 219.

Strom. K.M., Strom, S.E. and Yost, J. 1971, Ap. J. <u>165</u>, 479.

Strom, K.M., Strom, S.E. and Grasdalen, G.L. 1974a, Ap. J. <u>187</u>, 83.

Trefftz, E. 1952, Z. Naturf. <u>7a</u>, 99.

Turner, B.E. 1969, Astron. J. <u>74</u>, 985.

Webster, W.J. and Altenhoff, W.J. 1970, Astrophys. Lett. <u>5</u>, 233.

Weizsäcker, C.F., von, 1948, Z. Naturf. <u>3a</u>, 524.

Wynn-Williams, C.G., Becklin, E.E. and Neugebauer, G. 1974, Ap. J. <u>187</u>, 473.

Wynn-Williams, C.G. and Becklin, E.E. 1974, PASP <u>86</u>, 5.

Recent Theoretical Work on Star Formation

R. Kippenhahn and W. Tscharnuter

Max-Planck-Institut für Physik und Astrophysik,
München, FRG

We shall review recent results which have been obtained in the
theory of star formation. Although many important contributions to
this subject have been made by Hayashi's school (1966) as well as by
other authors such as Bodenheimer (1968), a big breakthrough came in
with Larson's classic paper (1969) on the collapse of a one solar mass
cloud.

This is what he obtained: Starting with an initial density of
$10^{-19}$ grams per $cm^3$, the cloud collapses almost in free fall (Fig. (1)).
Due to the influence of pressure, however, the motion is non-homologous.
The cloud becomes more dense in the center, which makes the free fall
there even faster. Consequently, the density peak in the center builds
up. After 400 000 years a core is formed, because the dense central
material has become opaque, and the free fall is stopped by the pressure
(Fig. (2)). Matter still falls on top of this core, the surface of which
is a shock front. When dissociation of hydrogen sets in, the core becomes unstable, a second collapse in the core interior begins and even-

tually a second core will form (Fig. (3)).

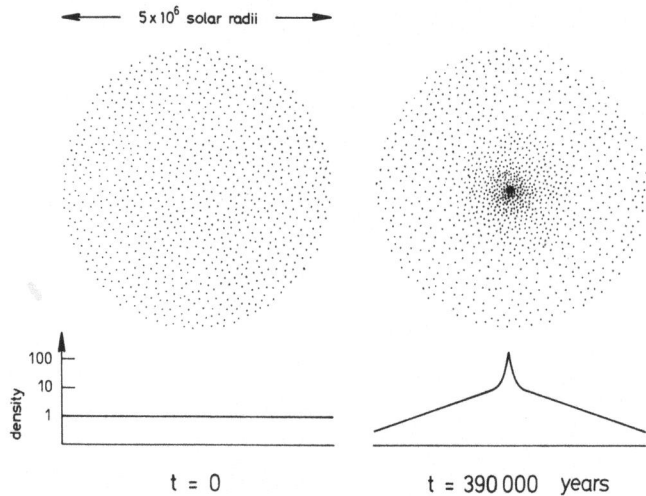

$5 \times 10^6$ solar radii

density

100
10
1

t = 0          t = 390 000  years

Fig. 1   First phases of Larson's homogeneous cloud.
         The density is measured in units of the initial
         density of $10^{-19}$ g/cm$^3$.

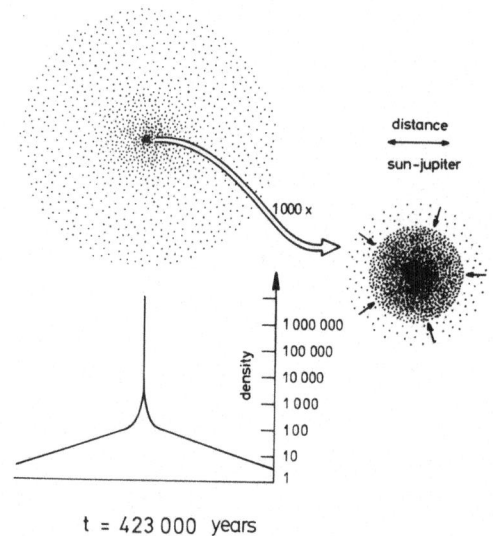

distance

sun-jupiter

1000 x

density

1 000 000
100 000
10 000
1 000
100
10
1

t = 423 000  years

Fig. 2   The formation of the first core after the central
         region became optically thick.

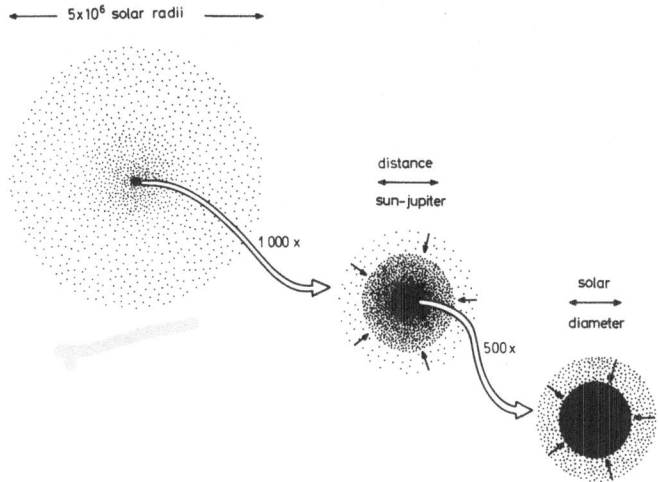

Fig. 3  Dissociation of hydrogen causes a collapse of
the core and the formation of a second core.

After all the outer material has fallen on it, this core will form the
star. We shall come to the details of this process later, but first we
shall discuss some basic physical problems of cores with infalling
envelopes.

I. Basic Problems

There are several time scales involved: The free fall time of the
initial cloud which for a long time is also the characteristic time for
the infall of the envelope.

$$t_{ff \text{ envelope}} = (\frac{3\pi}{32G\rho_o})^{\frac{1}{2}} = 2.1\text{x}10^3 \, \rho_o^{-\frac{1}{2}} \quad \text{c.g.s.} \tag{1}$$

(G gravitational constant, $\rho_o$ initial mean density of the cloud).

After a core has formed, its free fall time

$$t_{ff \text{ core}} = (\frac{3\pi}{32G\rho_{core}})^{\frac{1}{2}} = 2.1\text{x}10^3 \, \rho_{core}^{-\frac{1}{2}} = 4.4\text{x}10^3 R^{\frac{3}{2}} M^{-\frac{1}{2}} \tag{2}$$

($\rho_{core}$ mean density of the core, M, R mass and radius of the core) is much shorter than the free fall time scale of the envelope.

The characteristic time scale with which the core grows in mass is given by

$$t_{accr} = M/\dot{M} \quad , \tag{3}$$

where $\dot{M}$ is the rate with which the core receives mass from the envelope.

The core adjusts thermally within its Kelvin-Helmholtz time scale (see Appendix 1)

$$t_{KH\ core} = \frac{9G}{64\pi^2 ac} \kappa \frac{M^3}{R^5 \bar{T}^4} = 2.4 \times 10^{-6} \kappa \frac{M^3}{R^5 \bar{T}^4} \tag{4}$$

(a is the constant of radiation pressure, c velocity of light, $\kappa$ opacity, T temperature, bars indicate mean values over the core).

It might also be of interest to consider the thermal time scale of a thin shell such as the transition layer at the surface of the core, where the temperature continuously varies from the hot compressed region to the cool region in front of the shock (see Appendix 2)

$$t_{KH\ shell} = \frac{\gamma}{4(\gamma-1)} \frac{\beta}{1-\beta} \frac{\Delta\tau}{c} 1 = \frac{\gamma}{4(\gamma-1)} \frac{\beta\kappa\rho}{(1-\beta)c} 1^2 \tag{5}$$

($\beta$ ratio of gas pressure to total pressure, $\Delta\tau$ optical thickness of the shell, 1 geometrical thickness of the shell).

The free fall time scale of the core is the time in which hydrostatic equilibrium is adjusted. If $t_{accr} \gg t_{ff\ core}$ then the core is in hydrostatic equilibrium in its interior. It is then similar to a star, except that the surface condition is that of a shock front which matches the core with the infalling envelope. Within the accretion time scale the hydrostatically adjusted core changes its properties. If $t_{KH\ core} \gg t_{accr}$ then the core is compressed adiabatically by the infalling matter. When $t_{accr} \gg t_{KH\ core}$ (say, for instance, if almost no

mass is falling on top of the core), the core is thermally adjusted and
may behave more like a single star without a free falling envelope and
it might contract like a pre-main sequence star until it reaches
hydrogen burning. The slope of the evolutionary track of the core
center in a temperature-density diagram is different for the two
cases (Fig. (4)).

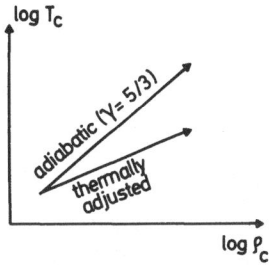

Fig. 4 The thermal evolution of a compressed or a contracting body in
the adiabatic case and in the
thermally adjusted case. In the
latter the contraction is more
homologous and the temperature
increase is less.

One can consider the core as a starlike object with special boundary
conditions and there are many analogies to normal stars. There is also
an analogy between the core in our case and cores of evolved stars.
Whereas, for instance, cores in stars sometimes have to be fitted to
radiative envelopes, here a core has to be fitted to a solution of
infalling material. When we come to numerical results we shall discuss
some properties of the cores from the point of view of classical
stellar model construction techniques.

We first formulate the virial theorem for the core. Since it has a
finite surface pressure, a surface term appears there.

The velocity of the infalling material is

$$v_e = \left(\frac{2GM}{R}\right)^{\frac{1}{2}} \qquad (6)$$

and its kinetic pressure is therefore

$$P_{kin} = \rho_e v_e^2 = \frac{\dot{M}}{4\pi R^2} v_e = \frac{\sqrt{2GM}}{4\pi} \dot{M} R^{-\frac{5}{2}} \qquad (7)$$

(the subscript e indicates the values in front of the shock). The virial
theorem is thus given by

$$3 \frac{\mathcal{R}}{\mu} \bar{T} = \alpha \frac{GM}{R} + (2GMR)^{\frac{1}{2}} \frac{\dot{M}}{M} \tag{8}$$

($\mathcal{R}$ gas constant, $\mu$ mean molecular weight). The dimensionless parameter $\alpha$ is given by

$$\alpha = \int_{o}^{M} \frac{M_r \, dM_r}{r} \cdot \frac{R}{M^2} \tag{9}$$

where $M_r$ is the mass within radius r. The numerical value of $\alpha$ depends on the density distribution of the core. It is e.g. 3/(5-n) for a polytrope of index n, it is 0.6 for the homogeneous sphere (n=0) and it is between 0.6 and 1.11 if the core is isothermal (see Appendix 3). For fixed mass and for a fixed accretion rate one can plot a mean-temperature-radius diagram for the core (Fig. (5)).

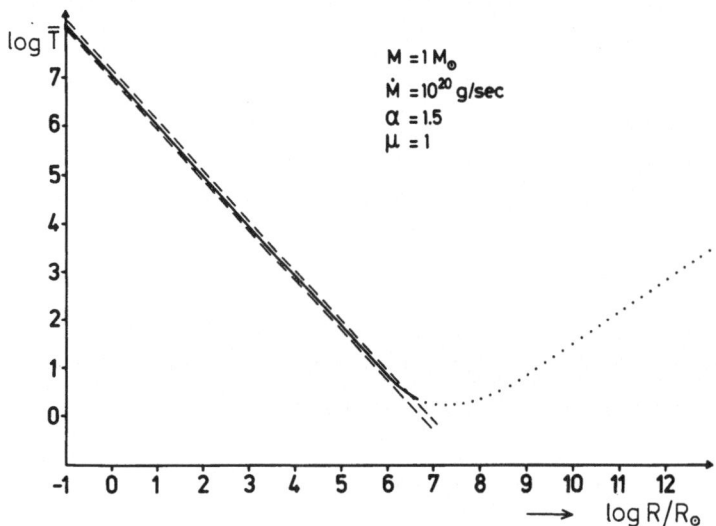

Fig. 5  The mean-temperature-radius diagram for a core of
1 $M_\odot$ imbedded into a free falling envelope with a
given accretion rate. The solid line gives the branch
where the core is dominated by gravity, the dotted
line where the surface pressure dominates the core.
The latter is physically unrealistic. The upper dashed
straight line gives a limit against non-spherical perturbations (see text and Appendix 4). The lower is a
limit for isothermal cores; strictly isothermal cores
must lie below the lower dashed line and since the solid
line does not, no isothermal cores are possible (see
text and Appendix 3).

On the left hand side the core is dominated by gravity and the finite surface pressure is only a small perturbation. On the right hand side, where the curve is dotted, gravity is negligible, but the internal pressure of the core is dominated by the infalling material. These cores are for several reasons physically unrealistic. One is that the shock conditions cannot be fulfilled there. The two broken lines indicate two limits: the upper curve is the stability limit which has the same slope as the equilibrium curve at a height which depends somewhat on the detailed temperature structure near the shock wave. Models below this curve are stable against nonradial perturbations (Appendix 4).

Although mechanically the cores behave like normal stars, since the surface pressure is relatively small, they are different thermally. Their main energy source is the kinetic energy of the infalling material which is released at the surface. Therefore, they are more similar thermally to stellar cores surrounded by shell sources. Similar to the Schoenberg-Chandrasekhar theorem in stellar structure theory, where isothermal cores are to be fitted to radiative envelopes, one can prove that isothermal cores cannot be in mechanical equilibrium with free falling envelopes (Appendix 3).

The new mechanical boundary condition makes the core a bit more stable against spherically symmetric perturbations. Whereas normal stars cannot become unstable when $\gamma = c_p/c_v < 4/3$, the upper bound is slightly diminished by the stabilizing kinetic pressure at the surface (see Appendix 5).

For a given M and a given accretion rate $\dot{M}$ in the diagram of Fig. 5 there is still a one parametric set of solutions possible with the radius of the core or the mean temperature as the free parameter. This free parameter in normal stellar models has to be fixed by a thermal boundary condition, the Stefan-Boltzman law which connects the luminosity with the effective temperature. There is no atmosphere theory up to now for this type of object ; something which certainly has to be developed. If it were to exist then we would know the temperature at an optical depth of say $\tau=2/3$. We then could fix the radius. We therefore can see that core mass and accretion rate fixes the core

mechanically and thermally for a given chemical composition. This is in
effect the Vogt-Russell theorem for this type of core . Certainly, this
is only true to an approximation since the infalling envelope is not
neccessarily transparent to the outgoing radiation and there might be
some back-scattering.

As long as the accretion time scale is short compared to the Kelvin-
Helmholtz time scale of the core, the luminosity of the core is determined by the inflow of kinetic energy during the accretion process

$$L = \frac{G\dot{M}M}{R} \tag{10}$$

since there is no time for the core to absorb energy. If the Kelvin-
Helmholtz time scale is short compared to the accretion time, the
luminosity is given by the virial theorem.

Another consequence which we can use from normal stellar evolution
theory applies to the onset of hydrogen burning. If in an adiabatic compression due to the increase of the core mass the core is pushed up to a
temperature at which hydrogen burning starts, then the temperature does
not increase further. This is due to the thermostatic property of nuclear
reactions in nondegenerate matter.

II. Spherical Collapse

After this general discussion we come to solutions of the spherically symmetric case. Larson's one solar mass computation has been repeated by Appenzeller and Tscharnuter (1975) with several improvements
but with no qualitative differences in the results. In Fig. 7 we see
the evolution of the center of the object. After an isothermal contraction the core became optically thick. After one free fall time the first
core is formed which then under the pressure of the infalling material
is adiabatically compressed. When molecular hydrogen dissociates, the
adiabatic gradient and $\gamma$, the ratio of the specific heats is lowered.
Consequently, there is a second collapse which is stopped when all the
hydrogen is dissociated. The second core being formed is again compressed adiabatically. This adiabatic compression goes on until the

infalling envelope is exhausted so that the accretion time scale exceeds the Kelvin-Helmholtz time scale of the core.

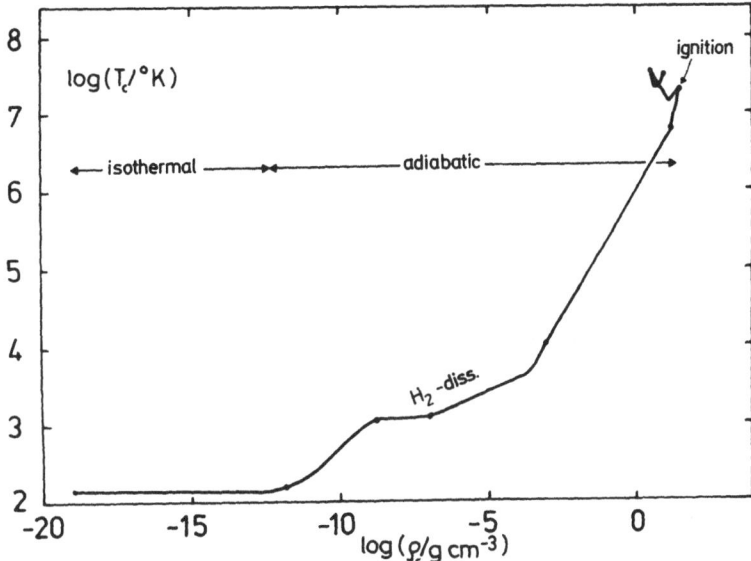

Fig. 6  The central evolution of a 60 M$_\odot$ cloud
        (Appenzeller and Tscharnuter).

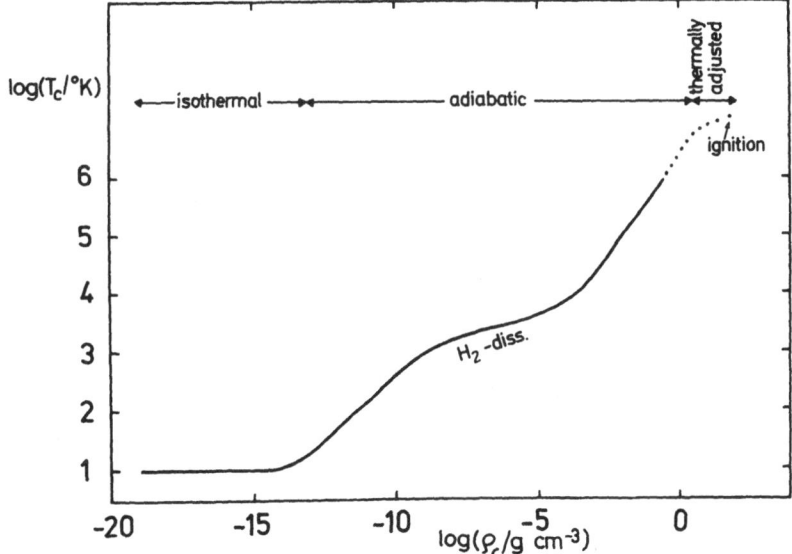

Fig. 7  The central evolution of a 1 M$_\odot$ cloud (Appenzeller and
        Tscharnuter). The dotted line is an extrapolation, indicating that after the adiabatic compression a phase of
        thermally adjusted contraction brings the center to ignition.

Then, a thermally adjusted contraction continues until the temperature of hydrogen burning is reached. The new calculations made by Appenzeller and Tscharnuter (1975) used the diffusion approximation at the surface of the core and therefore might describe the thermal behaviour of the core's surface better than Larson's approximation. But whenever shock fronts are treated with computer programs which cannot manage discontinuities, the shock front has to be smeared out by an artificial viscosity. With this, new uncertainties are introduced. One must be aware that the shock is normally broadened over a distance which by far exceeds the size of the core surrounded by the shock. Although one can show that no isothermal solution is possible, it is surprising that in the numerical solutions of Appenzeller and Tscharnuter (1975) isothermal cores are not a poor approximation, as one can see in Fig. 8.

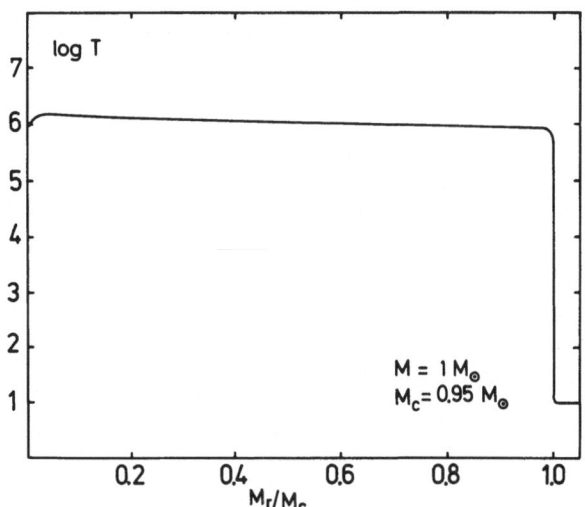

Fig. 8  Temperature as a function of mass for a cloud of 1 $M_\odot$ where 95 % is already in the core (Appenzeller and Tscharnuter).

If, however, the temperature is plotted versus the radius instead of versus the mass, this nice picture changes somewhat, as can be seen in Fig. 9.

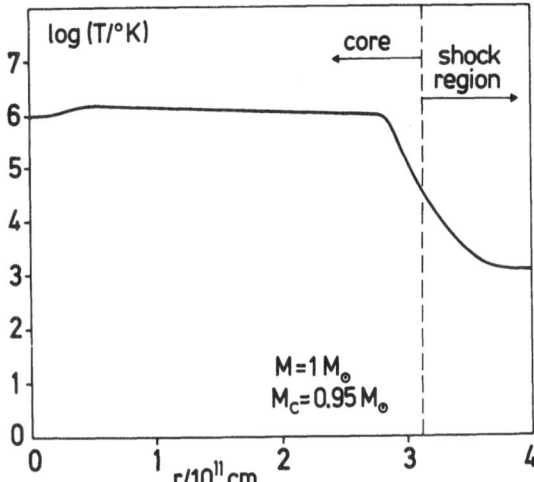

Fig. 9    Temperature as a function of radius for the same
case given in Fig. 8. The shock region is broadened
by artificial viscosity.


It is difficult to predict what the appearance of the object would
be as seen from the outside. Larson has plotted a Hertzsprung-Russell
diagram for the evolution of the core. Appenzeller and Tscharnuter (1975)
have determined an effective temperature and have plotted HR diagrams.
The result is seen in the Fig. 10 and Fig. 11, starting from the moment
the second core has formed. Since at the beginning the optical depth 2/3
is far outside in the opaque envelope, only infrared radiation comes
from there. In forming the second core, which itself is still optically
thin (because the opacity is very low after the dust has melted), a
brilliant radiative heat wave travels outwards and increases the infrared luminosity of the envelope. When the core becomes optically thick,
its luminosity is reduced since only the surface can radiate outwards
and the infrared luminosity drops again. The more mass collected by the
core, the higher the potential energy of the mass falling onto it, and
therefore the kinetic energy which the infalling matter delivers to the
surface of the core increases with time. This energy will be radiated
away and the models show an increasing luminosity during most of the
accretion phase. But finally the envelope is exhausted and the luminosity
drops. At the same time, the optical depth 2/3 approaches the accretion
shock, because the envelope is becoming more transparent.

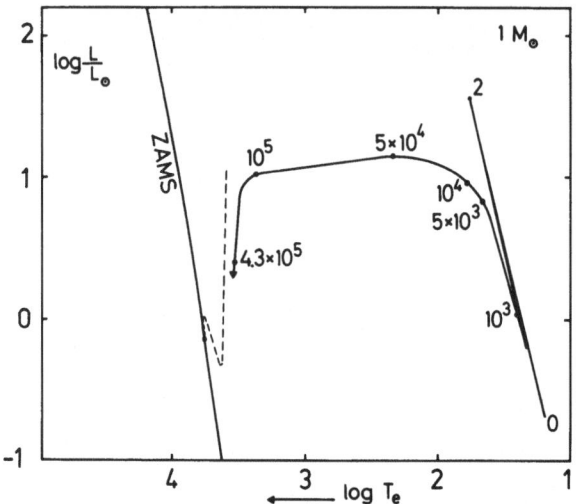

Fig. 10 The evolution of a 1 M$_\odot$ cloud in the infrared HR-diagram. The numbers along the curve give the age in years after the formation of the first core. The broken line gives the hydrostatic pre-main sequence evolution of a 1 M$_\odot$ star (K. v. Sengbusch, 1968).

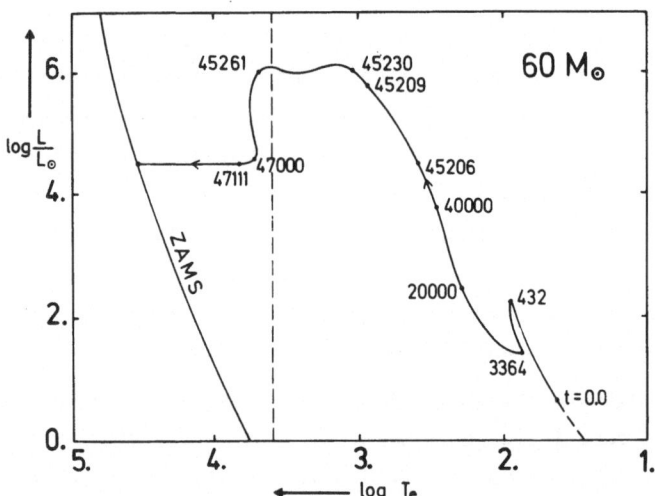

Fig. 11 The evolution of a 60 M$_\odot$ cloud in the infrared HR-diagram as in Fig. 10. The vertical line indicates the Hayashi line.

What is the behaviour of different masses? Appenzeller and
Tscharnuter (1974) have computed a 60 solar mass object. Its behaviour
in the center is given in Fig. 6. Again, we have an isothermal contraction, the formation of the first core, dissociation and hence contraction, leading to the formation of a second core being adiabatically compressed. The core is thereby heated to a central temperature at which
hydrogen starts to burn. It is well known that nuclear burning in a nondegenerate star acts in a manner to prevent a further increase in the
temperature. The computations show that, at the same time, matter at the
surface of the core is accelerated outwards and the motion of the envelope is reversed. In the case computed by Appenzeller and Tscharnuter
(1974) the 60 $M_\odot$ object leaves a core of only 17 solar masses. We must
confess that we do not completely understand what causes this inversion
 of the motion.

But in principle we suspect the following. If the core is adiabatically compressed up to the temperature of hydrogen burning, and if compression continues, then strong non-homologous motions in the region of
nuclear burning must take place in order to avoid too high an energy
production. These motions might still be small compared to the velocity
of sound there, and therefore hydrostatic equilibrium might still be a
good approximation for the interior.Near the surface, however, inertia
terms may become important and matter could be accelerated outwards. The
situation is a bit similar to the helium flash in degenerate stellar
cores, where the danger of an overproduction of energy also exists, since
a degenerate core is secularly unstable. In the case of the helium flash
the interior can adjust hydrostatically but at the surface inertia
terms can become important.

Due to the continuing adiabatic compression, the danger of an overproduction of energy is also inherent. Therefore expanding motions in
the region of nuclear burning are necessary in order to cool the central
region. This may be responsible for the outward motion.

Whether this explanation is true or not has to be checked by experiments with a program which includes inertia terms. If this explanation were true, it would mean that a star gaining mass at the surface

with an accretion time $\tau_{accr}$ can keep this mass only if $\tau_{accr} > \tau_{KH}$.
If, on the other hand, $\tau_{accr} \ll \tau_{KH}$, then matter will be blown off from
the surface in order to avoid overheating and the rate of infalling material is decreased until the accretion time scale and the Kelvin-Helmholtz time scale are of the same order of magnitude.

### III. Collapse of Rotating Clouds

Rotation certainly plays an important role in star formation, as is
indicated by the existence of binary stars and by the planetary system.
If during the collapse of a cloud the angular momentum is conserved, the
centrifugal force in the equatorial plane grows faster than gravity and
the collapse will be slowed down, possibly stopping the motion there or
even reversing it. One could try to estimate qualitatively how the formation of a shock front which surrounds a core is influenced by rotation.
If the rotation is very slow, then, as in the nonrotating case, a core
is formed with a density of about $10^{-13} g\ cm^{-3}$. If the rotation is faster
at the beginning, then the motion in the equatorial plane will be slowed
down or stopped before the core becomes optically thick. At the same
time matter is falling freely along the axis of rotation. When the density of the central region attains the value $10^{-13} g\ cm^{-3}$, a core is
formed and the infalling material forms a shock front. In this case,
however, the shock front forms two polar caps, whereas near the equator
of the core there is no such front, since there is no infalling material.
A rough overview is given in Fig. 12. A cloud starting at an initial
density $\rho_i = 10^{-19} g\ cm^{-3}$ with a certain initial angular velocity $\omega_i$
will move upward in the diagram during its collapse and will either
first encounter the line $\rho = 10^{-13} g\ cm^{-3}$ or the line $\chi = 1$ where $\chi$ gives
the ratio of centrifugal to gravitational force at the equator. In the
first case the core will be surrounded by a shock front; in the second
case it will be sandwiched between two shock fronts in the polar regions.

During the collapse the infalling rotating matter might become turbulent. Indeed, the Reynolds number can become rather large (see Appendix 7). Then due to the high turbulent viscosity angular momentum might
be exchanged in the collapsing cloud. One can, however, show that the
time scale $\tau_t$ during which angular momentum is redistributed by tur-

bulent viscosity is long compared to $\tau_{ff}$ (see Appendix 7).

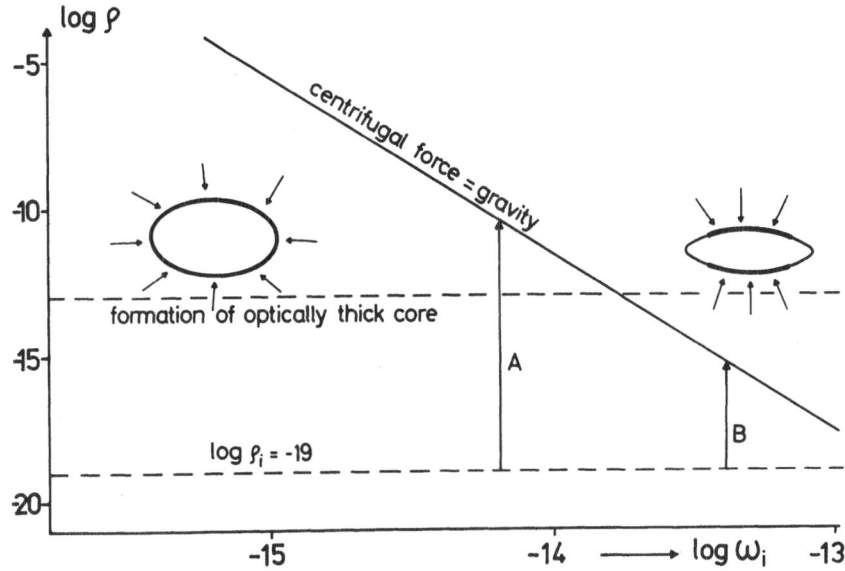

Fig. 12   Schematic picture of the evolution of a rotating cloud
starting with an initial density $10^{-19}g$ $cm^{-3}$ and an
initial angular velocity $\omega_i$ . A collapsing cloud
moves vertically upwards to higher densities. Qualitatively, two cases are possible. In the case A  the
core becomes optically thick before centrifugal force
becomes important. In case B  centrifugal force becomes
important before the central core is formed. In the first
case the core qualitatively behaves like in the non-rotating case. In particular the shock front surrounds the
central body. In the second case, however, a shock is
formed only in the polar region of the flattened central
object.

Only when $\chi$ becomes of the order one during the collapse and when the
matter is hindered to fall further inwards towards the rotating nebular
disk will turbulent friction become important. Since collapse calculations including rotation in the simplest case already require nonsta-
tionary two-dimensional hydrodynamics - axial symmetry instead of
spherical symmetry - the demands on computer time and storage as well
as on numerical techniques are very strong. For this reason, only a
few general results with relatively modest intrinsic accuracy are
available at present. Unfortunately, in contrast to the spherical
models, they are controversial in their basic contents.

The first to attack this very difficult problem was Larson (1972). He started with his usual protostellar cloud of one solar mass which is assumed to rotate uniformly (as a rigid body) with a given constant angular velocity. The unexpected outcome of his calculations was the growing of a ring-like instability. It was interpreted to be very important with respect to the fragmentation problem, because selfgravitating rings would be unstable against non-axisymmetric perturbations. Such a ring would tend to break up, thereby transforming spin angular momentum into orbital angular momentum, a mechanism which would widely resolve the so-called "angular momentum problem". At the same time it could explain the fact that so many young stars are members of binary or even multiple systems.

To reexamine these results Tscharnuter (1975) applied a new numerical method developed by Fricke, Möllenhoff & Tscharnuter (1975) to rotating protostellar clouds of Larson's type. The new calculations clearly show that one has to be very careful in interpretating numerical models obtained by finite difference methods using grids which are too çoarse. On the other hand, because of the vast amount of computer time necessary, the number of gridpoints must be kept as low as possible. Having taken into account all these difficulties the following general outline of rotating protostellar clouds can be drawn:

Starting with a small amount of angular momentum the collapse proceeds almost spherically. Because it also proceeds in a very non-homologous manner, centrifugal forces will become important only in the innermost regions of the cloud. So the general outcome is a rapidly spinning optically thick core. If the temperatures rise above 2000 K, the dust melts and molecular hydrogen dissociates. The collapse of the core thus proceeds further, but now the core containing a few percent of total mass has a relatively large amount of angular momentum. Consequently, the collapse becomes highly non-spherical. Due to local conservation of angular momentum the matter of high specific angular momentum is hindered to fall into the center. The accretion of matter at the center is performed solely by matter of low specific angular momentum, i.e. by matter coming from the neighborhood of the axis of rotation.

Starting with a large amount of angular momentum, however, the collapse is highly non-spherical from the very beginning. Thus the accretion at the center is taken over mainly by matter of low specific angular momentum. After some time this mass flow is stopped either because there is no more mass of low angular momentum or pressure effects become important. From this point on turbulent friction may influence the further evolution of the core considerably. In either case, starting with a small or a large amount of total angular momentum, a strong central condensation forms, whereas the envelope continues to fall and to flatten. (Fig. 13). Although ring instabilities have never been observed in Tscharnuter's calculations, it cannot, of course, be stated that they would actually never occur. But it seems to be quite clear that their significance is not as great as was originally expected.

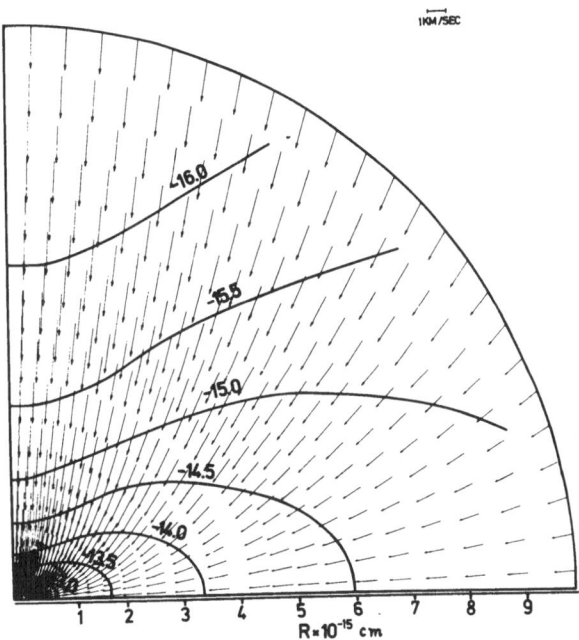

Fig. 13  The collapse of a rotating 60 M$_\odot$ cloud. The solid lines indicate density contours, the numbers give the logarithms of the density, the velocity field is given by arrows.

## Appendix 1   The Kelvin-Helmholtz Time Scale of a Core

The gravitational energy of the core is

$$\left| E_G \right| = \alpha \, \frac{GM^2}{R} \; .$$

(A1 1)

Let L denote the luminosity with which the core being left alone would cool off. From the equation of radiative transport we can estimate

$$L = \frac{64 \pi^2 ac}{9\bar{\kappa}} \, \frac{\bar{T}^4 R^4}{M}$$

(A1 2)

and

$$t_{KH \; core} = -\frac{\left| E_G \right|}{L} = \frac{9G}{64 \pi^2 ac} \, \bar{\kappa} \, \frac{M^3}{R^5 \bar{T}^4}$$

(A1 3)

where we have replaced $\alpha$ by 1.

## Appendix 2   The Kelvin-Helmholtz Time Scale of a Shell of
## Geometrical Thickness 1.

From the equation of radiative transport

$$F = -\frac{4ac}{3\kappa\rho} \, T^3 \, \frac{dT}{dr}$$

(A2 1)

and from

$$\frac{1}{r^2} \, \frac{d}{dr} \, (r^2 F) \approx \rho c_p \, \frac{\partial T}{\partial t} \approx \rho c_p \, \frac{T}{t_{KH \; shell}}$$

(A2 2)

we conclude that

$$t_{KH \; shell} = \frac{c_p \rho T l}{F} = \frac{3 c_p \rho^2 T l^2 \kappa}{4 acT^4} \; .$$

(A2 3)

With

$$\frac{aT^4}{3} = \frac{1-\beta}{\beta} \frac{\mathcal{R}}{\mu} \rho T \qquad \frac{c_p \mu}{\mathcal{R}} = \frac{\gamma}{\gamma-1} \tag{A2 4}$$

we obtain

$$t_{KH\ shell} = \frac{\gamma}{4(\gamma-1)} \cdot \frac{\beta}{1-\beta} \frac{\kappa\rho l^2}{c} = \frac{\gamma}{4(\gamma-1)} \frac{\beta}{1-\beta} \frac{l\Delta\tau}{c} \tag{A2 5}$$

where $\Delta\tau = \kappa\rho l$ is the optical thickness of the shell.


Appendix 3   Theorem:

It is not possible to fit a gravity dominated, isothermal core into a free falling envelope.

By "gravity dominated" we mean cores for which the surface pressure term in the virial theorem is negligible compared to the gravity term. We first determine the dimensionless quantity $\alpha$ (see Eq. (9)) for an isothermal gaseous sphere as a function of the distance from the center. The isothermal gaseous sphere is given by the solution of the Emden equation

$$\frac{1}{z^2} \frac{d}{dz} (z^2 u) = e^{-u} \tag{A3 1}$$

with $n = 0$ at $z = 0$. The radius from the center is proportional to $z$ while the density and pressure are given by

$$\rho(z) = \rho(0) e^{-u(z)} \quad , \quad P = \frac{\mathcal{R}}{\mu} T\rho(0)^{-u(z)}$$

and

$$\alpha(z) = \frac{1}{z^3 u'^2} \int_0^z z^3 u' e^{-u} dz$$

where primes indicate derivations with respect to $z$. In addition, the homology variable

$$V = \frac{\rho}{P}\, \frac{GM_r}{r} \tag{A3 3}$$

can be expressed by

$$V = -zu' \qquad . \tag{A3 4}$$

The numerical solution of Eq. (A3 1) gives $\alpha$ and $V$ as functions of $z$ as shown in Fig. 14.

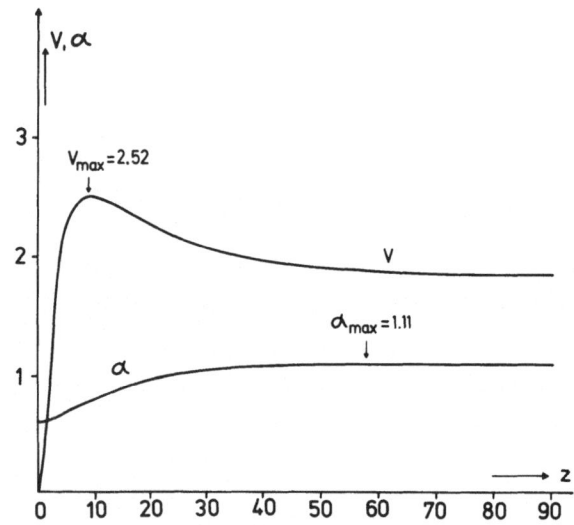

Fig. 14   The functions $V$ and $\alpha$ as defined in Eq. (A3 2) and (A3 3) as obtained from integration of the isothermal Emden equation.

One can see that

$$0 \le V \le 2.51 \quad , \quad 0.6 \le \alpha \le 1.11$$

With these properties of the isothermal gaseous sphere we can prove the theorem:

An isothermal core is a part of the (infinite) isothermal sphere. Therefore at the surface

$$V = \frac{\rho_i}{P_i} \frac{GM}{R} = \frac{\mu_i}{\Re T_i} \frac{GM}{R} < 2.51 \tag{A3 5}$$

where the subscript i indicates the values on the inner side of the boundary and therefore

$$\overline{T} = T_i > \frac{\mu_i GM}{2.51 \Re R} \quad . \tag{A3 6}$$

On the other hand we find from the virial theorem

$$\overline{T} = \frac{\alpha GM \mu_i}{3 R} + \frac{\sqrt{2GMR\mu}}{3 \Re} \frac{\dot{M}}{M} \quad . \tag{A3 7}$$

On the left hand side of Fig. 5, where the second term on the right hand sides in Eqs. (8) and (A3 7) can be neglected, Eq. (A3 6) and (A3 7) are compatible only if $\alpha > 3/2.51 = 1.195$. This is not possible for an isothermal core since $\alpha$ is always smaller there than 1.11.

# Appendix 4   Stability against nonspherical perturbations

At the surface of the core the kinetic pressure of the infalling material and the gas pressure of the core are in equilibrium

$$P_{kin} = \rho_e v_e^2 = P_i \quad . \tag{A4 1}$$

Here, the gas pressure of the infalling material as well as the kinetic pressure below the shock front have been neglected.

We now consider a nonspherical perturbation as indicated in Fig. 15. The surface of the core may be depressed by a small distance dr in a certain region. The falling material arrives with a higher velocity at the bottom of the depression. On the other hand, since the region below the shock is hydrostatic to a high degree of a accuracy, the gas pressure is also higher there. Stability demands that the gas pressure in the depression is more enhanced than the kinetic pressure of the infalling material, so that the shock front will be brought back into the original configuration.

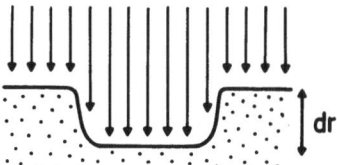

Fig. 15  Nonspherical perturbation of the surface of the core.

The increase of the kinetic pressure is

$$dP_{kin} = d(\rho_e v_e^2) = \rho_e v_e^2 \left(\frac{d\rho_e}{\rho_e} + 2 \frac{dv_e}{v_e}\right) \qquad . \tag{A4 2}$$

From the continuity equation

$$\frac{d\rho_e}{\rho_e} + \frac{dv_e}{v_e} + 2 \frac{dr}{R} = 0 \tag{A4 3}$$

and from

$$\frac{dv_e}{v_e} = -\frac{g}{v_e^2} dr \tag{A4 4}$$

we obtain

$$\frac{dP_{kin}}{P_{kin}} = \frac{dv_e}{v_e} - 2 \frac{dr}{R} = -\left(\frac{g}{v_e^2} R + 2\right) \frac{dr}{R} \qquad , \tag{A4 5}$$

whereas on the other hand

$$\frac{dP_i}{P_i} = -\frac{g\rho_i dr}{P_i} \qquad . \tag{A4 6}$$

In equilibrium we have $P_i = P_{kin}$ and since stability demands $|dP_{kin}| < |dP_i|$ we find

$$g \frac{\rho_i}{P_i} > \frac{g}{v_e^2} + \frac{2}{R} \tag{A4 7}$$

or

$$\frac{\rho_i}{P_i} > \frac{1}{v_e^2} + \frac{\rho_i}{P_i} \frac{2H_P}{R} \qquad (A4\ 8)$$

where we have introduced the pressure scale height

$$H_P = \frac{\mathcal{R} T_i}{g\mu_i} = \frac{P_i}{\rho_i g} \qquad . \qquad (A4\ 9)$$

The core envelope solution is therefore stable against nonspherical perturbations as long as the isothermal velocity of sound $v_{is} = (P_i/\rho_i)^{\frac{1}{2}}$
below the shock fulfills the condition

$$v_{is} < v_e (1-2HP/R)^{-\frac{1}{2}} \qquad . \qquad (A4\ 10)$$

Since normally $H_P \ll R$, the condition for stability is $v_{is} < v_e$.

If the temperature $T_i$ near the surface is given by $T_i = \xi\bar{T}$, and
if $v_e$ is expressed by Eq. (6), we obtain as stability condition

$$\frac{\mathcal{R}}{\mu_i} \bar{T} < \frac{2GM}{\xi R} \qquad (A4\ 11)$$

which, as long as $\xi$ is assumed to be roughly a constant, is a $45°$-line
in Fig. 5. Models below this line are stable.

Our discussion of stability deals with a simplified case only. If
a dip of the type as sketched in Fig. 15 occurs there is also no pressure balance in horizontal direction. Since the kinetic pressure acts
mainly in radial direction the dip is filled from the neighbouring regions by horizontal motions with velocities of the order of $v_{is}$. This
might also stabilize in the case of $v_{is} > v_e$. But for $v_{is} < v_e$ this
stabilizing effect might be too slow. But in this case - as we have
seen - the dip will be removed by vertical motions.

## Appendix 5  Stability against Spherically Symmetric Perturbations

For sake of simplicity we shall assume in the following the ratio
of the specific heats $\gamma = c_p/c_v$ to be constant throughout the core. For
normal stars the condition of dynamical stability is $\gamma > 4/3$. For cores
in equilibrium with a free falling envelope we have to rediscuss this
condition. We consider the equilibrium at the surface, again neglecting
the kinetic pressure inside and the gas pressure outside the core. Then
equilibrium demands

$$P_i = P_{kin} = \rho_e v_e^2 \qquad\qquad (A5\ 1)$$

If we rapidly change the radius R of the core by a small amount dr << R
then the variation of the internal density can roughly be discribed by
$\rho \sim R^{-3}$ and the internal pressure $P_i \sim \rho_i^{\gamma} \sim R^{-3\gamma}$ and we find

$$\frac{dP_i}{P_i} = -3\gamma\frac{dR}{R} \ . \qquad\qquad (A5\ 2)$$

For the kinetic pressure we obtain from Eq. (A4 5)

$$\frac{dP_{kin}}{P_{kin}} = \frac{5}{2}\frac{dR}{R} \ , \qquad\qquad (A5\ 3)$$

where the accretion rate $\dot{M}$ as determined by the envelope is assumed to
be constant. There is stability if $\left|dP_i\right| > \left|dP_{kin}\right|$ or if

$$\gamma > \frac{5}{6} \ . \qquad\qquad (A5\ 4)$$

We must keep in mind that a similar consideration must be made for
each layer in the core. Whether the criterion $\gamma > 4/3 = 1.333$ for
normal stars holds for the cores we are discussing here or the
condition $\gamma > 5/6 = 0.8333$ or a condition somewhere in between depends on the importance of the surface pressure for the mechanical
structure of the star. In the interesting region of the equilibrium
curve of Fig. 5 the surface pressure term is relatively unimportant for
the mechanical equilibrium of the star. We therefore conclude that
$\gamma > 4/3-\varepsilon$, $(0 < \varepsilon << 1)$ is the right criterion,

i.e. the stability condition is almost the same as for normal stars, but, due to the stabilizing effect of the new surface condition, $\gamma$ can go a little bit below 4/3 and the core may still be dynamically stable.

Appendix 6:    Derivation of the Critical Line in Fig.12

We assume a homogeneous spherical cloud of radius R and density and uniform angular velocity $\omega$. When the cloud contracts spherical symmetrically, then the conservation of mass and angular momentum demands $\rho \sim R^{-3}$ and $\omega \sim R^{-2} \sim \rho^{\frac{2}{3}}$. If the subscript i indicates initial values, we then have

$$\frac{\omega}{\omega_i} = (\frac{\rho}{\rho_i})^{\frac{2}{3}} . \tag{A6 1}$$

During the contraction both gravity and centrifugal acceleration at the equator increase, but the latter grows faster. We now want to estimate at which stage of contraction the centrifugal force at the equator becomes equal to the gravitational force. For a very rough estimate we still use the formulae for the spherically symmetric case ignoring the fact that the cloud becomes oblate before that stage is reached. With this approximation the gravitational and the centrifugal acceleration at the equator are given by

$$g_G = \frac{GM}{R^2} , \quad g_c = \omega^2 R \tag{A6 2}$$

and their ratio $\chi$ is unity

$$\chi \equiv \frac{g_c}{g_G} = \frac{3\omega^2}{4\pi G\rho} = 1 \tag{A6 3}$$

or

$$\omega^2 = \frac{4\pi G}{3} \rho . \tag{A6 4}$$

We now eliminate $\omega^2$ with the help of Eq. (A6 4) and obtain

$$\rho(\chi=1) = \left(\frac{4\pi G}{3}\right)^3 \rho_i^4 \omega_i^{-6} = 2.18 \times 10^{-20} \rho_i^4 \omega_i^{-6} \qquad \text{(A6 5)}$$

and with $\rho_i = 10^{-19}$ we find

$$\log \rho(\chi=1) = -6 \log \omega_i - 95.66 \qquad . \qquad \text{(A6 6)}$$

This is the critical line, plotted in Fig. 12.


## Appendix 7   Turbulent Friction in a Collapsing Rotating Cloud

If L is the mixing length and l a length characteristic for the angular velocity $\omega$ of a cloud of radius R, we can estimate

$$l \sim \omega \left| d\omega/dr \right|^{-1} \qquad . \qquad \text{(A7 1)}$$

Then the Reynolds number Re and the turbulent (kinematic) velocity $v_t$ and the turbulent viscosity $\nu_t$ are given by

$$Re = \frac{vl}{\nu_m} , \quad v_t = v\frac{L}{l} \quad \frac{\omega RL}{l} , \quad \nu_t = v_t L = \frac{\omega RL^2}{l} \qquad \text{(A7 2)}$$

($\nu_m$ molecular viscosity).

If we express the ratio $\chi$ of centrifugal to gravitational acceleration in terms of the free fall time $\tau_{ff} = (3\pi/32G\rho)^{\frac{1}{2}}$ and in terms of the rotational period $\tau_R = 2\pi/\omega$, we obtain

$$\chi = 32 \frac{\tau_{ff}^2}{\tau_R^2} \quad \text{or} \quad \tau_{ff} = \frac{\pi\sqrt{\chi}}{2\sqrt{2}\omega} \qquad . \qquad \text{(A7 3)}$$

The time scale during which turbulent friction influences the angular velocity distribution is therefore

$$\tau_t = \frac{l^2}{v_t L} = \frac{l^3}{\omega RL^2} = 2\sqrt{2} \frac{\tau_{ff}}{\pi\sqrt{\chi}} \frac{l^3}{RL^2} \qquad \text{(A7 4)}$$

and

$$R_e = \frac{vl}{\nu_m} = \frac{R\omega l}{\nu_m} = \frac{l^2}{L^2} \frac{\nu_t}{\nu_m} \qquad . \tag{A7 5}$$

In our picture the mixing length is still a free parameter. It seems to be reasonable (Lynden-Bell and Pringle (1974)) to take as a rough estimate

$$\nu_t \approx 10^{-3} \omega R^2 \tag{A7 6}$$

and if we compare this with our formula (A7 2) we obtain

$$L^2 = R1 \times 10^{-3} \qquad . \tag{A7 7}$$

Thus, we find for the Reynolds number and for the time scale of turbulent friction

$$Re = 10^3 \frac{1}{R} \frac{\nu_t}{\nu_m} \quad , \quad \frac{\tau_t}{\tau_{ff}} = \frac{2\sqrt{2}}{\pi\sqrt{\chi}} \frac{l^2}{R^2} \times 10^3 \tag{A7 8}$$

For all reasonable conditions of a rotating prestellar cloud $\nu_t/\nu_m \gg 1$. Therefore the motion will be turbulent as long as $l/R$ is not too far from the order of one and $\tau_t \gg \tau_{ff}$. This means that during the collapse turbulent friction will not redistribute the angular momentum. An exception will occur when during the collapse $\chi$ becomes of the order one and matter in the equatorial plane is hindered in falling inwards. Turbulent friction will then become important in the rotating nebular disk.

We thank Mr. K.H. Winkler for careful reading of the manuscript and Dr. H. Yorke for checking the English.

## References

Appenzeller, I., 1972, Mitt. d. Astron. Gesellschaft 31, 39

Appenzeller, I., Tscharnuter, W., 1974, Astron. & Astroph. 30, 423

Appenzeller, I., Tscharnuter, W., 1975, to be published in A & A

Bodenheimer, P., 1968, Ap. J. 153, 483

Fricke, K., Möllenhoff, C., Tscharnuter, W., private communication

Hayashi, C., 1966, Ann. Rev. Astron. & Astroph. 4, 171

Larson, R.B., 1969, M.N.R.A.S. 145, 271

Larson, R.B., Starrfield, S., 1971, Astron. & Astroph. 13, 190

Larson, R.B., 1972, M.N.R.A.S. 156, 437

Lynden-Bell, D., 1962 Proc. Cambr. Phil. Soc. 50, 709

Lynden-Bell, D., Pringle, J.E., 1974, M.N.R.A.S. 168, 603

Sengbusch, K.v., 1968, Zeitschr. f. Astroph. 69, 79

Tscharnuter, W., 1975, Astron. & Astroph., in press

# COCOON STARS

F.D. KAHN

(Sterrewacht, Leiden; on leave from Manchester University)

This is only a brief account. The main points of the lecture are dealt with in a recently published paper (Astronomy and Astrophysics $\underline{31}$, 149, 1974).

There are three important time scales which govern the formation and evolution of stars. They are $t_{acc}$, the accretion time, $t_{KH}$, the time required for a star to relax to the main sequence (the Kelvin-Helmholtz time), and $t_{evol}$, the time required for a star to evolve away from the main sequence. For stars near the top of the main sequence $t_{KH}$ is quite short, and much shorter than $t_{acc}$ in reasonable models. It may therefore be assumed that, at any stage, an accreting star actually has the same internal structure as a star of the same mass lying on the main sequence, provided the mass is large enough. But the radiation from the star is considerably changed by its having to pass through the surrounding interstellar dust. The star is therefore observed as an infra-red object. Its luminosity is, of course, not altered, only the distribution of radiant energy in the spectrum.

Radiation pressure builds up considerably in the circumstellar region during the accretion of gas and dust, because of the trapping of infra-red radiation there. The inflow of material is thus slowed down. In extreme cases the speed is reduced so much that the most refractory dust grains (graphite, probably) are not carried far enough in to reach a place where they get warm enough to melt. This quickly results in the formation of a very opaque layer at a distance of $10^{14}$ or $10^{15}$ cm from the star.

Radiation pressure then increases still further in the space enclosed by this surface. This effect will soon reverse the flow, so that accretion should come to an end. But it seems likely that the cocoon becomes unstable during the re-expansion phase, and fragments. No good description is yet available of this part of the process.

Clearly the present mechanism sets a limit to the mass of material which can accrete onto a star. The limit depends on the luminosity to mass ratio L:M, on the opacity of the interstellar mixture, on the temperature at which the most refractory grains melt, and on the rate at which mass is falling into the cocoon from outside. It appears that typical accretion rates are of the order of $10^{22}$ gm/s, for stars with a mass of the order of 30–40 solar masses. This gives typical accretion times $t_{acc}$ of the order of $10^{13}$ s, or 3– 4 x $10^5$ years. If $t_{acc}$ were much smaller or much larger, the radiation resistance would be too strong, and the most refractory grains would then not reach their melting surface. It is clear that $t_{acc}$ lies comfortably between $t_{KH} \sim 3 \times 10^4$ years, and $t_{evol} \sim 3 \times 10^6$ years. This is as it should be. A further useful result is that there is an upper limit to the ratio L:M which a star can attain in this way. The ratio seems to be about $10^4$ erg/gm s; it varies inversely as $\kappa_o$, the interstellar opacity in the ultraviolet. In regions of space where the heavy elements are less abundant, $\kappa_o$ should be smaller, and it should thus be possible for stars to form there with larger ratios of luminosity to mass.

There is a dust-free space inside an accreting cocoon. But the gas density within it is so large that the central star can only maintain an H II region of very small radius. Self-absorption would be severe for any radio emission coming from such a region. It therefore seems that the compact H II regions, which radio astronomers observe, cannot be associated with stars which are still gaining mass. Instead it is probable that the compact regions form once the associated star has stopped accreting, and its cocoon has begun to break up.

The calculations for this model are based on the approximation that dust and gas move together. This is reasonable enough in any region where the radiation field is infra-red, since the optical cross-section of the dust grains is small at long wavelengths. Even quite a slow drift speed $u_D$ through the gas then results in a large enough drag to balance the force due to radiation pressure.

However the approximation becomes less good in those regions where the dust grains are directly illuminated by the central star. Here an appreciable drift speed can arise. It will be largest for the grains nearest the star; further away a grain will be partly shielded and will drift more slowly. The result must be a partial separation of gas from dust, since this process clearly squeezes out the gas. The effect is not very important while the accretion flow continues. But once the cocoon breaks up the dust grains will be pushed away from the star more strongly than the gas. The space around the star therefore gradually fills up with gas from which the dust has been removed.

This separation process needs to be investigated further, since the problem of the abundance of dust in H II regions is important. It has a considerable bearing on the structure of compact H II regions, and of the associated infra-red and maser sources.

# The Early Stages of HII Regions

E. Krügel

Universitäts-Sternwarte
Göttingen, W-Germany

## Abstract

A model for W3(A) is described and the results of the
calculations for the dynamical evolution of a compact HII
region with a dust front are discussed.

## 1. Introduction

Let us define the early stage of an HII region not in terms of age, but of size and density. We can do this because HII regions due to their high internal gas pressure expand with a velocity of some 10 km/s. So if we follow their evolution backwards in time we end up with compact HII regions. Rather arbitrarily we define them by having a mean electron density $n_e \gtrsim 10^4$ cm$^{-3}$ and a diameter of less than 0.5 pc. Observations of several such objects are discussed in detail in the present volume. In this paper we try a theoretical approach. In section 2 a model for W3(A) is constructed on the basis of the observational data. In section 3 we discuss the results of dynamical calculations for the initial evolution of an HII region.

## 2. A Model for W3(A)

One of the most conspicuous features of compact HII regions is the fact that they emit the bulk of their energy in the IR. It is caused by heating of the dust within, and to a smaller extent, outside the HII region. For an HII region of constant density $n \approx 10^4$ cm$^{-3}$ the emission above $100\mu$ comes from the cool dust around the ionized nebula, at $20\mu$ from the dust in the HII region, at $10\mu$ only from the very hot grains near the star, and below $5\mu$ free-free and bound-free transitions from the gas are the main emission mechanisms. An increase in density shifts the maximum of the IR emission towards shorter wavelengths, because the radius of the HII region shrinks and the grains are closer to the star and therefore hotter. In addition, the optical depth of the dust in the HII region rises and more Lyman continuum radiation is directly absorbed by grains. A particuliarly fine example of IR emission from a compact HII region is W3(A), the observational data of which are discussed in the review by Mezger and Wink (this volume). In this section extensive reference is made to their paper.

Using the wealth of IR and radio measurements P. Mezger and I started building a model for W3(A). It should
reproduce all observational data and give insight into the
structure of W3(A). The method of constructing such a model
is described in another paper (Krügel, 1975). We assume
that the IR luminosity of the HII region equals the luminosity of the exciting star and infer from the theory of model atmospheres the stellar flux in the Lyman-continuum.
The electron density in the ring of W3(A) is about $8000\text{cm}^{-3}$.
Input parameters of the model which have to be chosen
suitably to match the observations are the density distribution in the inner region, the dust to gas ratio, the absorption coefficients of the dust in the UV, and its absorptivity in the IR. The mass ratio of dust to gas
$m_d/m_g$ should be of the order $10^{-2}$. A much lower value
of $m_d/m_g$ , i.e. dust depletion, can safely be excluded
because of the strong absorption of the Lyman continuum
radiation inside the HII region. The absorptivity $Q_\lambda$ of
the dust in the IR, which is needed for the calculation of
the IR spectrum, is approximated by a power law $Q_\lambda \propto 1/\lambda$ .
A $1/\lambda^2$ - dependance would shift the peak of the IR spectrum to much shorter wavelengths (Krügel, 1975).

All constant density models with $n_e = 8000 \text{ cm}^{-3}$, as
has been determined for the ring, yielded near IR fluxes
some two orders of magnitude in excess of the observed
ones and peaks at around $30\mu$ . Therefore, the inner regions of W3(A) have to be very rarified. The alternative
that W3(A) has a constant density, but that its near IR
radiation is transformed to longer wavelengths in the obscuring matter in front of the HII region is not acceptable,
because the visual extinction towards W3(A) is only 14 mag
and the optical depth at $\lambda \gtrsim 10\mu$ should be less than
unity.

The existence of the dense shell with $n_e = 8000$ cm$^{-3}$ has important consequences on the interpretation of two quantities derived from radio measurements:

a) the ratio of R of the volume of the HeII zone to that of the HII zone weighted by the square of the proton density

b) the ratio $N_c'/N_c$ of Lyman continuum photons absorbed by the gas to those emitted by the star.

For W3(A) one obtains R = 0.65, and $N_c'/N_c$ = 0.32. The mass absorption coefficients of the dust in the Lyman continuum have been derived by Mezger et al. (1974). As an average over a sample of 16 HII regions they obtain $K_D^{He}$ = 500 cm$^2$/g for $\lambda < 504$A, and $K_D^{H}$ = 100 cm$^2$/g for $504 < \lambda < 912$A. With the particular geometry of W3(A) where a tenuous inner region is surrounded by a dense shell, both $K_D^{He}$ and $K_D^{H}$ have to be about five times higher than stated by Mezger et al. in order to reproduce the observed low values for R and $N_c'/N_c$ . So either the dust particles in the shell are of a different kind and absorb much more effectively or the dust to gas ratio is some five times higher than in "usual" HII regions. Because $K_D^{He}/K_D^{H}$ is still required to be around five in order to explain R = 0.65, we favour the latter explanation.

The spectrum of the model which fits the observational data with sufficient accuracy is shown in Fig. 3 of the paper by Mezger and Wink (this volume). The absorptivity of the dust in the IR was taken to be $Q_\lambda = 2\pi a/\lambda$ . The grain radius a cancels out in the further calculations. The dust to gas ratio used was $m_d/m_g$ = 0.1. For the central light source we assumed an O5 star with a luminosity $L_* = 3.3 \ 10^{39}$ erg s$^{-1}$ and $N_c = 5.0 \ 10^{49}$ photons/s.

The geometric structure of the model is depicted in Fig. 1. The abscissa is the normalized distance from the centre. The extension of the HeII zone is marked by $r_{HeII}$.

The atom density in the inner region is only 40 cm$^{-3}$ and
rises steeply towards the shell. The grain temperature in
the shell varies between 80K and 50K. In the neutral gas
outside the radiation with $\lambda >$ 912A is the only heating
mechanism. The two other curves in Fig. 1 show two maps
at 20$\mu$ and 100$\mu$ ; the ordinate of the brightness temperature is on the left in an arbitrary logarithmic scale.
There is little variation across the nebula at 100$\mu$ , only
a small bump at the ring. The profile at 20$\mu$ agrees qualitatively with the 20$\mu$ map of Wynn-Williams et al.
(1972): there is a broad dip in the middle and a rise towards the edge by a factor of two and a somewhat greater
increase at the centre. For a constant density model, on
the other hand, the surface brightness at 20$\mu$ in a cut
across the nebula would decline rapidly towards the edge.

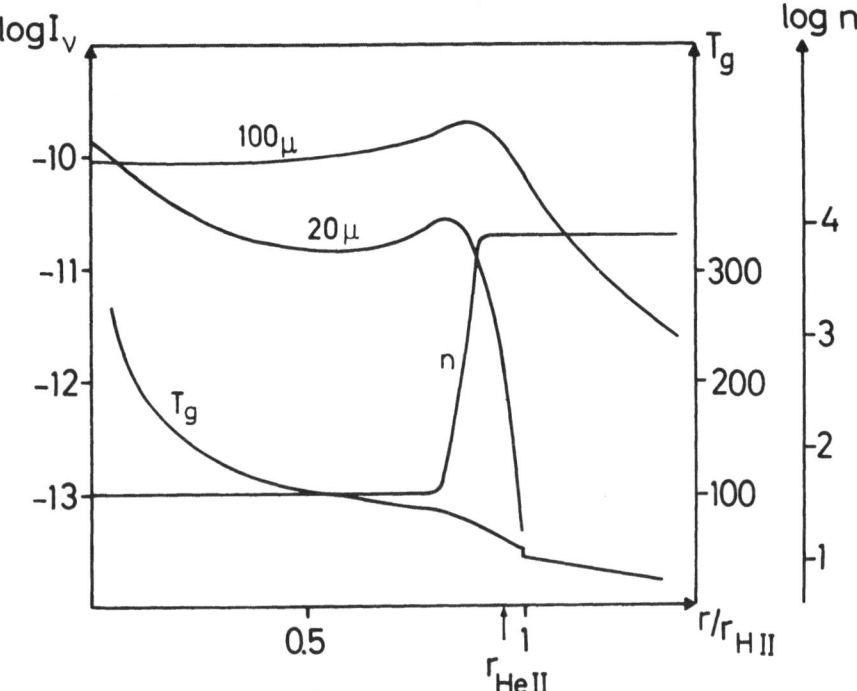

Fig. 1   A cross cut through the model for W3(A). The abscissa is the distance from the centre normalized
to the radius of the HII region. The arrow marks
the border of the HeII region. The left ordinate
is the surface brightness in an arbitrary logarithmic scale. The right ordinates give the grain
temperature and the hydrogen density.

The dust plays a decisive role in compact HII regions
not only in the transport of radiation, but also in the
dynamics. To investigate this in more detail is the goal
of the next section.

## 3. Dynamical Evolution

Pioneering works in the study of the dynamical evolution of an HII region were made by Mathews, Lasker, Vandervoort, Kahn, Goldsworthy. A star is placed in a homogeneous
medium at rest and suddenly "switched on". It still evolves
towards the main sequence. An ionization front (IF) of type

R rapidly propagates outward without accelerating the gas
noticably. When the IF nears the Strömgren radius, a shock
begins to form in the HII region, crosses the IF, and then
travels ahead of it. During this phase the front changes
to type D. The gas at the edge of the HII region expands
with a velocity of some 10 km/s, while the inner parts are
still at rest. Therefore, a density minimum forms behind
the IF and gradually spreads inwards until it reaches the
centre. Then the HII region expands with constant density.
It is bounded by shocked neutral material. For a very compact HII region this stage is reached after only a few
thousand years. Mathews (1969) also included radiation
pressure on the dust. It speeded up the velocity of the IF,
but the overall effect of the dust on the dynamics was
small. His calculations were very illustrative and included
many details. Their main shortcoming was the lack of realistic initial conditions. To improve such models one has
to know more about the structure of the protostellar cloud
into which the HII region expands.

Calculations of the collapse of a protostar have been
carried out by a number of authors (see review by Kippenhahn in this volume) with varying initial conditions. They
yielded similar results for the development of the core
where gravitation is well in control of the evolution. But
the numerical models are certainly not very reliable for
the envelope, because we do not know what a protostellar
cloud looks like when a central condensation begins to
form and because the unknown boundary conditions enter critically into the calculations, to say nothing of the
possible effects of magnetic fields or rotation. Hopefully,
the analysis of molecular line emission from dense massive
clouds, where we expect the conditions to be favourable
for star formation will tell us more about the physical
state of the cloud from which stars form.

In 1967 Davidson and Harwit outlined the evolution of

an HII region by introducing the notion of a cocoon star.
Their cocoons are essentially different from those discussed by Kahn (this volume): they are further away from
the star, not stationary, and the drift velocity of the
dust through the gas is not negligible. Davidson and Harwit did not use a more refined model for the structure of
the protostellar cloud, but first argued that dust plays a
fundamental role in the dynamics. When a massive protostar
approaches the main sequence its surface temperature is low
and it cannot ionize the ambient gas, but it has already
acquired its full luminosity. The radiation pressure acts
on the dust and drives it at high speed ( $\gg$ 10 km/s) outwards without dragging along the neutral gas, because friction between the gas and dust by elastic collisions is weak
The dust piles up in a front until its density there is so
high that coupling between the gas and dust becomes very
effective. This dust front is optically thick at visible
wavelengths and hides the star from the optical observer.
Fig. 2 shows the result of a 30 $M_\odot$ star in a cloud with a
density $n$ = $10^5$ cm$^{-3}$ by Davidson and Harwit. The distance
of the dust front and the extension of the HII region are
given as a function of time. At the beginning the dust
front moves faster, but the HII region will eventually
catch up with it. The bottom solid line gives the equilibrium Strömgren radius when expansion is not taken into account. The arrows on the distance scale show the mass interior to that radius.

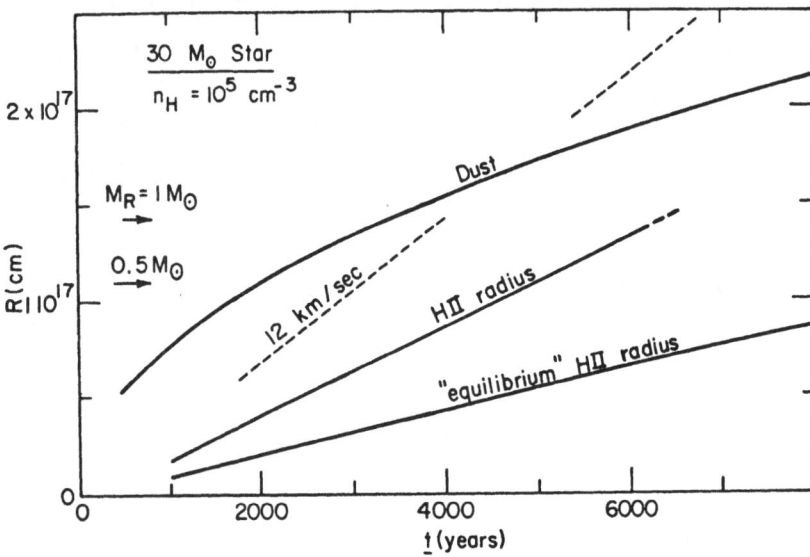

Fig. 2   The evolution of a cocoon by Davidson and Harwit
(1967) Description is given in the text.

Stimulated by the ideas of Davidson and Harwit H.
Yorke and I calculated numerical models for such an evolution with a more complete set of equations. To allow for a
dust front to form we used a two component model where the
equations of motions were solved seperately for the dust
and the gas. They were linked by a friction term. The optically thick dust front takes up the whole momentum $L/c$
of the stellar radiation. Its drift velocity $w$ through the
gas is given by

$$\frac{L}{c} = M_D \frac{w}{t_s}$$

$M_D$ is the mass of the dust front which contains all the
dust that has been swept up and $t_s$ is a slowing down time

which can be approximated in neutral gas of 50K by

$$t_s = 2 \times 10^{-5} \frac{a}{\rho}$$

$a$ is the grain radius, $\rho$ the gas density. Typical drift
velocities are around 1 km/s. It is difficult to estimate
the width of the dust front. Fortunately, the optical depth
and the number of collisions a gas atom undergoes while
traversing through the dust front is independent of it. The
drift velocities are low in any case and the results were
not sensitive to the value chosen. We assumed a width of
$5 \cdot 10^{14}$ cm. The dust to gas ratio is about a factor of ten
higher in the dust front than in the medium ahead of it.
Fig. 3 shows a calculated model. The exciting star will be
of type O6 when it arrives on the main sequence. The Lyman
continuum flux is one third of its final value. The velocity and the gas density are plotted over the distance
from the centre. The dashed line gives the initial $1/r$
density profile, the gas was then assumed to be at rest.
The dust free HII region is already expanding at constant
density $\rho = 10^{-18}$ g/cm$^3$. The IF is of type D. Ahead of it
there is a shell of shocked gas. Although it is very thin
it contains three times more mass than the HII region. The
density in the shell decreases slightly. The shock is assumed to be isothermal. The dust front has arrived at a
distance of $6.4 \cdot 10^{16}$ cm. Up to this point the nebula is
cleared from grains. The mass of the dust front is $1.2 \cdot 10^{31}$ g
and it drifts with a velocity of 0.8 km/s. It drags the gas
along at supersonic speed. This explains the broad density
minimum around $5.2 \cdot 10^{16}$ cm. The gas enters the dust front
with high relative velocity and leaves slowed down, but
very compressed. The hydrogen number density inside is
about $2 \ 10^{8}$ cm$^{-3}$.

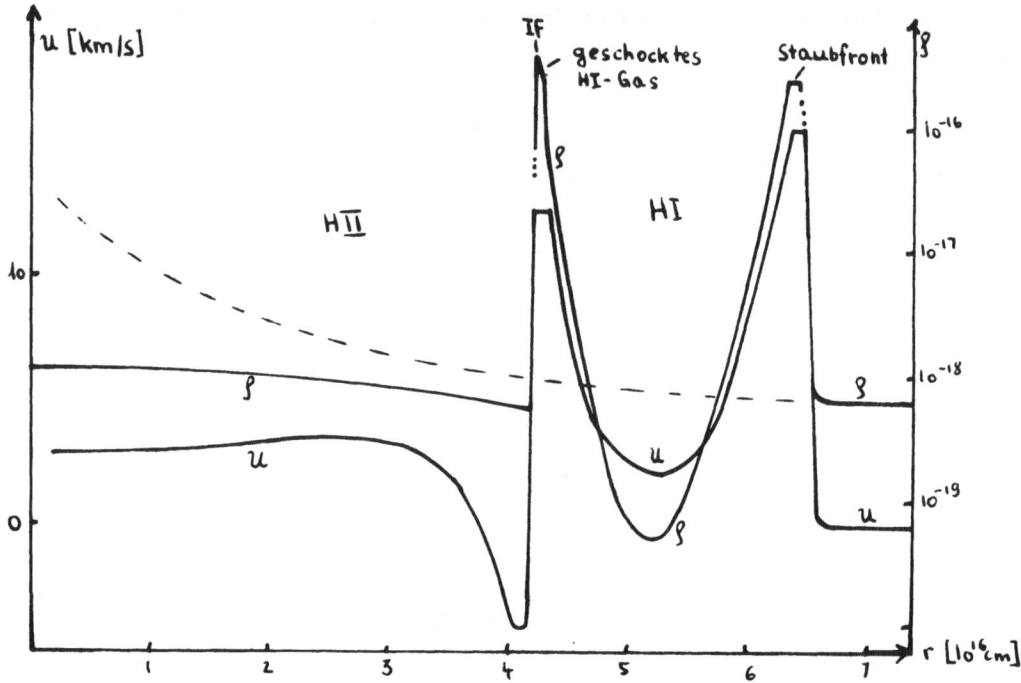

Fig. 3   The run of the density $\rho$ and the gas velocity $u$
in a model protostellar cloud. The star has a luminosity of $1.5 \cdot 10^{39}$ erg s$^{-1}$, but has not yet reached
the main sequence. The dashed line shows the ini-
tial density profile. IF marks the ionization front.
Further description is given in the text.

In the further evolution the HII region grows in mass.
When it has eaten up the surrounding shocked neutral (dust-
less) shell it rapidly spreads through the rarified gas
outside until it reaches the dust front. The grains there
are exposed to energetic Lyman continuum radiation and are
(probably) charged by the photoelectric effect. The ambient
gas is ionized, and the grains are effectively frozen into
the gas by Coulomb forces. When the compressed material in
the dust front is ionized its pressure rises and it expands.
At this stage we ran into numerical difficulties. The details of the evolution depend sensitively on the initial
velocity and density distribution. Let us therefore just

qualitatively describe the appearance of an HII region at
a stage where the IF has not yet caught up with the dust
front, and the star has not yet reached the main sequence:

In the optically thin radio regime the observer sees
a very compact HII region. The few remaining grains absorb
the trapped $L\alpha$ -photons and are therefore rather hot. So
the ionized region emits effectively at, say, $20\mu$ . Its
size at this wavelength is the same as that of the radio
source. Because the HII region is dust depleted most of
the radiation is absorbed further out in the dust front.
This explains why the IR spectrum is rather "normal",
peaking around $100\mu$ . The extension of the source at $100\mu$
is determined by the position of the dust front and much
greater than at $20\mu$ . The total IR flux from the shell
and the core yields the luminosity $L_*$ of the star. The IR
luminosity from the dust depleted HII region alone equals
$N_c \cdot h\nu_\alpha$ , where $h\nu_\alpha$ is the energy of the $L\alpha$ -photons,
and $N_c$ the number of Lyman continuum photons emitted by
the star.

Although the star is hidden behind a dust shell one
can show from the observational data that it is still e-
volving towards the main sequence by the following argument: Assuming for the moment that the star of luminosity
$L_*$ is on the main sequence. Then its Lyman continuum output $N_c$ is greater than the measured $N_c'$ (number of Lyman
continuum photons absorbed by the gas). So $(N_c - N_c')$ Lyman continuum photons are absorbed by dust in the ionized
gas and the HII region cannot be dust depleted. The energy of these $(N_c - N_c')$ Lyman continuum photons plus the
energy of the trapped $L\alpha$ -photons are emitted from the
dust in the HII region at rather short IR wavelengths,
( $\lambda \lesssim 30\mu$ ). In this case the bulk of the total IR radiation stems from the HII region and the spectrum of the
whole nebula would consist of two distinct components: a
very "hot" one peaking around $20\mu$ from the core and a

weaker one at $100\mu$ from the shell.

It is interesting to note that the above described picture seems to apply to W3(OH). Mezger and I are at present trying to work out the details.

At a later stage, when the inner side of the dust front is being ionized the situation resembles that of W3(A). The star is already on the main sequence, and the inner zones of low density are surrounded by a shell of high gas density and high dust to gas ratio. IR measurements of compact HII regions with higher resolution will help to decide whether this crude model is realistic.

## REFERENCES

Davidson, K., Harwit, M.   1967, Ap.J., 148, 443
Goldsworthy, F.   1961, Phil. Trans.R.Soc. London, A253, 227
Kahn, F.   1954, BAN 12, 456
Kahn, F.   1975, this volume
Kippenhahn, R.   1975, this volume
Krügel , E.   1975, submitted to Astron. & Astrophys.
Lasker, B.   1967, Ap.J., 143, 700
Mathews, W.   1965, Ap.J., 142, 1120
Mathews, W.   1969, Ap.J., 157, 583
Mezger, P., Wink, J.   1975, this volume
Mezger, P., Smith, L., Churchwell, E.   1974, Astron. &
      Astrophys. 32, 269
Vandervoort, P.   1964, Ap.J. 139, 869
Wynn-Williams, C.G., Becklin, E.E., Neugebauer, G. 1972,
      Monthly Notices Roy. Astron. Soc., 160, 1.

# MODEL CALCULATIONS OF DUSTY STRÖMGREN SPHERES

TEIJE DE JONG, FRANK P. ISRAEL and ALEXANDER G.G.M. TIELENS

Sterrewacht,
Leiden, the Netherlands

## ABSTRACT

We compare parameters derived from the observed radio and near-infrared flux densities of a sample of 21 compact H II regions with theoretical parameters calculated for dusty Strömgren spheres. Preliminary results of this comparison suggest that the dust inside compact H II regions is quite often depleted by large factors, consistent with evaporation of the mantle material of core-mantle particles.

## 1. INTRODUCTION

Recently a large body of infrared observations of compact H II regions has become available. At the same time detailed radio maps of compact H II regions have been obtained by means of aperture synthesis

techniques. By comparing radio and infrared data of H II regions with
model calculations of dusty Strömgren spheres one can learn something
about the spatial distribution and the physical properties of the
interstellar dust (de Jong, 1975).

In order to allow detailed model fitting one should be certain
that the infrared and the radio data pertain to one and the same source.
Thus high spatial resolution of a few arc seconds is required so that
at present one is restricted to using aperture synthesis radio observations and near-infrared observations. On the one hand this is a
disadvantage since most sources emit most of their energy at longer
infrared wavelengths, typically at about 100μ. On the other hand at
near-infrared wavelengths one studies the hot dust that is generally
confined to the H II region so that a comparison of radio and near-
infrared data is more meaningful.

There are several ways in which one can approach the model fitting
process. Either one tries to match the spectrum of one single H II
region in detail or one tries to extract statistical information from
observations of a group of H II regions by fitting observed parameters
of that group of H II regions. In this paper we report some preliminary
results following the latter approach.


## 2. THE OBSERVATIONS

From the wealth of presently available infrared and radio
observations of compact H II regions we have selected sources according
to the following criteria:
(i) the radio and infrared observations should have comparable spatial
    resolution,
(ii) the radio and infrared observations should pertain to the same
     source.

In Table 1 we give the source names, the adopted distances, the
radio diameters and the 20μ and the 6 cm flux densities of 21 sources
obeying the above criteria. We chose to use 20μ fluxes rather than 10μ

Table 1    Parameters of H II regions

| Source | D(kpc) | d(pc) | $S_{20\mu}$(f.u.) | $S_{6cm}$(f.u.) | u(pc cm$^{-2}$) | $S_{20\mu}/S_{6cm}$ | Ref. |
|---|---|---|---|---|---|---|---|
| W3A-IRS1 | 3.0 | 0.58 | 2000 | 31 | 93 | 65 | 1,2 |
| B-IRS2 | | 0.25 | 200 | 10 | 64 | 20 | 1,2 |
| C-IRS4 | | 0.025 | 300 | 0.6 | 25 | 500 | 3,2 |
| OH-IRS8 | | 0.024 | 200 | 0.62 | 25 | 320 | 4,3 |
| DR 21 N | 3.0 | 0.062 | 38 | 0.82 | 28 | 46 | 5,6 |
| M 17 N | 2.1 | 2.9 | 31000 | 220[++] | 141 | 141 | 7,8 |
| S | | 2,3 | 51000 | 230[++] | 143 | 220 | 7,8 |
| E | | 1.95 | 2400 | 37[++] | 78 | 65 | 7,8 |
| NGC 7538-IRSI | 3.5 | 0.031 | 200 | 0.12 | 16.1 | 1670 | 9,2 |
| IRS2 | | 0.187 | 700 | 1.40 | 37 | 500 | 9,2 |
| IRS3 | | 0.054 | 71 | 0.02 | 8.9 | 3600 | 9,2 |
| IRS4 | | 0.55 | 45 | 0.8 | 30 | 56 | 10,2 |
| W43-G30.8-0.0 A | 7.1 | 0.60 | 100[+] | 0.54 | 43 | 185 | 11 |
| B | | 0.60 | 180[+] | 0.46 | 40 | 390 | 11 |
| C | | 0.83 | 275[+] | 1.02 | 53 | 270 | 11 |
| D | | 0.83 | 140[+] | 0.58 | 44 | 240 | 11 |
| G45.5+0.1 | 9.7 | 0.84 | 750 | 4.4 | 106 | 170 | 12,13 |
| W51-G49.5d-IRS2 | 7.8 | 0.42 | 1500 | 11 | 124 | 136 | 14,2 |
| S156A | 3.5 | 0.16 | 300 | 0.80 | 30 | 375 | 10,15 |
| S157B | 3.5 | 0.034 | 25 | 0.2 | 19.1 | 125 | 10,16 |
| H2-3 | 5.0 | 2.6 | 2050 | 28[++] | 126 | 73 | 17,18 |

[+]extrapolated from 12.6µ;  [++]extrapolated from 1.94 cm

References to Table 1:

1. Wynn-Williams (1971)
2. Wynn-Williams et al. (1972)
3. Wynn-Williams (1975)
4. Harten (1975)
5. Harris (1973)
6. Wynn-Williams et al. (1974)
7. Gordon and Williams (1971)
8. Lemke and Low (1972)
9. Martin (1973)

10. Israel (1975)
11. Pipher et al. (1974)
12. Matthews et al. (1975)
13. Zielik et al. (1974)
14. Martin (1972)
15. Cohen and Barlow (1973)
16. Neugebauer (1974)
17. Rubin (1970)
18. Becklin et al. (1974)

fluxes since the effects of the observed 10μ resonance in the dust
material would unnecessarily complicate the model fitting process.
We also list the excitation parameter u, calculated from the 6 cm
flux densities according to the relation $u = 14.2 \ (SD^2)^{1/3}$ pc cm$^{-2}$
where D is the distance in kpc and S is the 6 cm flux density in flux
units, and the 20μ to 6 cm flux ratio $S_{20\mu}/S_{6cm}$. In the last column we
give references to the papers where the radio and the infrared
observations of each individual source are described.

It is hard to judge how strongly our sample is affected by
selection effects. We believe that for diameters between about 0.1 and
1 pc our sample is representative. Many of the larger sources have not
been studied in the near-infrared because of instrumental limitations.
Many of the smaller sources may escape detection at radio wavelengths
because of radio self-absorption, or they may not yet have been observed
in the near-infrared.

## 3. THE MODEL

We have calculated infrared and radio spectra of dusty Strömgren
spheres, taking into account the effect of the dust on the size of the
Strömgren sphere and including both stellar photons and recombination
line photons to heat the dust. Given the complexity of such calculations
we have idealized the problem by making the following assumptions:
(i)   the gas density is uniform and taken to be equal inside and
      outside the Strömgren sphere,
(ii)  the dust particles are spherical and of equal size, a = 1500Å.
      We allow for destruction of the dust inside the H II region by
      defining a mass depletion factor δ such that the depleted dust
      particles are decreased in size to a = 1500 $\delta^{1/3}$ Å.
(iii) the number density of the dust particles, $n_d = 1.2 \times 10^{-12} \ n_H$,
      follows from the assumption that the undepleted dust has a mass
      density of 1 gr cm$^{-3}$ and that the dust to gas ratio of the
      undepleted dust is 1 percent by mass,

(iv)   the ionization equilibrium in an H II region with internal dust
       is treated according to the analytical relations of Petrosian,
       Silk and Field (1972),

(v)    the dust inside the Strömgren sphere is heated by stellar photons
       and by Lyman α photons. The dust outside the Strömgren sphere is
       heated by stellar photons longwards of the Lyman limit and by
       Balmer continuum photons emitted inside the Strömgren sphere.

(vi)   the efficiency of the dust particles for absorption of these
       photons is taken to be $Q_{abs}$ = 1. The efficiencies for emission
       and extinction of the dust particles for infrared photons are
       calculated from Mie theory,

(vii)  the dust material is characterized by a solid with one infrared
       resonance, either at 10μ or at 50μ. The strength and the width
       of the resonance at 10μ matches the 10μ absorption in silicates
       (Pollack, Toon and Khare, 1973). The complex refractive index
       of such material m = n + ik is calculated from classical
       dispersion theory. The resonance at 50μ is shifted in wavelength
       but taken to have the same strength and relative width as the
       resonance at 10μ.

Our model is rather similar to that discussed by Wright (1973)
but it has several new features. In the first place we calculate dust
temperatures using computed absorption efficiencies that are much more
realistic than $Q_{abs} \propto \lambda^{-1}$. This results in a very different temperature
distribution of the dust. The representation of the dust material by
a solid with one resonance at a variable wavelength has the advantage
over the use of real solids (like silicate and ice) that one can more
easily sort out the effect of varying the dust material. Secondly we
include the radio and infrared free-free emission of the H II region
adopting an electron temperature $T_e = 10^4$ K. Finally, we take into
account the circumnebular extinction by dust particles outside the
H II region. We rather arbitrarily assume that the thickness of the
circumnebular shell is equal to the Strömgren radius of the H II region
since the Strömgren radius is approximately equal to the scale height
of the circumnebular shell if the gas density in the circumnebular
cloud falls off with radius like a power law distribution.

## 4. MODEL FITTING

We have chosen to fit the excitation parameter u and the flux density ratio $S_{20\mu}/S_{6cm}$ of the sample of the H II regions in Table 1. The excitation parameter is sensitive to the amount of internal dust since a certain fraction of the Lyman continuum photons is absorbed by the dust. The ratio of the 20μ and 6 cm flux densities is sensitive to the dust temperature and may thus provide some clues about the dust material.

In Figures 1 and 2 we have plotted the u and $S_{20\mu}/S_{6cm}$ values of our sample of H II regions against d, the source diameter. The observed excitation parameters rise with increasing diameter and the ratios of the 20μ and 6 cm flux densities drop with increasing diameter. This behaviour is a priori to be expected because for a certain exciting star the dust competes more effectively with the gas for Lyman continuum photons when the H II region decreases in size (increasing density) and the temperature of the dust inside the H II region increases when the H II region increases in size. The positions of the points plotted in Figures 1 and 2 depend on the adopted distances of the H II regions. Since $u \propto D^{2/3}$ and $d \propto D$ the points in Figure 1 would slide along a line with slope 1/3 and the points in Figure 2 would slide along horizontal lines if the actual distances of the H II regions would be different.

In order to compare the observations with the models we have calculated diameters, 20μ flux densities and 6 cm flux densities of dusty Strömgren spheres varying the electron density ($n_e = 10 - 10^5 cm^{-3}$) the spectral type of the exciting star (04, 09), the dust material (10μ, 50μ resonance) and the depletion factor ($\delta = 1,10^{-2}$). For an 04 star we adopted $L = 1.42 \times 10^6 L_\odot$, $S_H = 9.4 \times 10^{49} s^{-1}$ and $< h\nu_{Lyc} > = 24$ eV and for an 09 star $L = 5.4 \times 10^4 L_\odot$, $S_H = 1.19 \times 10^{48} s^{-1}$ and $< h\nu_{Lyc} > = 18.0$ eV (Cesarsky, Cesarsky and de Jong, 1975). The curves in Figures 1 and 2 are constructed from the model calculations by plotting the excitation parameters and the flux density ratios against the diameters of the dusty Strömgren spheres. We computed the excitation parameters from the model flux densities using the same relation as used for the computation of the excitation

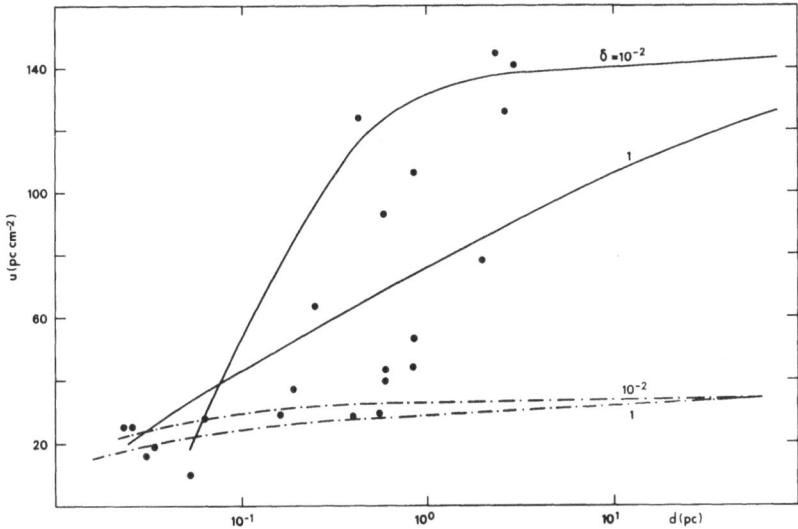

Fig.1  Excitation parameters u in pc cm$^{-2}$ plotted against the radio
source diameters in pc. The points correspond to the observed H II
regions in Table 1. The curves are computed for dusty Strömgren
spheres with dust depletion factors $\delta$ = 1,10$^{-2}$, excited by 04
stars (solid lines) and 09 stars (dashed-dotted lines).

parameters in Table 1.

It is evident from the curves in Fig. 1 that the effect of depletion of the internal dust shows up most clearly for Strömgren spheres
excited by 04 stars. In undepleted dusty Strömgren spheres excited
by 04 stars with diameters less than 10 pc the dust optical depth
becomes greater than unity so that a large fraction of the Lyman
continuum photons is absorbed by the dust. The sharp drop in u at small
diameters for dust depleted Strömgren spheres excited by 04 stars is
due to self-absorption of the radio continuum at 6 cm. The curve
representing the depleted 04 star case intersects with the undepleted
case because the electron density in a dust depleted Strömgren sphere
is larger than in an undepleted Strömgren sphere with the same diameter
resulting in larger radio continuum optical depths. Figure 1 suggests
that in at least 30 percent of the sources the dust must be strongly
depleted. Furthermore it appears that the determination of the spectral
types of exciting stars from observed excitation parameters is

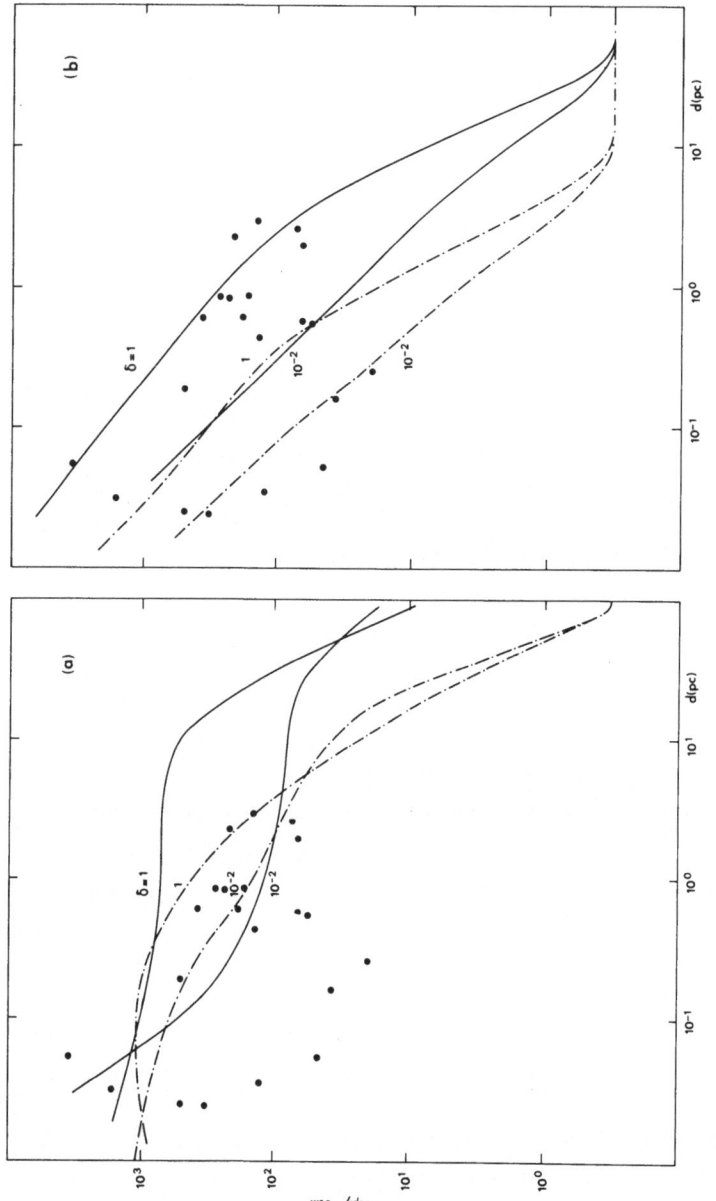

Fig.2   Ratios of the 20 μ to 6 cm flux densities plotted against the radio source diameters in pc.
The points correspond to the observed H II regions in Table 1. The curves are computed for dusty
Strömgren spheres with dust depletion factors δ = 1,10⁻² and with dust material characterized by
a resonance at 10 μ (Fig.2a) and at 50 μ (Fig.2b). The solid and dashed-dotted lines have the
same meaning as in Fig.1.

unreliable for H II regions larger than about 0.1 pc in diameter and completely meaningless for H II regions smaller than about 0.1 pc in diameter, unless some additional information about the amount of interstellar dust is available. This conclusion still holds if the turnover point of the radio spectrum is known so that the effect of radio self absorption can be properly taken into account.

In Figure 2 we show the $S_{20\mu}/S_{6cm}$ curves for dusty Strömgren spheres with different exciting stars, different depletion factors and different dust material. Several features of these curves should be discussed. In the first place, for any combination of exciting star, dust depletion factor and dust material the weighted mean temperature of the dust increases with decreasing diameter. Secondly, in the case of depletion the increase in temperature of the dust particles (smaller particle size) partly offsets the decrease in mass of the emitting dust. Thirdly, in the depleted 04 star case the effect of radio self-absorption shows up in the rise of the curves at small diameters. Finally at large diameters the dust becomes very cool and free-free emission of the ionized gas takes over at 20μ so that the flux density ratio approaches 0.32, the ratio of the free-free Gaunt factors at 20μ and 6 cm.

From the curves in Fig. 2 it is clear that the 20μ to 6 cm flux density ratios are very sensitive to the adopted dust material. Dust particles consisting of 10μ material have smaller emission efficiencies at long wavelengths than dust particles consisting of 50μ material and are therefore appreciably hotter. Comparing the calculated curves with the observed points one might be tempted to conclude that the observed flux density ratios can be best explained by dust material with a resonance at about 50μ. However, that conclusion would be too simpleminded since most probably the dust material and the dust depletion factor are not independent variables, as we shall discuss below.

The most plausible mechanism to destroy the dust is evaporation. Dust grains evaporate if the dust temperature rises above the sublimation temperature of the dust material. Both the sublimation temperature and the dust temperature depend on the assumed dust material. A priori one would expect that the 50μ material (ice) is less strongly bound than the 10μ material (silicate). Consistent with the concept of

core-mantle particles, we envisage that depletion corresponds to
evaporation of the 50μ mantle material at about 100 K, while the cores
consisting of 10μ material survive temperatures up to about 1500 K.
Then the transition between the depleted dust and the undepleted dust
is determined by the dust temperature distribution so that there is no
a priori reason why it should coincide with the boundary of the
Strömgren sphere as we have implicity assumed in our definition of the
dust depletion factor. In particular around the smaller Strömgren
spheres the circumnebular dust will attain rather high temperatures.
This causes the dust to be depleted far outside the Strömgren sphere and
it leads to a large drop in the calculated flux density ratios in Fig.2.

These complications make it rather difficult to draw firm
conclusions about the dust material and the dust depletion factor from
a comparison of the calculated and observed flux density ratios in
Fig. 2. However a few sources can still be discussed on an individual
basis. Take for instance the North and South components of M 17.
According to Fig. 1 these sources are probably excited by 04 stars
while the dust inside is strongly depleted. At the same time the
observed flux density ratios in Fig. 2a can be explained by emission
of depleted 10μ material, consistent with the picture of core-mantle
particles.

In order to resolve the present ambiguity in the interpretation
of Fig. 2 we clearly need to improve our models by coupling the dust
material and the dust depletion in a self-consistent manner through the
dust temperature.

REFERENCES

Becklin, E.E., Frogel, J.A., Kleinmann, D.E., Neugebauer, G., Persson,
    S.E., Wynn-Williams, C.G. 1974, Ap. J. 187 487.
Cesarsky, C.J. Cesarsky, D.A., de Jong, T. 1975, in preparation.
Cohen, M., Barlow, M.J. 1973, Ap. J. (Letters) 185, L37.
Gordon, M.A., Williams, T.B. 1971, Astr. and Ap. 12, 120.
Harris, C.S. 1973, MNRAS 162, 5P.
Harten, R. 1975, in preparation.
Israel, F.P. 1975, in preparation.
de Jong, T. 1975, in "Proceedings of the Second European Regional
    Meeting in Astronomy", in press.
Lemke, D., Low, F.J. 1972, Ap. J. (Letters) 177, 53.

Martin, A.H.M. 1972, MNRAS 157, 31

Martin, A.H.M. 1973, MNRAS 163, 141.

Mathews, H.E., Winnberg, A., Goss, W.M., Habing, H.J. 1975, in
    preparation.

Neugebauer, G. 1974, private communication.

Petrosian, V., Silk, J., Field, G.B. 1972, Ap. J. (Letters) 177, 69.

Pipher, J.L., Grasdalen, G.L., Soifer, B.T. 1974, Ap. J. 193, 283.

Pollack, J.B., Toon, O.B., Khare, B.N.T. 1973, Icarus 19, 372.

Rubin, R.H. 1970, Astr. and Ap. 8, 171.

Wright, E.L. 1973, Ap. J. 185, 569.

Wynn-Williams, C.G. 1971, MNRAS 151, 397.

Wynn-Williams, C.G. 1975, private communication.

Wynn-Williams, C.G., Becklin, E.E., Neugebauer, G. 1972, MNRAS 160, 1.

Wynn-Williams, C.G., Becklin, E.E., Neugebauer, G. 1974, Ap. J. 187, 473.

Zeilik, M., Kleinmann, D.E., Wright, E.L., 1975, preprint.

# INFRARED OBSERVATIONS OF HII REGIONS AT $\lambda < 40\mu$

C. G. WYNN-WILLIAMS

Mullard Radioastronomy Observatory,
Cavendish Laboratory,
Cambridge, England

## ABSTRACT

Most infrared sources found at $\lambda < 40\mu$ and associated with HII regions appear to fall into one of three classes which may be loosely referred to as the reddened O stars, the compact HII regions and the protostellar objects. However   recent observations indicate that the distinction between the last two types is blurred.

REDDENED O STARS

Highly obscured exciting stars have been found in W3 (Wynn-Williams, Becklin and Neugebauer 1972) and H2-3 (Becklin et al 1974). In both these cases the estimated visual extinction in front of the star is about 15 magnitudes and the star is of spectral type around O5. Grasdalen (1974) claims to have found the exciting star of NGC 2024, and proposes that it is a B0 supergiant of $1.6 \times 10^6$ $L_\odot$ hidden behind 32 magnitudes of visual extinction. However this proposed luminosity is about thirty times greater than the total measured infrared luminosity of NGC 2024 at all wavelengths, so in this case there is a difficulty in explaining how 97% of the starlight is able to escape from such a highly obscured region. Elsässer and his colleagues (elsewhere these proceedings) have found new hidden stars associated with W3 and M17 using the extreme photographic infrared.

COMPACT HII REGIONS

Only comparatively compact HII regions can be studied at $\lambda < 40\mu$, since it is necessary at these wavelengths to reduce background noise by the use of beamswitching and small diaphragms. Unlike those at $\lambda < 40\mu$ the present observations are not usually diffraction limited so that it is possible to obtain resolution of a few seconds of arc, comparable to that available from aperture synthesis radio telescopes. In almost all cases where a compact HII region is visible on aperture synthesis maps infrared emission from heated dust has been detected at 10-20 microns. Most of these results have been summarised by Wynn-Williams and Becklin (1974). In general the infrared source has the same shape and size as the ionized region but among different compact HII regions there is a wide variation in the ratio between $20\mu$ and 5 GHz flux density. The similarity between radio and infrared maps can be explained most naturally if the dominant heating mechanism for the dust is absorption of resonantly-trapped Lyman-α photons, and if it is assumed that most of the emission at $\lambda > 20\mu$ comes from dust in the neutral gas exterior to the ionised region. In some cases, however, energetic considerations indicate that direct absorption of Lyman-continuum photons by dust grains must be occurring; the relative importance of these two mechanisms is still controversial.

## PROTOSTELLAR OBJECTS

The third important type of infrared source consists of compact objects such as W3-IRS5 and the Orion BN source, which have no detectable radio emission. These objects, which are in general less than 2 arc sec in diameter, have an energy distribution resembling a black body at a temperature of a few hundred K modified by a strong 10-$\mu$ "silicate" absorption. Objects of this type with luminosities in the range 300 to $3 \times 10^4$ $L_\odot$ are known; they are often associated with OH or $H_2O$ masers. As summarized by Wynn-Williams (1974) there are strong grounds for identifying such objects with protostars or "cocoon" stars of the type described by Appenzeller and Tscharnuter (1974) or Kahn (1974). These objects consist of a compact core, probably nuclear burning, onto which matter is rapidly accreting. The outer layers of the infalling envelope are at a few hundred K, and produce the infrared emission.

Interest has recently been centering on a group of sources which have the same infrared properties as the protostellar objects but which have weak radio emission. The objects NGC 7538 IRS1 (Wynn-Williams, Becklin and Neugebauer 1974), AFCRL 809-2992 (Merrill and Soifer 1974) and, possibly, W3(OH) (Wynn-Williams et al 1972) are of this type. Careful measurement of the relative angular sizes of the radio and infrared emitting regions are currently being made in an attempt to establish whether or not they represent an evolutionary link between infrared protostars and compact HII regions.

## REFERENCES

Appenzeller, I., and Tscharnuter, W. 1974, Astron. Ap., 30, 423.
Becklin, E. E., Frogel, J. A., Kleinmann, D. E., Neugebauer, G.,
    Persson, S. E., and Wynn-Williams, C.G. 1974, Ap. J., 187, 487.
Grasdalen, G. L. 1974, Ap. J., 193, 373.
Kahn, F. D. 1974, Astron. Ap., 37, 149.
Merrill, K. M., Soifer, B. T. 1974, Ap. J. (Letters), 189, L27.
Wynn-Williams, C. G. 1974, Talk given at Regional Meeting of the IAU,
    Trieste. To be published in Pub. Ital. Astron. Soc.
Wynn-Williams, C. G., and Becklin, E. E. 1974, Pub. Astron. Soc.
    Pacific, 86, 5.
Wynn-Williams, C. G., Becklin, E. E., and Neugebauer, G. 1972, Mon.
    Not. R. Astr. Soc., 160, 1.
Wynn-Williams, C. G., Becklin, E. E., and Neugebauer, G. 1974, Ap. J.,
    187, 473.

# INFRA-RED OBSERVATIONS AT $\lambda > 40$ μm OF COMPACT HII REGIONS

R.E. JENNINGS

University College London

## ABSTRACT

A brief account of the techniques used at $\lambda > 40$ μm is followed by photometric data, maps of IR regions and spectral data. In particular the $He^{+}/H^{+}$ ratio is discussed and new features on maps of W3 and the Galactic Centre are presented.

## INTRODUCTION

Due to attenuation by the earth's atmosphere, infra-red astronomical observations at wavelengths beyond twenty microns are generally made by getting either partially or completely above the atmosphere, using aeroplanes, balloons, rockets, etc. To date no satellite measurements have been published but such vehicles are planned. However, in all these vehicles the size of the telescope and the corresponding angular resolution in the far infra-red has been severely limited compared with ground based observations at shorter wavelengths. It has so far not been possible to match the increase in wavelength with a corresponding increase in the primary mirror - in fact the reverse is true, the largest primary mirrors that it has been possible to get above the atmosphere have diameters of order one metre only. Consequently many of the astrophysical models proposed in this wavelength range have simple geometries, when it is known from measurements at shorter (and longer) wavelengths that this is probably not the case. However, even with these relatively small telescopes it has already been possible to make many useful measurements during the last few years, measurements which have shown that a large number of objects, in particular sources associated with HII regions, radiate most of this energy at wavelengths beyond forty microns. This fact alone has been sufficient stimulus for the rapidly increasing effort that is taking place at the present time to make more detailed observations in the far infra-red. It must also be mentioned that all is not lost as regards angular resolution because it is possible, from high dry sites, to use ground based telescopes at 34, 350 and 450 µm, but unfortunately these atmospheric windows are rather poor and there are no good ones until a wavelength of a few mm is reached. In addition it is hoped that the space shuttle will provide an observatory equipped with a relatively large telescope well outside the earth's atmosphere.

## TECHNIQUES

Before discussing the measurements that have been made of the HII regions in the far infra-red, it is appropriate briefly to mention the techniques that are employed. The detector, often a semiconductor such as doped germanium cooled to about 2 K, is capable of detecting a signal

of $10^{-13}$ watts, but has falling on it a background signal from the telescope itself which can be $10^6$ times greater. This background signal
raises a number of problems; in particular it introduces noise due to
the statistical fluctuation in the number of photons arriving at the
detector. By careful design of the telescope optics and suitable choice
of field of view, every effort is made to keep this noise below the inherent noise of the detector. To overcome drifts in the background some
form of spatial modulation is generally used, either by relatively rapid
movement of the telescope through the object or more usually by modulat-
ing one of the mirrors in the optical system. Often the secondary mirror
is chosen and made to follow a square wave at a frequency of a few tens
of Hz, the corresponding signal being phase sensitively detected. Even
when the telescope has no object in either beam a level will be produced
due to slight imbalance in the two fields of view, but this level will
be constant over short periods of time so that the change when an object
passes through the beam can be observed. In this way fluxes smaller by
many orders of magnitude than that from the background can readily be
detected, the frequency components of fluctuations in the background level
being much lower than the signal frequencies. What is effectively being
observed is the spatial differential of the intensity distribution of
the source. This has the disadvantage that for extended objects a
gradual change in intensity gives rise to a small slowly changing signal
level which can easily be confused with a zero level drift.

As a slight digression, it is probably worth remarking that small
fields of view and correspondingly small angles of chop make a detection
system relatively insensitive to large sources of low surface brightness.
This is of particular interest in connection with the 'large' source
R. Weiss reports 'seeing' during a recent balloon flight to determine
the uniformity of the 3 K background radiation. For these measurements,
he was using a $20^o$ F.O.V. Similar arguments might apply to the recent
observations of Friedlander, Goebel and Joseph (1974).

The observations of HII regions ($\lambda > 40$ μm) will now be discussed
under the headings of photometry, mapping and spectroscopy.

PHOTOMETRY

The great interest in photometry at these long wavelengths came
when it was realised in the late '60s that many HII regions could radiate
much of their energy at wavelengths around 100 µm, this energy being that
of the central star re-radiated from cool dust in and around the HII
region. Two experimental programmes which confirmed these ideas both
used very modest 12 inch telescopes. Hoffmann et al (1971b) made use of a
balloon to lift their telescope above the atmosphere and carried out a
partial survey of the sky, principally in the galactic plane, while F.J.
Low and his collaborators (see Harper & Low, 1971) relied upon a Lear
Jet as a suitable vehicle from which to make their observations. Other
programmes started up at about this time, rockets were employed in
addition to the aeroplane and balloon, until today the photometry data
which is available is surprisingly extensive. For this review, it was
initially considered useful to prepare a list of all the photometric
measurements made to date, but the list proved to be too long. At the
present time over 300 separate far infra-red measurements have been made
on over a hundred objects, a large proportion of which are HII regions.
Included among these observations are ground based measurements at 350
µm by Joyce et al (1972) and Harper et al (1972).

Rocket observations of HII regions using a helium cooled telescope
(18 cm) have been made by the Cornell group (Soifer et al, 1972, 1973), who
observed the Galactic Centre region, M8, NGC 6357 and NGC 1499. The use
of a four colour photometer system enabled the colour temperature of the
sources to be determined.

A multiband photometer has also been used on the Lear Jet to observe
HII regions (Harper, 1974). The objects measured include the Galactic
Centre and M8, and, in addition, W3, NGC 2024, M42, M17 and upper limits
obtained for the supernova remnants M1 (the Crab) and Cas A. The upper
limit for M1 is very similar to that quoted by the UCL group (Furniss et
al, 1972), who have made a series of photometric measurements from
balloons (Emerson et al, 1973, Furniss et al, 1974). Other photometric
measurements in five bands between 30 and 196 microns made from a balloon
borne telescope are reported by Olthof (1974).

Correlation of the infra-red flux with the radio-continuum flux has always been important for HII regions and initially it was popular to plot one against the other when, to limited accuracy, a proportional relationship appeared to exist. This proportionality was 'encouraged' by the fact that two objects, identical in all respects but at different distances would automatically fall on the line. However, the basic proportionality was still to be explained and an attempt was made to do this in terms of the resonantly scattered Lyman α in the HII region eventually being absorbed by any dust grains present, the energy being re-radiated in the infra-red. In this way it was easy to establish a simple proportionality between the radio-continuum intensity and the infra-red. Unfortunately the infra-red exceeds the Lyman α energy by large factors and it is necessary to consider other processes, in particular direct absorption of the continuum radiation by dust in and around the HII region. Following Petrosian et al (1972) the infra-red luminosity is given by

$$L_{IR} = L_\alpha + (1-f)\langle h\nu \rangle_{Lyc} S + L_{\nu < Lyc}\left(1 - e^{-\tau_o'}\right)$$

where f is the fraction of Lyman continuum radiation absorbed by the gas. It turns out that to a good approximation $f = e^{-\tau}$ where $\tau$ is the dust optical depth in the ionised region. $\tau_o'$ is the effective optical depth for $h\nu < 13.6\text{eV}$ in the HII and surrounding HI region.

For particular cases one can get a good estimate of the value to be assigned to the terms in this equation. $L_{IR}$ is the total luminosity as measured in the infra-red and is quite a good estimate of the luminosity of the exciting star. Then, knowing the luminosity of the exciting star, a continuum radio flux measurement at a sufficiently high frequency will enable f to be calculated. A value for Lα, the Lyman α luminosity, can be obtained from the radio measurements; Lα is usually sufficient to supply the 1 - 25 μm IR flux. In this way the overall energy balance for the object can be investigated.

With this model in mind, the total far infra-red luminosity has been plotted, in Figure 1, against the flux of Lyman continuum photons (S) as deduced from the radio measurements for a set of photometric observations made with the University College London 40 cm balloon borne telescope system. The integrated value of the infra-red luminosity has been deduced from the peak signal and the infra-red size, if known, or else the radio size and a correction of 1.48 has been applied to allow for the limited bandwidth.

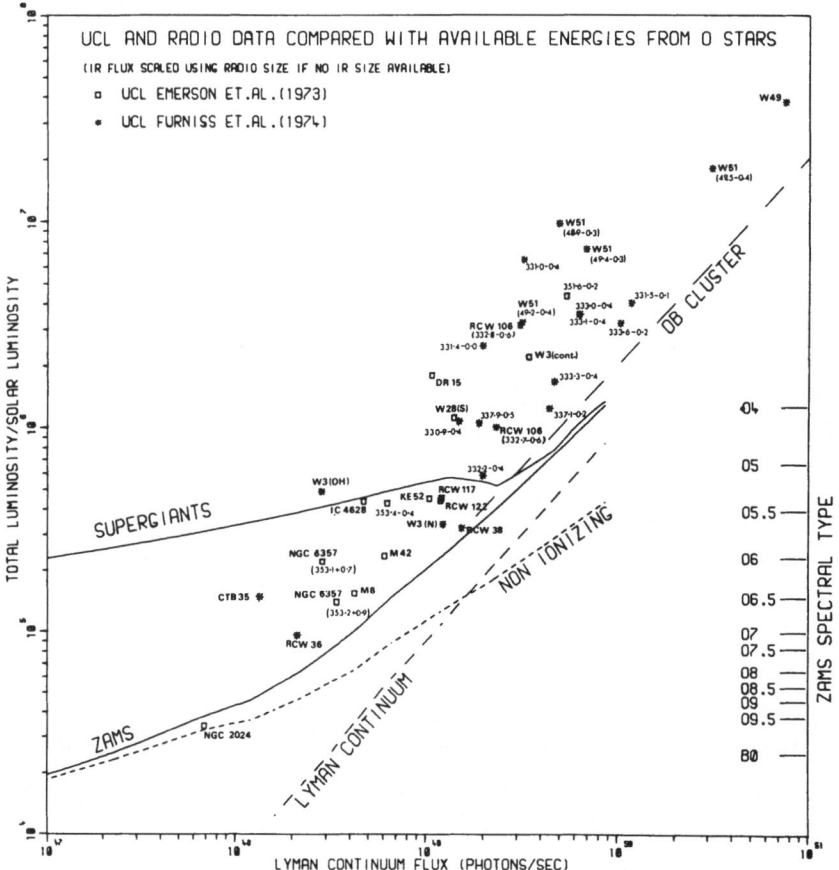

Fig. 1   Total Infra-Red Luminosity plotted against the Flux of Lyman Continuum Photons as deduced from Radio Observations

A plot of this type is likely to provoke severe criticism but before mentioning the Zero Age Main Sequence curve it is perhaps worth considering the following. The IR luminosity is a reasonable estimate of the stellar luminosity – at least it gives a lower limit for the stellar luminosity – while the Lyman continuum flux can only be less – not more – than the flux from the star without any dust absorption. Thus the points tend to be to the left and down a little relative to their position on the main sequence. Nevertheless it is possible to draw a boundary curve such that all the objects lie above it - and this curve would in fact be in quite good agreement with the set of stellar parameters as plotted (Panagia, 1973). These parameters are based on the absolute magnitude (Conti & Alschuler, 1971) and temperature (Conti, 1973) scales derived for a set of homogeneous observations of O type stars and also on the model atmosphere calculations of Auer and Mihalas (1972). NGC 2024 does in fact fall below the line but this measurement is thought to be low, particularly when compared with that of Harper (1974).

Accepting the diagram at its face value, and assuming the energising star is on the main sequence, the dust optical depth in the HII region corresponds to the horizontal distance the object is to the left of the ZAMS line. (Fraction absorbed by gas $\sim e^{-\tau}$).

All three components of W3 have been observed and are shown on the plot. The main continuum source appears to have a large optical depth but this is deceptive, as it is known to have many components including IRS5 which alone could account for 25% of the IR flux and, assuming it to be a dust embedded protostar (Aitken & Jones, 1973) would give no continuum flux. The remaining luminosity is accounted for by the two O5 stars identified by Wynn-Williams, Becklin & Neugebauer (1972), and Beetz, Elsässer & Weinberger (1974). However, the corresponding Lyman continuum flux is more than twice that required to account for the radio flux at 2 cm (Schraml & Mezger), so that some absorption in the HII region appears likely. However, it is dangerous to be too specific in explaining such a complex region in such broad terms.

Of the other components, W3 (OH) has a large IR luminosity, and very little radio flux, implying a large optical depth indicative of a young region in which the HII region has not developed fully. W3 (N) has also been observed and here the optical depth is less.

Another object of interest is G 333.6 - 0.2, with an IR luminosity of 3 x $10^6$ $L_\odot$ and a radio flux of 84 Jy, one of the brightest radio sources measured at 5 GHz by Shaver & Goss (1970). This HII region is surrounded by locally obscuring dust (Becklin et al, 1973) and therefore the measured luminosity approximates closely to the true luminosity of the stars energising the region. It is seen that the U.V. optical depth is small and that in this case the bulk of the IR radiation probably originates in the obscuring dust shell which surrounds the ionised gas. This picture is borne out by the ground based observations of Aitken & Jones (1974), who show that the absence of a silicate feature can be explained by the obscuring cloud modifying any emission features from the central HII region. The optical depth of the inner source is found to be $0.02$ at 10.4 μm corresponding to 0.1 to 0.25 in the visible. This compares quite well with the relatively small optical depth as deduced from the far infra-red observation ($\sim 0.4$ for the Lyman continuum).

The variation in dust absorption between the visible and U.V. is not well known. Any significant variation in the absorption by dust of photons capable of ionising hydrogen and not helium and those with sufficient energy to ionise helium will affect the observed ratio of He and H recombination lines. This is discussed below.

## RELATIONSHIP BETWEEN THE OBSERVED HELIUM ABUNDANCE AND THE IR EXCESS

Churchwell, Mezger and Huchtmeier (1974) presented values of $<N(He^+)/N(H^+)>$ for 39 galactic HII regions obtained from observations of Helium and Hydrogen recombination lines. Values of this ratio were found to lie between 0.06 and 0.12 for spiral arm HII regions and were less than or equal to 0.02 for the giant HII regions in the galactic centre. They suggested that the present abundance of Helium in our Galaxy, $\frac{N(He)}{N(H)}$, was 0.10 and that the smaller values of $<N(He^+)/N(H^+)>$ were due to the contraction of the $He^+$ stromgren sphere relative to the $H^+$ sphere due to greater absorption of Helium photons by dust. In cases where there is significant absorption of ionising photons the value of $<N(He^+)/N(H^+)>$ would be expected to be smaller and the corresponding value of the Infrared excess ($\frac{\text{I.R. Luminosity}}{\text{Lyman } \alpha \text{ Luminosity}} -1$) to be larger. This possible correlation was tested by Churchwell et al using the infra-red data for 17 sources obtained by a number of different observers. The result was not completely conclusive but showed that a correlation, as suggested above, was possible. As pointed out by the authors, it is difficult to combine IR data from different telescopes with various fields of view and spectral bandwidth. In particular, the values for G 348.7 - 1.0 and W31

are strikingly different from the proposed relationship. Recent values
for these objects (W31 - Olthof (1974) and G 348.7 - 1.0 - Emerson et
al (1973) are in better agreement.

The problems of using I.R. data obtained with different beam sizes
etc. can be overcome if a set of measurements obtained with one telescope
system can be compared with a corresponding set of Radio continuum
measurements, obtained with the same size beam. Fortunately, such a
'pair' exists between the U.C.L. data and the measurements at 5 GHz of
Shaver and Goss, both effectively using a H.P.B.W. of 4 arc min. Making
the assumption that this 4 arc min beam measures all, or the same
fraction, of the total Radio and Infra-red flux, peak fluxes rather
than integrated fluxes can be used to calculate the I.R. excess. This
assumption leads to smaller errors than would be introduced by allowing for different beam sizes.

Using the expression for the number of Lyman Continuum photons
(Rubin, 1968) with $T_e$ =10000 K and assuming that each Lyman continuum
photon gives rise to one Lyman α photon, the I.R. excess is given by

$$\text{I.R. excess} = 12.2 \times 10^{10} \left( \frac{\text{Peak I.R. flux } (wm^{-2})}{\text{Peak Radio flux } (Jy)} \right) - 1$$

In this expression the I.R. flux has been increased by a factor of 1.48
to allow for energy outside the passband (40 - 350 μm) assuming a source
colour temperature of 80 K.

In Figure 2 the values of ionised helium abundance (Churchwell et
al, 1974) against Infra-red excess is shown for the following 21
sources:

Emerson et al (1973)      ▢      Reading left to right Fig. 2.

NGC 6357 (G 353.2 + 0.9), M8, NGC 6357 (353.1 + 0.7), M42, RCW 117,
RCW 122, NGC 2024.

Furniss et al (1974)   ✷

RCW 38,  G 333.6 - 0.2,  G 331.5 - 0.1,  G 333.3 - 0.4,  W 51,  W 49,
and G 333.0 - 0.4,  G 333.1 - 0.4.

Emerson et al.  Preliminary Southern Hemisphere Data   ✕

RCW 49,  RCW 57 (G 291.6 - 0.5),  G 298.9 - 0.4,  RCW 57 (291.3 - 0.7),
G 316.8 - 0.1,  G 305.4 + 0.2.

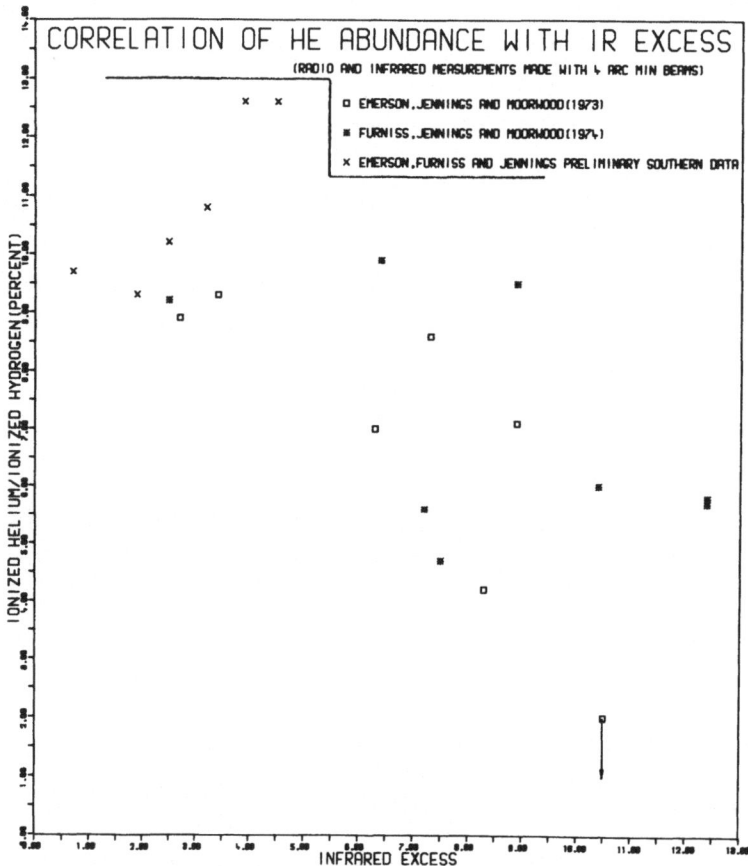

Fig. 2  Ratio of Ionised Helium to Ionised Hydrogen as a function of
the Infra-red excess

## Accuracy

$$\frac{N(He^+)}{N(H^+)}$$ Standard Error $\sim \pm 1$ or 2% (see Table 2 Huchtmeier et al 1974)

I.R. Excess Standard Deviation $\sim \pm 20\%$ of (I.R. Excess + 1)

At least qualitatively, Figure 2 shows a correlation between the Helium abundance and the I.R. excess, the spread of points being compatible with the accuracy of the measurements. Assuming as a first approximation that a linear relationship exists, the least squares line has a slope of $-0.54.10^{-2}$ and cuts the abundance axis at 11.4 per cent, a result compatible with the galactic helium abundance. It is questionable whether NGC 2024 should be included in this plot as it is not clear what is the exciting star and whether sufficient radiation is available short of 504 Å to ionise the helium (Harper, 1974). If a line is fitted ignoring NGC 2024 it has a slope of $-0.47.10^{-2}$ and cuts the abundance axis at 11.2 per cent. For the remainder of the objects it is reasonable to assume that the exciting star is of a sufficiently early type.

Assuming that such a correlation does exist, the explanation is not straightforward and very many different parameters are involved.

## MAPS OF HII REGIONS

A number of contour maps have been produced during the last few years at a resolution of a few arc minutes with relatively small telescopes. One of the first was by Hoffmann et al(1971a) of the Galactic Centre region at a wavelength of 100 μm and with a H.P.B.W. of 12 arc mins. By comparison of this map with the corresponding radio map, the remarkable agreement which is usually found between the infra-red and radio contours of thermal HII regions can be demonstrated.

A map of the galactic centre region at somewhat higher resolution (5.75' x 6.50') is shown in Figure 3. This was obtained by the University College London group by making a raster scan consisting of 50 cuts across the galactic plane during a balloon flight. Again, comparison with a radio map at comparable resolutions shows correlation between the majority of the features - of particular interest, of course,

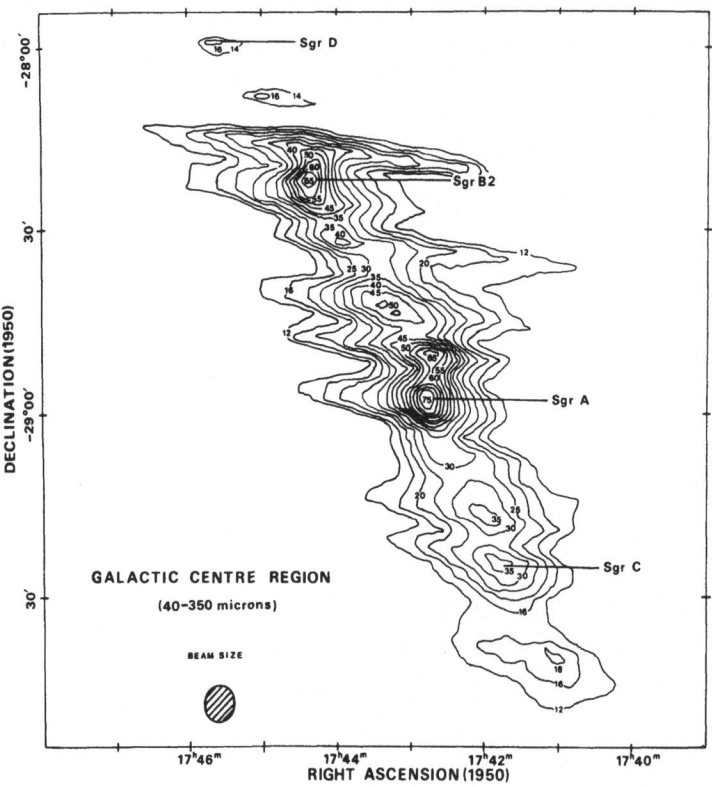

Fig. 3. Map of the Galactic Centre Region

       Wavelength Band:   40 - 350 μm
       Beam Size:          5.75 arc min. x 6.5 arc min.

are those features which do not correlate, as, for example, the arc to
the north of Sgr A, which does not extend round to the east as far on
the I.R. map as it does on the radio, due, it is thought to this being
a non-thermal radio source (Downes - Private Communication).

     The Galactic Centre region consists of a collection of compact
sources superimposed on a background which is more extended than in the
radio.  Figure 4 shows a single cut through Sgr B2 at a resolution of
3.75 x 4 arc min, and shows that the central peak is comparable in width

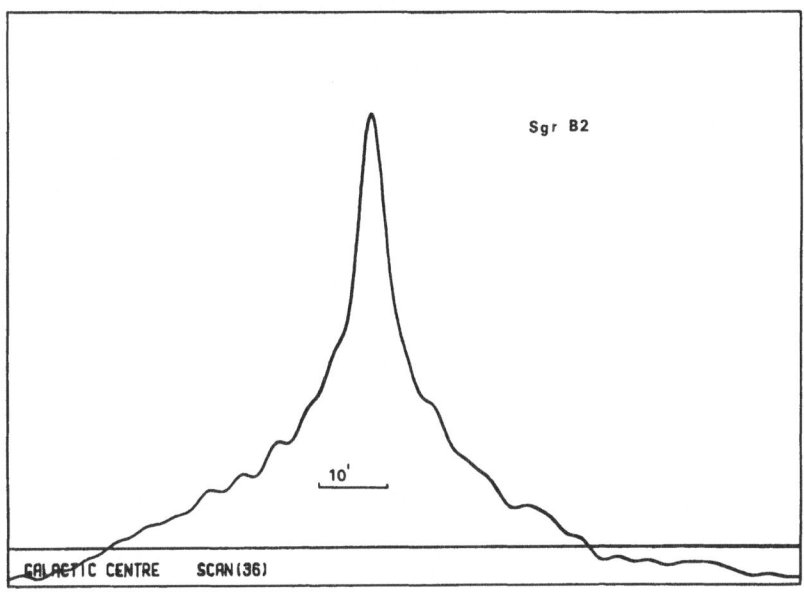

Fig. 4.  Single Cut Through Sgr B2

       Wavelength Band:    40 - 350 μm
       Angular Resolution: 3.75' along scan, 4' across

to the 3 arc min found by Kaptizky and Dent (1974) at 15.5 GHz. Rieke
et al (1973) at 350 μm showed that this peak was 1 arc min in R.A.
and 2 arc min in declination.  More recently Righini, Simon, Joyce and
Gezari (1975) mapped this region at 350 μm and an angular resolution
of 56 arc seconds.  Their contour map is shown in Figure 5 on which the
compact HII regions detected interferometrically by Martin & Downes
(1972) have been marked.

The Orion nebula has been mapped by Ade et al (1974) at 1.4 mm, by
Harvey et al (1974) at 1 mm and in the 2 cm $H_2CO$ line, by Gezari et al
(1974) at 350 μm, by Soifer and Hudson (1974) at 400 μm, at 91 μm by
Harper (1974) from the Lear Jet and at 69 μm by Fazio et al (1974) from
a balloon borne 102 cm telescope.  This latter map was able to separate

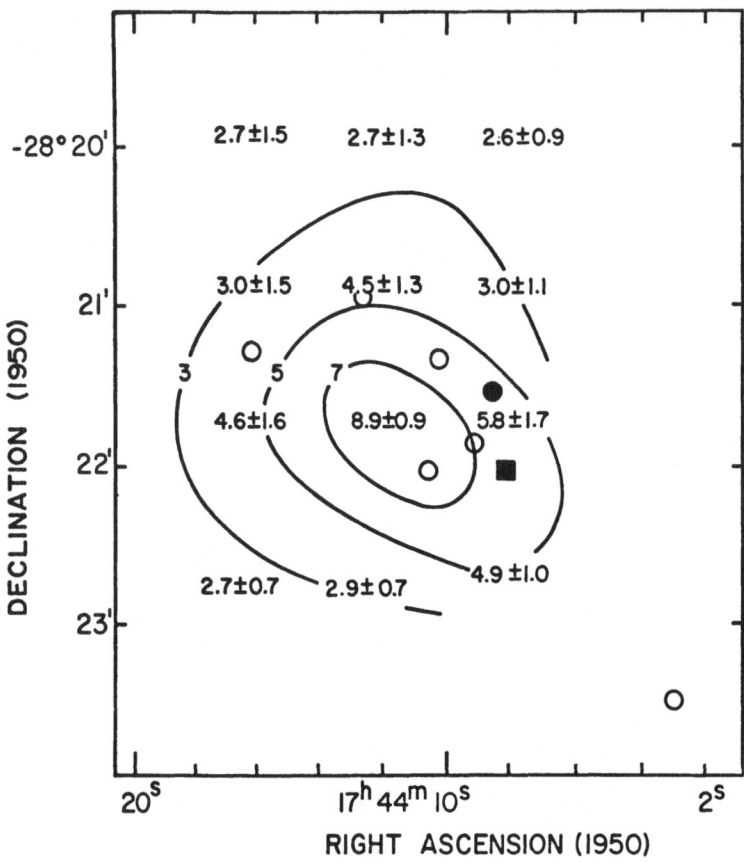

Fig. 5.   350 μm of Central Part of Sgr  B2 at 56 sec resolution
Flux unit:      $10^{-23}Wm^{-2}Hz^{-1}$ in the beam
Filled circle:  Peak Position at 350 μm by Rieke et al (1973)
Filled square:  Peak 3.5 mm. position (Hobbs et al, 1971)
Open circles:   Compact HII regions (Martin & Downes, 1972)
(from Righini et al, Ap. J. (Letters), 195, L77)

out two main components, one centred on the Kleinmann Low nebula and the second on an emission peak near $\theta^2$Ori.  On the same balloon flight Fazio et al obtained a map of W3 which showed the continuum and OH sources and in addition two small sources not previously observed.  A map of the same area (Figure 6) has also been obtained by the University College London group (Furniss et al, 1974), in which, as well as the continuum and OH sources, the northern source G 133.8 + 1.4 has been included.  This map shows an extension of the continuum source to the east as well as to the

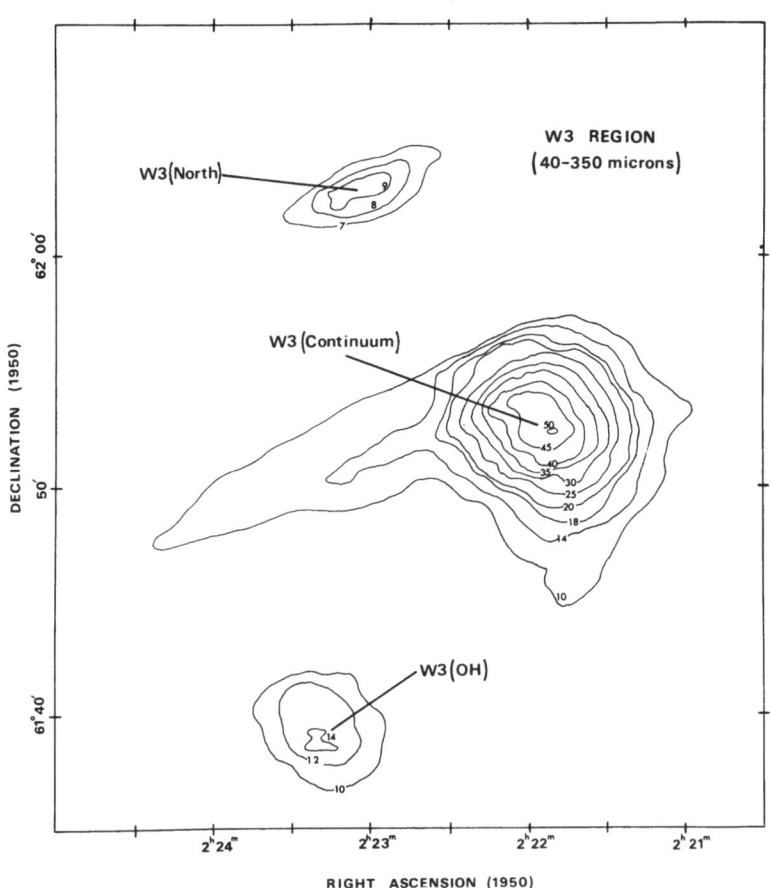

Fig. 6. Map of W3

       Wavelength Band:    40 - 350 μm
       Beam size:         5.5 arc min x 5.5 arc min.
       Contours normalised to 50 for peak value of W3 continuum.

south. The eastern extension coincides with the optically visible nebula
IC 1795, and also with a region in which CO emission has been detected
(Wilson et al, 1974). Fig. 6 is at a lower resolution than the map due
to Fazio et al and the sensitivity is not sufficient to show the two
small sources which they detected. It differs slightly from the
preliminary map presented at the 8th ESLAB Symposium.

SPECTROSCOPY

Spectroscopic measurements in the far infra-red are in their
infancy and to date the spectral resolution has been insufficient for the
detection of lines.  However, the shape of the continuum of a number of
sources has been determined as for instance, in the case of the Orion
nebula, by Erickson et al (1973), using a Michelson interferometer and
by Ward & Harwit (1974), using a cooled grating, both sets of observations being made from the Lear Jet.  Measurements such as these are
sufficient to begin to construct models for which possible temperature
distributions of the dust and corresponding emissivity can be determined.
In the case of W51 a low resolution spectrum of the principal component,
G 49.5 - 0.4, has been obtained (Alvarez et al, 1974), and showed some
evidence for the radiating particles being composed of ice.

Mention should also be made again here of the careful multifilter
observations of Harper (1974) which enabled the colour temperature of a
number of sources to be found.

It is confidently expected that observation of lines will be made
in the near future, both from NASA's C 141 aircraft and also from balloon
borne systems, which have the advantage of getting above a lot more of
the water vapour in the atmosphere.

CONCLUSIONS

As yet only a small beginning has been made in the study of
astronomical objects at $\lambda > 40$ $\mu$m.  New techniques are being developed
all the time, better platforms are becoming available and the quality of
the data is steadily improving.  By the 1980s, hopefully with super-
heterodyne techniques at long I.R. wavelengths and 2 - 3 m telescopes
in space, our understanding of these complex energy sources should be
very much more advanced.

ACKNOWLEDGEMENTS

It is a pleasure to thank all the members of the Infra-red Group
at University College London for their assistance over the last few
years and in particular Jim Emerson for his enthusiastic help in pre-
paring this paper.

REFERENCES

Ade, P.A.R., Clegg, P.E. and Rather, J.D.G. 1974, Ap. J. (Letters),
189, L23.

Aitken, D.K. and Jones, B. 1973, Ap. J., 184, 127

Aitken, D.K. and Jones, B. 1974, M.N.R.A.S., 167, 11P.

Alvarez, J.A., Furniss, I., Jennings, R.E., King, K.J. and Moorwood,
A.F.M. 1974, Proc. 8th ESLAB Symposium 'HII Regions and the
Galactic Centre'.

Auer, L.H. and Mihalas, D. 1972, Ap. J. Suppl., 24, 193.

Becklin, E.E., Frogel, J.A., Neugebauer, G., Persson, S.E. and Wynn-
Williams, C.G. 1973, Ap. J. (Letters), 182, L125.

Beetz, M., Elsasser, H. and Weinberger, R. 1974, Astr. and Ap., 34, 335.

Churchwell, E., Mezger, P.G. and Huchtmeier, W. 1974, Astr. and Ap.,
32, 283.

Conti, P.S. 1973, Ap. J., 179, 181.

Conti, P.S. and Alschuler, W.R. 1971, Ap. J., 170, 325.

Emerson, J.P., Jennings, R.E. and Moorwood, A.F.M. 1973, Ap.J., 184, 401.

Emerson, J.P., Furniss, I. and Jennings, R.E. Southern Hemisphere Data.
To be published.

Erickson, E.F., Swift, C.D., Witteborn, F.C., Mord, A.J., Augason, G.C.,
Caroff, L.J., Kunz, L.W. and Giver, L.P. 1973, Ap. J., 183, 535.

Fazio, G.C., Kleinmann, D.E., Noyes, R.W., Wright, E.L.,    Zeilik II,M.
1974, Ap. J. (Letters), 192, L23.                          and Low,F.J.

Friedlander, M.W., Goebel, J.H. and Joseph, R.D. 1974, Ap. J. (Letters),
194, L5.

Furniss, I., Jennings, R.E. and Moorwood, A.F.M. 1972, Nature, 236, 6.
      "            "                 "            1974, Proc. 8th ESLAB
Symposium 'HII Regions and the Galactic Centre.'

Gezari, D.Y., Joyce, R.R., Righini, G. and Simon, M. 1974, Ap.J.(Letters) 191, L33.

Harper, D.A. 1974, Ap. J., 192, 557.

Harper, D.A. and Low, F.J. 1971, Ap. J. (Letters), 165, L9.

Harper, D.A., Low, F.J., Rieke, G.H. and Armstrong, K.R. 1972, Ap. J. (Letters), 177, L21.

Harvey, P.M., Gatley, I., Werner, M.W., Elias, J.H., Evans II, N.J., Zuckerman, B., Morris, G., Sato, T. and Litvak, M.M. 1974, Ap. J. (Letters), 189, L87.

Hobbs, R.W., Modali, S.B. and Maran, S.P. 1971, Ap.J.(Letters), 165,L87.

Hoffmann, W.F., Frederick, C.L. and Emery, R.J. 1971(a), Ap.J.(Letters), 164, L23.

Hoffmann, W.F., Frederick, C.L. and Emery, R.J. 1971(b), ibid.,170,L89.

Joyce, R.R., Gezari, D.Y. and Simon, M. 1972, Ap.J.(Letters), 171, L67.

Kapitzky, J.E. and Dent, W.A. 1974, Ap.J., 188, 27.

Martin, A.H.M. and Downes, D. 1972, Ap.Letters, 11, 219.

Olthof, H. 1974, Astr. and Ap., 33, 471.

Panagia, N. 1973, Astron. J., 78, 929.

Petrosian, V., Silk, J. and Field, G.B. 1972, Ap.J.(Letters), 177, L69.

Rieke, G.H., Harper, D.A., Low, F.J. and Armstrong, K.R. 1973, Ap. J. (Letters), 183, L67.

Righini, G., Simon, M., Joyce, R.R. and Gezari, D.Y. 1975, Ap. J. (Letters), 195, L77.

Rubin, R.H. 1968, Ap. J., 154, 391.

Schraml, J. and Mezger, P.G. 1969, Ap.J., 156, 269.

Shaver, P.A. and Goss, W.M. 1970, Aust. J. Phys. (Astrophys Supplement No. 14), 133 - 196.

Soifer, B.T., Pipher, J.L. and Houck, J.R. 1972, Ap.J., 177, 315.

Soifer, B.T. and Houck, J.R. 1973, Ap. J., 186, 169.

Soifer, B.T. and Hudson, H.S. 1974, Ap.J.(Letters), 191, L83.

Ward, D.B. and Harwitt, M. 1974, Nature, 252, 27.

Wilson, W.J., Schwartz, P.R., Epstein, E.E., Johnson, W.A., Etcheverry,
R.D., Mori, T.T., Berry, G.G. and Dyson, H.B. 1974, Ap. J., <u>191</u>,
357.

Wynn-Williams, C.G., Becklin, E.E. and Neugebauer, G., 1972, M.N.R.A.S.
<u>160</u>, 1.

# COMPACT H II REGIONS, MAINLY RADIO OBSERVATIONS

H.J. Habing

Sterrewacht, Leiden,
The Netherlands

## ABSTRACT

Small dense clumps of ionized gas      exist during various stages
in the evolution of H II regions. They can be manifest as bright rims or
as very small, very compact H II regions surrounding new born stars.
Examples of both types of objects are given and some speculation is made
about how massive stars are formed.

## I. HISTORY AND QUESTIONS

At the end of the sixties radio observations showed, that H II regions sometimes contain very compact (diameter < 1 pc) components with

high surface brightness (emission measure $EM > 10^6 \, cm^{-6} \, pc$). The first completely convincing case became the isolated compact H II region DR 21, which Ryle and Downes (1967) had mapped with the Cambridge aperture synthesis telescope with a resolution of 25 arcsec. However, earlier observations with single dishes (e.g. Schraml and Mezger, 1969) had already done much to break the grounds. At that time the general conclusion was that compact H II regions are all very young, because their small sizes and high densities should lead to rapid expansion. This conclusion stimulated much enthusiasm because it might make it possible to study conditions of formation of massive stars.

Nevertheless, the enthusiasm appeared somewhat premature, because exactly the same way of estimating ages had failed already in the case of the Orion Nebula. This nebula was recognized as a compact H II region (avant-la-lettre!) in 1959 by Osterbrock and Flather. Dynamical models (Vandervoort, 1964) suggested that the nebula is at most 20,000 yr old. Such an age conflicts with the age of the Trapezium stars, which are definitely on the main sequence. Clearly there is a paradox here, which may have been overcome only recently (see section II.1).

From the beginning until now a few important questions have remained to be answered by the study of compact H II regions:
"1) How does a compact H II region come into being, how does it evolve, what is its ultimate fate?
2) What does this evolution tell us about the conditions during the formation of massive stars?"
I think that we can formulate at present an overall but qualitative and schematic answer to these questions. Even though many details and steps are still not understood it is possible to illustrate this schematic answer with numerous examples. I will quote only a few in order to remain within the boundaries of this Review.

This answer has become available mostly because of observational progress. Essential is the availability of infrared data, both high resolution, near-infrared measurements and low resolution, far-infrared observations. Since the topic of infrared astronomy has been dealt with extensively in the two preceding reviews by Wynn-Williams and by Jennings I will not enter into a general discussion of this point. A further major step has been the availability of a larger sample of H II regions studied with improved radiotelescopes. Finally it proves to be of increasing importance to <u>include optical observations</u>, because of the rich information content of optical spectra and of the high angular resolution available. I hope that you find this point proven at the end of my review.

Finally let me outline briefly where the progress in radioastronomical observations is to be found.

$1^o$. Angular resolution has been increased. For isolated and symmetrical sources the 3-element N.R.A.O.-interferometer reaches about 1 arcsec resolution at 3 cm. But otherwise the Cambridge 5 km array appears superior with a resolution of 2-3 arcsec at 6 cm. At Westerbork we reach 6 arcsec at 6 cm.

$2^o$. Sensitivity has been improved and, depending on the presence of confusing sources in the field, fluxes may be detected in Westerbork and at N.R.A.O. as low as 4-5 m.f.u.

$3^o$. Dynamic range has been increased. This is an important fact because very compact, weak sources that are sometimes found near less compact, stronger sources, appear to be of fundamental interest.

## II. <u>ANSWERS</u>

### II.1. <u>Two types of compact H II regions.</u>

Compact H II regions are characterized by small diameters and high emission measures, and consequently by high densities, $n_e \geq 10^4$ cm$^{-3}$. Since the electron temperature $T_e \approx 10^4$ K, the gas pressure in the compact H II region exceeds that of the general interstellar medium by at least a factor $10^4$. Gravitational attraction by the ionizing star can be ignored. Hence the observed ionized gas will rapidly disperse. However, what the observed situation will look like in the course of time depends on the boundary conditions. Two extreme situations have been met in reality. The <u>first</u> situation results when the ionizing star is located in the center of a density increase, because it was born there. Since the surrounding medium has a considerably lower density a rapid expansion will follow with an expansion time $t = 1/c_i$, where 1 is the diameter of the object and $c_i$ the sound speed in ionized gas. Typically $t \lesssim 10^4$ yr. This value is confirmed by model calculations made e.g. by Mathews (1969). The second extreme situation is found in the type of object usually called a "bright rim" (Pottasch, 1956; Mathews and O'Dell, 1969). The ionizing star is at some distance outside a dense neutral cloud and the ionization front is slowly eating into the cloud. The ionized gas flows away from the front. The D-type ionization front advances into the cloud with a speed of roughly $u = c_n^2 / c_i$, where $c_n$ is the sound speed in the neutral gas, say $c_n \sim 1$ km s$^{-1}$. The speed of the ionization front is so low that the cloud erodes only very slowly. Hence the observed situation can last for times equal to the main sequence life time of the ionizing star.

In both situations the appearance of the object can be quite similar: we see very dense, small blobs of ionized gas. The example of the Orion Nebula shows that it is meaningful to make a distinction between the two varieties of compact H II regions. In Section I I noticed already the old paradox between the main sequence position of the Trapezium stars and the dynamical age of the Orion nebula. The dynamical age was based, essentially, by taking the nebula to be an isolated, freely expanding object. However, Zuckerman (1973; and this Symposium) proposed a model in which the Orion Nebula is bounded on one side by a neutral cloud. Such a model is close to that of the bright rim variety; the situation is quasi stationary. Hence the nebula may be very old and the paradox is resolved. Another object of which the interpretation has recently given some trouble is NGC 2175 or S 252 – it probably is an old H II region (Grasdalen, 1975); Orion B or NGC 2024 may be another candidate. One thus requires additional criteria to see to what variety a specific compact H II region belongs. Unfortunately, simple criteria do not exist and I will proceed by using ad-hoc criteria. The only simple criterion with some general use is the flux density ratio S(20μm/S(6 cm), which most probably measures the temperature of the dust inside the ionized gas. High temperatures (and high ratio's) are expected only when the ionized gas is so close to the ionizing star that a genetic relation between the two may be assumed.

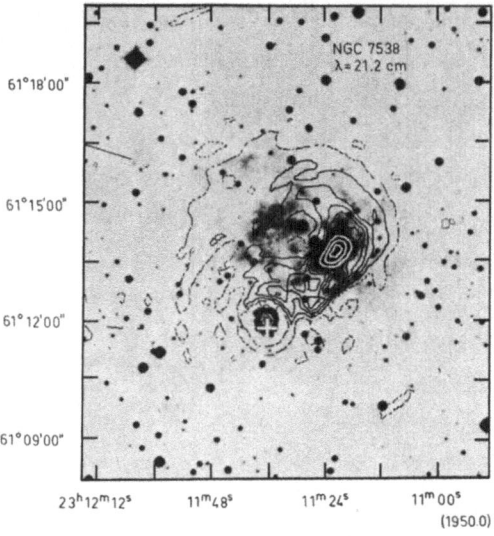

Figure 1. NGC 7538. Isophotes of the continuum radiation at 21 cm super-imposed on the blue photograph from the National Geographic Society - Palomar Sky Survey. The ringlike object to the south-east of the nebula (NGC 7538 G) containing the white cross is a point source in the radio-map and does not correspond to anything conspicuous in the optical photo-graph. The cross indicates the position of the OH maser. This figure is from Habing et al., 1972.

Figure 2. Isophotes of NGC 7538 G, the point source in figure 1 as mea-sured at 6 cm continuum with 10 times higher resolution (Martin, 1973).

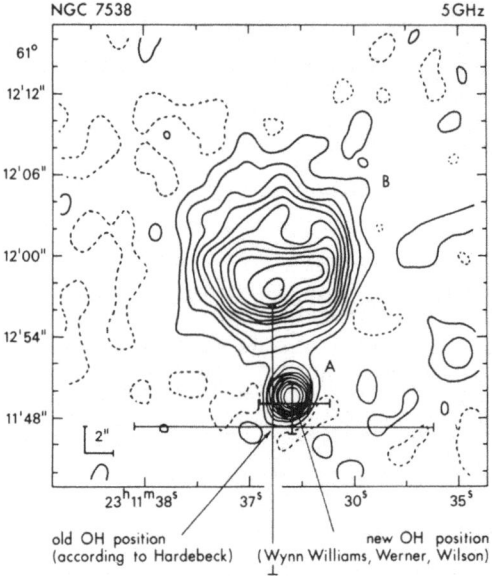

## II.2. Expanding compact H II regions.

Some expanding compact H II regions are shown in figure 2. Figure 1 gives the general area. A small, dense H II region, NGC 7538 or S 158, of about 5 pc diameter shows up almost equally well on a red Palomar Sky Survey print and a 20 arcsec radiomap at 21 cm. The nebula contains two exciting stars, one of type 07, the other, discovered recently by Gluzhkov et al. (1975), is probably of type 05. The radio map (Habing et al., 1972) shows the presence of a very compact radio source south of the nebula. Recent measurements with higher angular resolution made in Cambridge (Martin, 1973) have shown that this radio source consists of three compact H II regions with diameters varying from $6 \times 10^{16}$ to $4.5 \times 10^{17}$ cm. All three H II regions are intense near-infrared emitters, by which I mean that the ratio of flux-densities, $S(20\mu m)/S(6 cm)$, is much larger for these sources than for sources of larger diameter like, e.g., M 17. In addition one of the three sources coincides with a "main line" OH maser to within the observational accuracy of only a few arcsec, that is to within 0.1 pc. (This latter coincidence is by no means unique. Out of a sample of 14 main line OH masers, Habing et al. (1974) found 7 positional coincidences between a maser and an H II region of only a few arcsec in diameter ). Finally the three sources are embedded in an intense molecular cloud, with several of those molecular lines that usually indicate very high densities, notably $H_2CO$ in emission at 6 cm (Downes and Wilson, 1974). The fact that the H II regions coincide with such hot infrared sources strongly suggests (although it does not yet prove the point) that the H II regions have a powerful energy source inside. Hence they are probably very young. The presence of an OH maser makes it likely that the three sources are embedded in dense neutral material and the overall impression is that here one has three genuine cocoon stars together.

Several other cases of H II regions with diameters less than 0.1 pc and with probably the same nature are known - see for example table 1 in Habing et al. (1974). However, infrared and optical data are lacking for most of these. Let me give two more examples, not included in the table mentioned. The first example (S 157A) is shown in figure 3. Again the Hβ photograph and the radio map are quite similar. But there is one remarkable difference. On the red Palomar Sky print it looks as if in the lower right hand corner of S 157A there is a blackmark. But in the same corner in the radio map two small radio sources appear. They are called

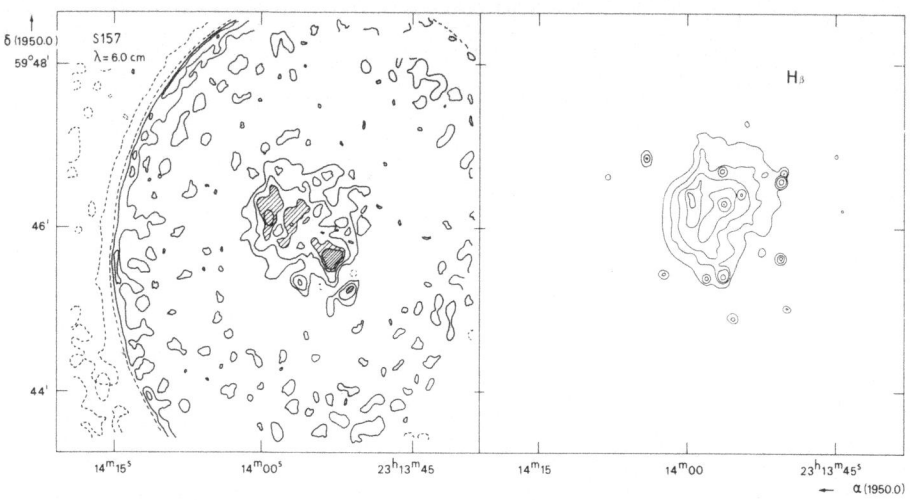

Figure 3. Isophotes of Hβ intensity and of the 6 cm continuum of S 157. The original Hβ photograph was taken at the Haute Provence Observatory, France. The 6 cm observations were obtained by F.P. Israel with the Westerbork Synthesis Radio Telescope.

S 157B;    recent observations at Cambridge (Wynn-Williams, private communication) and Westerbork show them to be about 2 arcseconds in size. Observations by Becklin and Neugebauer (private communication) indicate that S 157B is an unusually strong infrared source at 20 μm.

My third candidate for a very compact, but expanding H II region is K3-50, an object in Cygnus. K3-50 is a very red stellar object with a strong emission line spectrum. It coincides to within a few arcsec with a poorly mapped infrared source and with a complicated radiosource. Recently Israel (in preparation) has put forward an interesting interpretation of all the data (optical, infrared and radio) in which he distinguishes a very compact H II region of only 0.1 pc in diameter at the edge of a compact H II region of approximately 0.5 pc. The very compact H II region of 0.1 pc is totally obscured by $A_v = 50^m$ at least and probably contains a cocoonstar. Its thermal radio flux-density requires an O4 star. In the slightly larger H II region of 0.5 pc the exciting star is visible (K3-50), possibly through a hole in the surrounding dust. The star is reddened by $A_v = 7$ to $10^m$.

A fourth likely candidate for a cocoon star is G 69.54-0.98, a 1 arc-

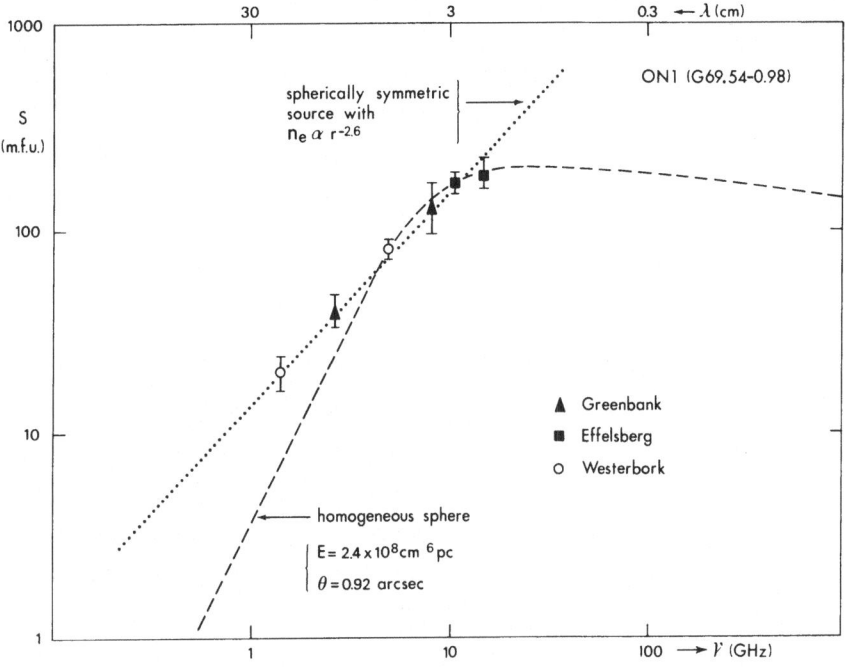

Figure 4. The spectrum of G 69.54-0.98. The N.R.A.O. measurements are from K. Lo (1974), the Effelsberg and Westerbork data were supplied by H. Matthews and A. Winnberg.

sec thermal radiosource coinciding with the OH maser ON 1. There is very
little known about the source. It is a weak near infrared source, but
this may mean that most of the energy is transformed into far infrared
emission.Its radio spectrum is remarkable (figure 4). The six well determined flux density values cannot be matched by the (Bremsstrahlung )
spectrum of a homogeneous sphere filled with electrons. The well-deter-
mined 11 and 21 cm points indicate that a straightline fit is much better.
A straightline can be explained by a model in which the electron density
decreases outward, in this case $n_e \propto r^{-2.6}$ (Olnon, 1975). There are more
sources known with a straightline spectrum (e.g. MWC 349) but they may
be evolved stars with large mass losses. To me the origin of the straight
line spectrum is a puzzle.

Finally, I would like to show another H II region called S 156 or IC 1470
(Israel, 1975, and (at optical wavelengths) Deharveng, 1974) and which
is a rather strong source at 10 and 20 $\mu$m (Cohen and Barlow, 1973). An
aperture synthesis map at 21 cm continuum shows that the 0.8 pc source
S 156 is embedded in an envelope of about 1.5 pc in size with $n_e$ (r.m.s.)
$\sim$ 600 cm$^{-3}$. Six centimeter observations show that S 156 itself has a
ring type structure with one very intense spot of approximately 10 arcsec
or 0.18 pc in diameter. The whole structure is quite similar to that in
H$\alpha$ and in NII , as is shown in figure 5. Also shown in figure 5 is the
position of the exciting star of spectral type 07, with $A_v$ = 3.6 and
V = 12.7. Mass estimates, which have considerable uncertainties, yield
about M(H II) $\lesssim$ 2 M$_o$ for the central nebula, and $\lesssim$ 20 M$_o$ for the envelope.

Rings as the one in figure 5 are quite frequently observed (Israel,
1975). The best known, and one of the most regular, is W 3A, which is
discussed in Mezgers' review of the W 3 area at this Symposium. It is
not clear at all how the rings come into existence. I am not aware of
dynamical models that explain their origin. It is conceivable that radiation pressure on grains has something to do with it. Reliable dynamic
models including grain/gas interaction hardly exist, and may be worthwhile to construct. An interesting early attempt has been made by
Mathews (1966) to explain the structure of the Rosette Nebula. But that
paper should not be taken as the last word on the subject. Apart from
the interest in the phenomenon itself, there is some interest from the
point of fragmentation of the initial cloud. How much matter was concen-

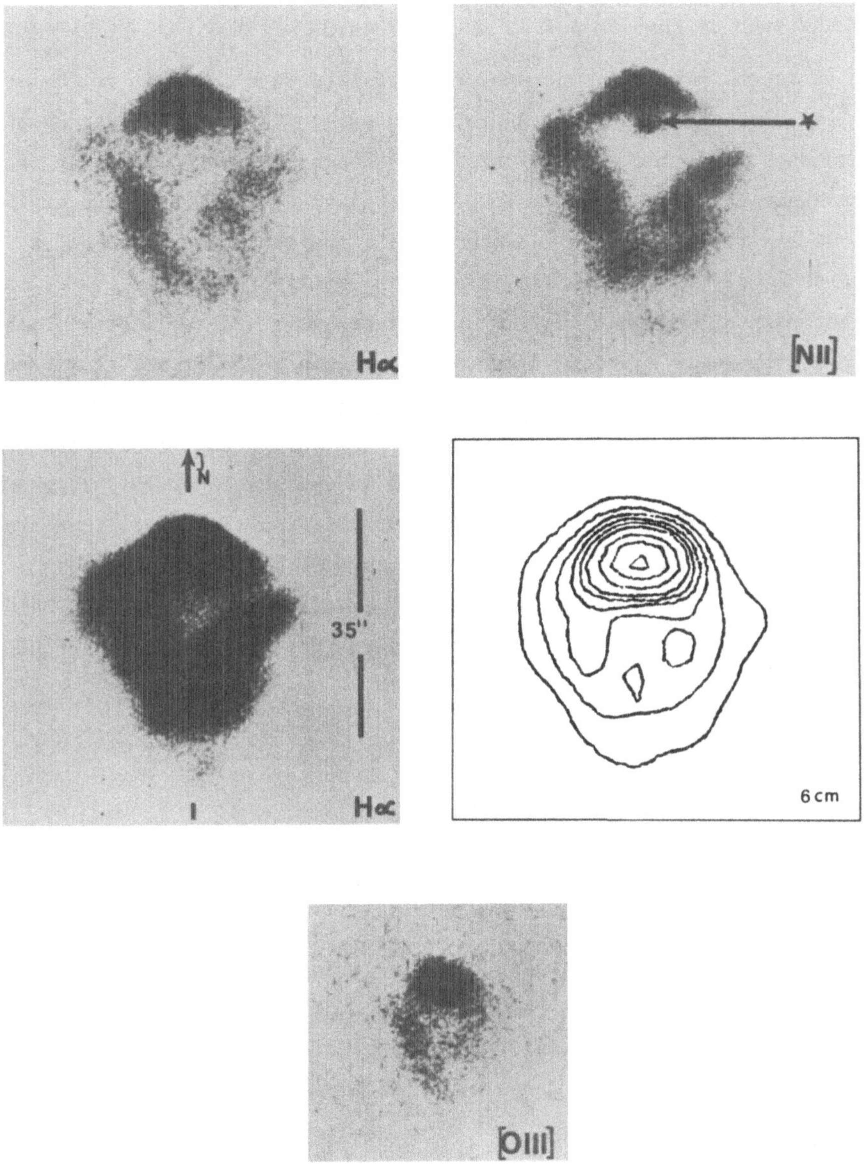

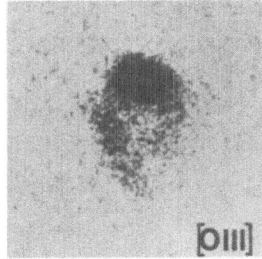

Figure 5. Photographs and 6 cm continuum isophotes of the ring of S 156. The optical data have kindly been supplied by Lise Deharveng (see Deharveng, 1974). The radiodata were obtained by F.P. Israel with the Westerbork Synthesis Radio Telescope.

trated near the star of S 156 when it was born? What was the distribution? And how much of what we see is actually swept up from the surroundings?

A second point to consider is the existence of the 1.5 pc, $n_e = 150$ $cm^{-3}$ envelope around S 156. Again, such envelopes are a quite common phenomenon – see for example the discussion of DR 21 by Harris at this Symposium. The envelope around S 156 requires at least one O8 star, but one O7 star, like the one detected, could ionize the envelope and S 156 simultaneously. I favour this latter possibility (because there is no direct sign of an additional O8 star whatsoever) and the interpretation that I propose is that the ring that we see in S 156 is the remainder of the condensation in which originally the O7 star was formed. The condensation formed by fragmentation of a general background cloud. Part of the background cloud we see illuminated in the envelope. This suggestion has two consequences: 1. There was a sort of final collapse, that started in a volume of less than the present size ($\approx$ 1 pc). 2. During this collapse most of the material ($\approx$ 20 $M_\odot$) ended up in the star and only very little ($\approx$ 2 $M_\odot$) was ejected back into the surrounding background.

## II.3 Bright rims.

In figure 6 a convincing case is given of globules ionized from the outside. The complex is referred to by various names: S 162, NGC 7635 or the "Bubble nebula". References to earlier studies may be found in Johnson (1974), and new studies by Cohen and Barlow and by Israel are forthcoming. In the blue photograph from Lick one can recognize three components: (i) the star, called BD+60$^\circ$ 2522; (ii) a spherical nebula (the Bubble); (iii) three dense blobs of ionized gas to the right of the star. The Bubble and the dense blobs of ionized gas are easily recognized in the 6 cm free-free continuum map. The star is not in the radiomap but it is probably located in a low-brightness hole in the radiomap. As far as one can tell, the star, of type Oef, is probably old and decaying (Conti, private communication). Mass loss from the star may explain the Bubble (Icke, 1973). What about the dense blobs? 6 cm radio-observations indicate r.m.s. electron densities of 4 x $10^3$ or higher (Deharveng, 1972; Israel, private communication), whereas 10 and 20 μm observations by Cohen and Barlow (private communication) indicate no detectable emission,

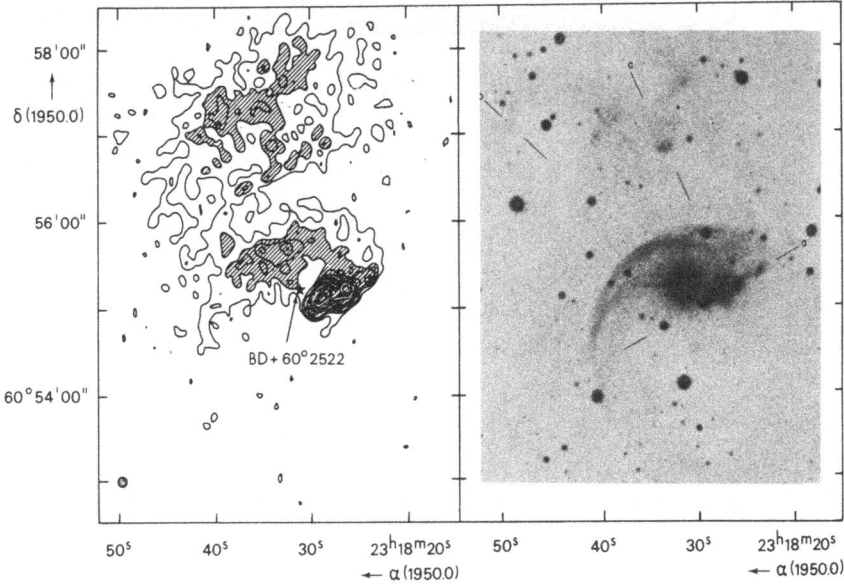

Figure 6. Isophotes of the 6 cm continuum radiation from NGC 7635 compared with a blue Lick photograph. The photograph, taken in 1934 by Stoy, has been discussed by Johnson (1974). The radio map has been obtained by F.P. Israel with the Westerbork Synthesis Radio Telescope.

however, such emission appears to come from the immediate surroundings from
the star, say, from within 10 arcsec. It therefore appears most likely
that the dense blobs have no internal energy source, but are being
ionized from the outside. This is then the same interpretation as was put
forward by Pottasch in 1956. The interpretation is strongly supported by
a 200" red photograph that shows the presence of small black globules
near the brightest emission – like in so many bright rims.

I consider NGC 7635 as an example of an "old" situation. The star is
probably evolving already off the main sequence, but at the same time it
is still interacting with its surroundings, most conspicuously with density inhomogeneities therein. Although these density inhomogeneities have
small scales, they are not necessarily short living. If we had only the
radio picture, and no optical and infrared information, we might have been
tempted to join the bandwagon and state that star formation is occuring.

## III. The formation of massive stars.

What have we learned about the formation of massive stars from the
study of compact H II regions? Let us follow the answer through conse-
cutive steps.

### III.1. The collapsing dark cloud.

It is a well established fact that compact H II regions frequently
coincide with dense molecular clouds (Zuckerman and Palmer, 1974). In
fact Dickinson et al. (1974) found surprisingly strong CO lines at the
peak positions of compact Sharpless regions. A well established but unpublished example of such a coincidence is given in figures 7 and 8.
Figure 7 is a map of the 21 cm continuum radiation of three compact H II
regions, together called G 75.8+0.4. Observations of the continuum radiation at 6 cm with higher angular resolution show the two main sources to
consist of smaller sources (Matthews et al., 1973). H 109$\alpha$ measurements
show that the two main sources in figure 7 have the same radial velocity.

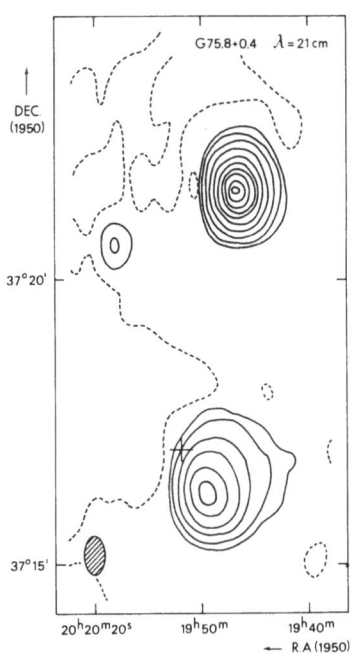

Figure 7. Isophotes of the 21 cm continuum radiation from G 75.8 measured with the Westerbork Synthesis Radio Telescope by Matthews, Goss, Habing and Winnberg (unpublished). The cross indicates the position of the OH maser ON 2.

Figure 8. Isophotes of the intensity of CO (J = 1 → 0) radiation (integrated over the line profile) measured by B. Baud (unpublished) with the Texas telescope in the direction of G 75.8 (see figure 7).

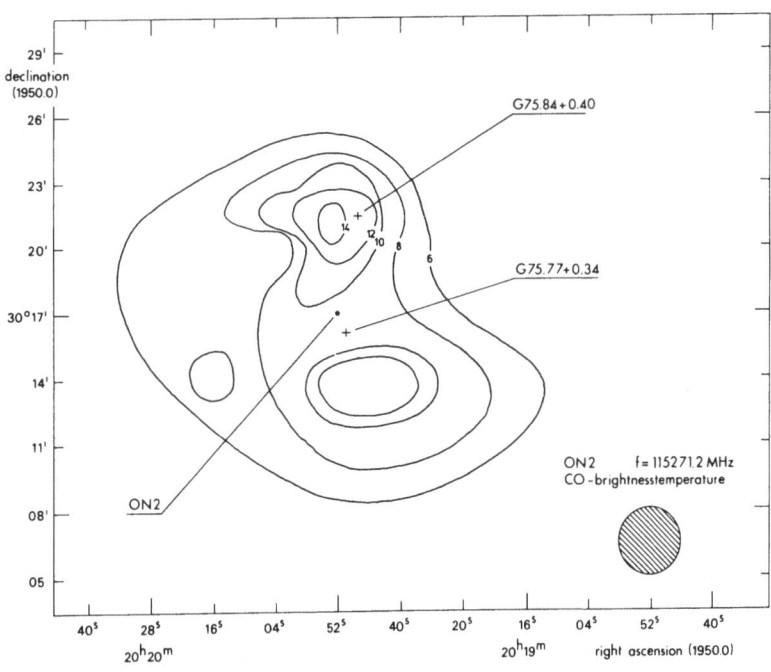

Note that the declination scale should read 37° 17' instead of 30° 17'.

Baud (unpublished) has made CO-line measurements with the Texas telescope and finds the three objects in figure 7 embedded in an extensive CO-cloud (figure 8). The cloud contains at least two nuclei, separated over 10 pc. This probably is a sign of fragmentation of the cloud as a whole. In this cloud two parts evolved faster and produced O-type stars which produced compact H II regions. It is of interest to point out that either the two main compact H II regions came into being independently from each other, or that they were both caused by the same phenomenon. In any case it can be excluded that the birth of one triggered the birth of the other. This follows from the distance between the two objects (9 pc) and their maximum age ($< 10^5$ yr). In 1962 Blaauw concluded that independent bursts of starbirth have occurred in the Orion region, with typical time intervals of $10^6$ yr. Here the time interval is practically zero.

Nevertheless, in other H II regions triggering may occur. In NGC 7538 (figure 1) the expansion of the presently visible nebula may have triggered the collapse of the three sources shown in figure 2. A similar sequence of events may have occurred in S 157 (figure 3), where the larger, visible nebula may have pushed the nearby dark object into collapse.

III.2. Further fragmentation.

Obviously fragmentation does not stop at scale sizes of about 10 pc. As I mentioned already each of the two main compact H II regions in figure 7 consists actually of smaller sources, indicating that fragmentation also occurs in smaller, denser parts, at the scale of 1 pc. Other outstanding examples of fragmentation on that scale are W 3 and DR 21 (see the discussions at this Symposium). It is quite striking that often the compact H II regions are embedded in a lower density envelope. I have noted this in Chapter II for the case of S 156. This suggests that the 1 pc fragment, that ultimately produced the ionizing star of S 156, became an isolated feature in a lower density background at some stage during the collapse.

What about the size of such a fragment that isolates itself from the background? Obviously it will shrink during the collapse, but is there a minimum size for the outer parts? The ring of S156 has a diameter of 0.8 pc. If the ring is a left-over of the collapsing, isolated fragment in S 156 this size is an upper limit, because the ring may have expanded. Let us therefore look for still smaller objects, of which I have described already several in section III.2. A common property of the three very compact H II regions near NGC 7538 (figure 2), the two in S 157B and the one near K3-50 is that they are invisible, whereas the associated, less compact H II regions are visible. This indicates that the very compact H II regions are still embedded in dust shells (cocoons) not much bigger than the H II regions themselves. For the region near K3-50 Israel estimates $A_v = 30^m$ to $50^m$ for the shell only! The shell size should be of the order of (4 to 10) x $10^{16}$ cm or (0.015 to 0.04) pc.

But is 4 x $10^{16}$ cm really the minimum size? There is some indication that this is the case. The observed sizes of OH masers are of the same order. A representative value is 1 arcsec at 3 kpc or 4 x $10^{16}$ cm (Harvey et al., 1974). A considerable fraction (> 50%) of the OH masers appears to coincide with very compact H II regions, but some do not. Habing et al. (1974) quote OH 205.1 - 14.1; Lo (private communication) suggests OH 0609+18 near Sharpless 255; and Harris (this Symposium) mentions DR 21 OH S. This may mean that the OH phenomenon does already occur before a detectable H II region emerges around the forming star. This will then be a still earlier phase in the formation chronology, possibly around the time when the collapse stops.

How much matter is contained in the "final" fragment? Again, I think, that the data at hand suggest an answer: probably less than twice the mass of the star that forms inside. The main argument here is that the measurement of the radio flux density of the compact H II region leads to an upper limit of the radiating mass of ionized gas. In the case of objects like S 156 where a compact source is embedded in an envelope, the mass of ionized gas in the compact source is at most of the order of the stellar mass. Since the envelope is well visible all around the source, it seems unlikely that large, hidden masses of neutral material are present. Hence at least, say, 50 percent of the matter in the fragment went into the star. Star formation at this stage of the game appears to be an efficient process.

The outcome of these speculations is quite interesting. It states that a "final" collapse takes place when a fragment of approximately twice the stellar mass has contracted to a sphere with radius less than

1 pc. This is not incompatible with the initial conditions used by Larson (see e.g. Larson, 1973) in his collapse calculations. Apparently these assumed conditions are met in reality. During the collapse the outer parts of the fragment never get closer to the star than about $10^{16}$ cm , again in agreement with the predictions.

### III.3. Final collapse.

If one wants to build a 30 $M_\odot$ star, one has to accumulate material onto a stellar core in a time shorter than the main sequence life time of such a star. Therefore one needs accretion rates $\dot{M} \gtrsim 10^{-5}$ $M_\odot$ yr$^{-1}$. Given the (speculative) initial conditions derived in the previous section one can calculate the maximum accretion rate under the optimized condition that no rotation and no magnetic braking occurs. The results (Larson and Starrfield, 1971) indicate that this maximum accretion rate is just fast enough to build massive stars. There are two consequences. 1. Rotational effects probably do not halt the collapse for a considerable amount of time. 2. Since the accretion time $M/\dot{M}$ is always larger than the Kelvin-Helmholtz time, massive stars form by moving upward along the main sequence, meanwhile growing in mass because of accretion (Kahn, 1974; this Symposium). What observations exist about this critical stage of accretion?

First, the accretion rate is so large, that the star will be surrounded by only a very small H II region, with a Strömgren radius of, probably, only a few times the stellar radius. H II regions of such diameters are too weak to be detected at radio wavelengths. But the objects should appear as strong, near-infrared souces. We thus expect that the precursors of the very small H II regions shown, for example, in figure 2, are small near infrared sources without radio continuum. A prime candidate for such an object is the infrared source IRS 5 in W 3 (Wynn-Williams, Becklin, Neugebauer, 1972). In the first place, its infrared spectrum and the strong silicate absorption band set it aside from the other infrared sources in that area. Secondly, all other infrared sources in the W 3 area have now been shown to coincide with radiocontinuum sources Harten (1975); Wynn-Williams, this Symposium, but IRS 5 remains undetectable as a radiosource. (This second argument is of modest value, actually, because it is very difficult to detect intrinsically weak sources so close to strong sources like W 3A and W 3B). Considering the overall properties of W 3, I conclude that IRS 5 is probably a precursor of a very compact H II region, and hence a prime candidate for an accreting main sequence star. Apart from IRS 5 one should consider the BN object as another candidate (Larson, 1973), because (1) its infrared continuum spectrum has the same form as IRS 5, and (2) its location inside the KL nebula and the dense molecular cloud OMC 1 suggest that stellar formation is underway. We really need an explanation of masers (OH, SiO) in order to understand the conditions in the accreting circumstellar nebula.

The final question is: what stops the accretion? Kahn (this Symposium) has shown that radiation pressure gradients in the accreting envelope may ultimately stop the inflow, after which an ionization front can speed away from the stellar surface and a very compact H II region results. It is questionable whether this effect will occur before the inflow rate drops for other reasons. From our speculation that compact H II regions contain hardly any neutral material the accretion appears to stop because of exhaustion of accreting material.

## ACKNOWLEDGEMENTS

My own thoughts on compact H II regions and the information that they contain about star formation arise mainly from discussions with my colleagues F.P. Israel, T. de Jong, P. Bedijn, F.M. Olnon, A. Winnberg, W.M. Goss and H.E. Matthews. Unpublished maps have been shown here that were measured with the Westerbork Synthesis Radio Telescope. This telescope is funded and operated through funds provided by Z.W.O., the Netherlands Organization for the Advancement of Pure Research.

174

## REFERENCES

Barlow, M.J., Cohen, M., Gull, T.R. 1974, Proceedings 8th ESLAB Symposium Frascati, June 1974 (ed. A. Moorwood).
Blaauw, A. 1962, Ann. Rev. Astron. Astrophys. 2, 213.
Cohen, M., Barlow, M.J. 1973, Ap. J. (Lett.) 185, L37.
Deharveng, L. 1972, 18th Liège Symposium on Planetary Nebulae, p. 357.
Deharveng, L. 1974, Astron. Astrophys. 35, 63.
Dickinson, D.F., Frogel, J.A., Persson, S.E. 1974, Ap. J. 192, 347.
Downes, D., Wilson, T.L. 1974, Ap.J. (Lett.) 191, L 77.
Grasdalen, G 1975, preprint.
Gluzhkov, Yu.I., Denisyuk, E.K., Karayagina, Z.V. 1975, Astron. Astrophys. (in press).
Habing, H.J., Israel, F.P., de Jong, T. 1972, Astron. Astrophys. 17, 329.
Habing, H.J., Goss, W.M., Matthews, H.E., Winnberg, A. 1974, Astron. Astrophys. 35, 1.
Harvey, P.J., Booth, R.S., Davies, R.D., Whittet, D.C.B., McLaughlin, W. 1974, Monthly Notices Roy. Astron. Soc. 169, 545.
Icke, V. 1973, Astron. Astrophys. 26, 45.
Israel, F.P. 1975, thesis Leiden University (in preparation).
Johnson, H.M. 1974, Astron. Astrophys. 32, 17.
Kahn, F.D. 1974, Astron. Astrophys. 37, 149.
Larson, R.B. 1973, Ann. Rev. Astron. Astrophys. 11, 219.
Larson, R.B., Starrfield, S. 1971, Astron. Astrophys. 13, 190.
Martin, A.H. 1973, Monthly Notices Roy. Astron. Soc. 163, 141.
Mathews, W.G. 1966, Ap. J. 144, 206.
Mathews, W.G. 1969, Ap. J. 157, 583.
Mathews, W.G., O'Dell, C.R. 1969, Ann. Rev. Astron. Astrophys. 7, 67.
Matthews, H.E., Goss, W.M., Winnberg, A., Habing, H.J. 1973, Astron. Astrophys. 29, 309.
Olnon, F.M. 1975, Astron. Astrophys. (in press).
Osterbrock, D.E., Flather, E. 1959, Ap. J. 129, 26.
Pottasch, S.R. 1956, Bull. Astron. Inst. Neth. 13, 77.
Ryle, M., Downes, D. 1967, Ap. J. (Lett.) 148, L17.
Schraml, J., Mezger, P.G. 1969, Astrophys. J. 156, 269.
Vandervoort, P.O. 1964, Ap. J. 139, 869.
Wynn-Williams, C.G., Becklin, E.E., Neugebauer, G. 1972, Monthly Notices Roy. Astron. Soc. 160, 1.
Zuckerman, B. 1973, Ap. J. 183, 863.
Zuckerman, B., Palmer, P. 1974, Ann. Rev. Astron. Astrophys. 12, 279.

# SELECTIVE ABSORPTION BY DUST IN H II REGIONS

Lindsey F. Smith

Max-Planck-Institut für Radioastronomie,
Bonn, FRG

## ABSTRACT

A summary is given of evidence supporting the hypothesis that the helium abundance is constant throughout the Galaxy and that variations in observed ionized helium abundance are due to selective absorption of the helium ionizing photons by dust in the H II regions. New support comes from mapping of the recombination lines in the Orion Nebula and from detection of $He^+$ in the region of the galactic center.

The absorption cross section per H-atom for photons $\lambda < 504$ Å increases by a factor of 5 from the solar neighbourhood to the galactic center. This probably reflects a change in the dust to gas ratio.

In the solar neighbourhood, the cross section per H-atom for $912 > \lambda > 504$ is less than the minimum predicted if the number of large grains per H-atom is the same as in the diffuse IS medium. A depletion of large grains in the H II regions by a factor of about 3 is needed to make the cross sections compatible with theory.

## INTRODUCTION

It is now reasonably widely accepted that dust selectively absorbs ionizing photons, in the sense that high energy photons - capable of ionizing helium - are more strongly absorbed than the less energetic photons capable of ionizing only hydrogen. This provides a <u>qualitative</u> explanation for the failure of earlier observers to detect $He^+$ in the galactic center region (see Churchwell et al. 1974). It also explains why density bounded H II regions only release low energy ($\lambda > 504$ Å) photons, thus limiting the degree of ionization of the diffuse inter-stellar (IS) medium in the way observed by the Copernicus satellite (Rogerson et al. 1973; Jenkins, private communication). It likewise explains the low excitation of the diffuse H II observed by Comte and Monnet (1974) between the spiral arms of M 33.

Extensive <u>quantitative</u> data on ionization of galactic H II regions is available from recombination line observations of He and H. Church-well et al. (1974) noted that the ratio of $He^+/H^+$ varies from one H II region to another and, in particular, confirmed that it is very low in the region of the galactic center. The $He^+/H^+$ ratio is the product of the total He/H abundance ratio (by number), called y, and R, the ratio of the volumes of the $He^+$ and $H^+$ regions weighted by the square of the proton density:

$$R = \int_{He^+} n_p^2 dV \; / \; \int_{H^+} n_p^2 dV$$

One or both of these quantities varies from one region to another. The maximum observed value of $He^+/H^+$ was 10%. Churchwell et al. (1974) argued that this value probably represents the total He/H ratio, that this is constant throughout the Galaxy, and that R varies from one region to another, principally due to selective absorption by dust. This paper summarizes the evidence that now supports these hypotheses.


## $y$ = CONSTANT

In spiral arm H II regions, the $He^+/H^+$ ratio varies from 0.06 to 0.1. The most accessible example of a spiral arm H II region is the Orion Nebula, which has $He^+/H^+$=0.09. We have mapped the recombination lines of He and H with the $2\overset{!}{.}6$ arc beam of the Effelsberg telescope at 6-cm and find that to the east and north-east, the $He^+$ ceases to be detectable ($He^+/H^+ < 0.04$) at a distance of $\sim5'$ arc from the center. Thus R<1, and we can confirm directly that $y \overset{\sim}{\sim} 0.1$ in this nebula.

In the galactic center region, $He^+$ was unsuccessfully searched for several times. Pauls et al. (1974) eventually observed a broad feature in the direction of Sgr A West, which they interpreted as blended H 109α, He 109α and H 137β lines. Repeated observations have confirmed the detection and interpretation (Pauls, private communication).

G0.7-0.0 (= Sgr B2), recently observed with the $2\overset{!}{.}6$ arc beam at 6-cm at Effelsberg by Mezger and myself shows $He^+/H^+ \overset{\sim}{\sim} 4\%$ compared to an upper limit of 2% set by Churchwell et al. (1974) with a 6' arc beam (140 ft., NRAO) and 4% by Mezger et al. (1970) with a 4' arc beam (210 ft., Parkes). G0.5-0.0 also shows $He^+/H^+ \overset{\sim}{\sim} 4\%$, compared to an upper limit of 1% set by Huchtmeier and Batchelor (1973) with a 4' arc beam (Parkes).

These detections are due to the smaller beam of the Effelsberg telescope: the H II regions are resolved, but the smaller, ionized helium spheres are still smaller than the beam and hence contribute relatively more to the observed signal. Thus, we again find that R<1. Although, no definite value for y can be derived in this case: y=0.1, everywhere, appears to be a reasonable hypothesis.

# R VARIES

Variations of R can be due to ionization by cool (spectral type later than 09) stars or to selective absorption by dust in the H II region. Churchwell et al. (1974) argued that, in the giant H II regions observed, the Salpeter initial luminosity function for young clusters should be a reasonable statistical approximation, and hence that the majority of ionizing photons comes from hot stars.

Specific demonstration of dust absorption is provided by the Orion Nebula. Here the exciting star is 06, hot enough to ionize He throughout the H II region. Yet, our mapping confirms that R<1. The discrepancy is most likely due to selective absorption of the ionizing photons by dust in the H II region.

# ABSORPTION CROSS SECTIONS

The critical parameter for the ionization structure of the nebula is the absorption cross section of the grains. Witt and Lillie (1973) derived this quantity from 3000-1500 $\overset{\circ}{A}$ from observations of the diffuse galactic background. Their results are shown in Figure 1 as $x\sigma^{abs}/x_1\sigma_v^{ext}$, the absorption cross section per H-atom, where x is the number of grains per H-atom, $\sigma^{abs}$ the cross section per grain and $x_1\sigma_v^{ext}$ is the extinction cross section at visual wavelengths. The Witt and Lillie observation applies to the diffuse IS medium in the solar vicinity. The absorption cross sections are related to the extinction cross sections via the albedo (= $\sigma^{scattering}/\sigma^{abs}$). The cross sections may be assigned absolute values since $x_1\sigma_v^{ext}=3.7 \ 10^{-22} cm^2 H\text{-atom}^{-1}$ for the diffuse IS material in the solar neighbourhood may be derived from the observed value of $N_H/A_V$ for the nearby stars (Jenkins and Savage, 1974).

Mezger et al. (1974) derived a mean absolute value for the absorption cross section per H-atom in the wavelength region 504-228 $\overset{\circ}{A}$ from observations of the recombination lines from giant H II regions ($N_c>3.5 \ 10^{49}$ photons sec$^{-1}$). An average was made over all observed regions, although it was noted by Churchwell et al. (1974) that $x\sigma_{He}$ (galactic center)

> 8.4 $x\sigma_{He}$ (spiral arm).

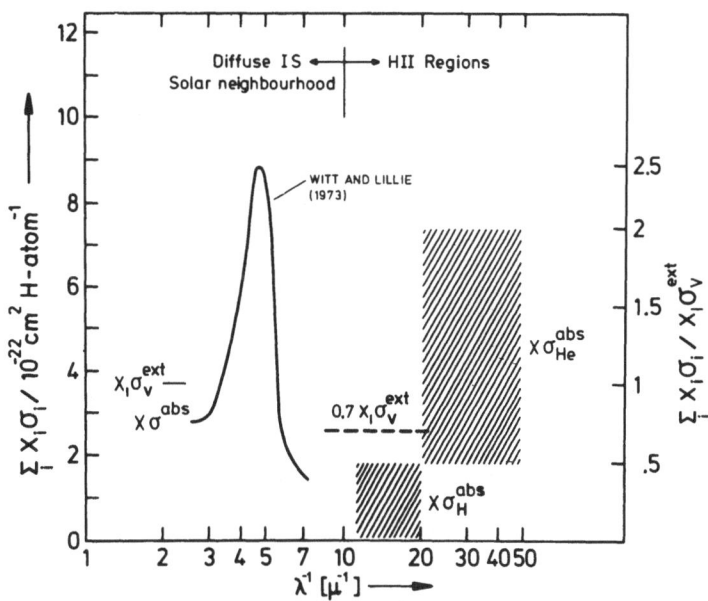

Fig. 1    The cross section of dust as a function of wavelength.
Above 1500 Å (6.7μ⁻¹) the data is from Witt and Lillie
(1973) and applies to the diffuse interstellar matter
in the solar neighbourhood. Below 912 Å (10.9μ⁻¹) the
data applies to the giant H II regions observed by
Churchwell et al. (1974).

Figure 2 shows the values of $x\sigma_{He}$ for individual H II regions
plotted against distance from the galactic center. These are derived
from eq. (24) of Mezger et al. (1974) in units of $3.7 \ 10^{-22}$ $(=x_1\sigma_v^{ext})$

$$\frac{a_o R^{1/3}-1}{a_o} \ \frac{x\sigma_{He}}{x_1\sigma_v} = \frac{\ln (\gamma/\gamma')}{3.4 \ 10^{-18}(N_c'n_e)^{1/3}}$$

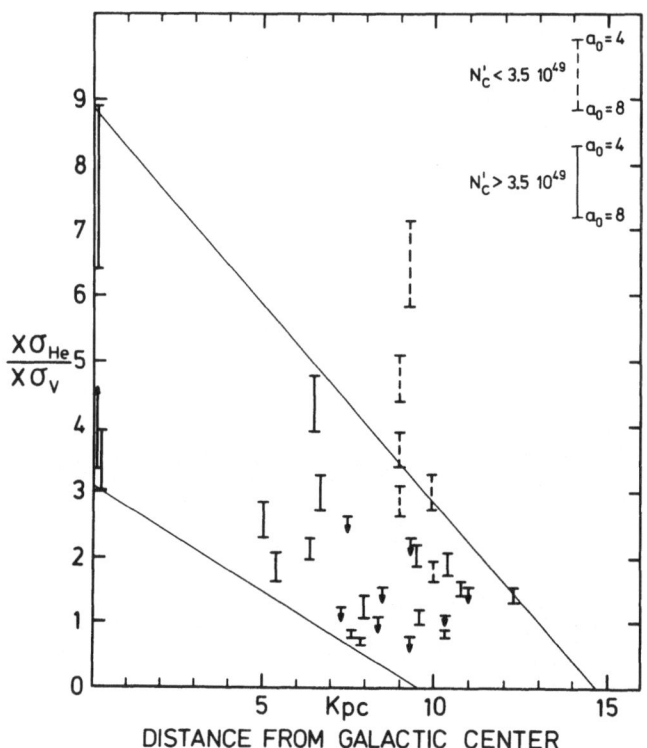

Fig. 2     The cross section of dust in the He$^+$ region to He-
           ionizing photons as a function of distance from
           the galactic center. The cross section is per H-
           atom and is in units of $x_1\sigma_v=3.7\ 10^{-22}$ cm$^2$ H-atom$^{-1}$,
           the extinction cross section for the diffuse inter-
           stellar matter in the solar neighbourhood.

Note that Mezger et al. (1974) used $x_1\sigma_v^{ext}=5.5\ 10^{-22}$ H-atom$^{-1}$ derived from the earlier paper of Savage and Jenkins (1972). $a_o=x\sigma_{He}/x\sigma_H$
and was shown by Mezger et al. (1974) to be greater than 4. The bars in
Fig. 2 allow for the range $a_o=4$ to 8. (Observational errors are not considered.) The data used are those given by Churchwell et. al. (1974)

except that R=0.2 and 0.25 are adopted for G0.5 and G0.7 based on the new detections and the old upper limits. (Derivation of these values will be explained in a later paper.)

The dashed bars refer to small H II regions $N'_c < 3.5 \ 10^{49}$ photons sec$^{-1}$, in which a single star may be responsible for the ionization. When this is a cool star(s), the point falls too high.

The giant H II regions show a clear increase of $x\sigma_{He}$ toward the galactic center. This is most likely due to an increase in the dust/gas ratio for H II regions nearer to the center of the Galaxy. This effect is certainly related to the dramatic increase in heavy element abundances observed toward the centers of external galaxies (Searle, 1971; E. Smith, 1975; Burbidge in this volume).

In the solar neighbourhood, $x\sigma_{He}/x_1\sigma_v^{ext}$ ranges from 0.5 to 2.0. A range of $a_o$ from 4 to 8 allows $x\sigma_H/x_1\sigma_v^{ext}$ between 0.06 and 0.5. These ranges are indicated by the hatched areas in Figure 1. The value of $x\sigma_H/x_1\sigma_v^{ext}$ now lies below the expected minimum, 0.7, from large grains alone. This expected minimum value follows from the theoretical prediction that, at short wavelengths, large grains, of any known substance, will have absorption cross sections equal to their geometric cross sections. At visual wavelengths, extinction is mainly due to the large particles and their efficiency is estimated between 1 and 2. Thus the absorption cross section in the ultraviolet due to these particles alone should be $x_1\sigma_v^{ext}/(1.5\pm0.5)$. Note that $x\sigma^{abs}$ at short wavelengths is also expected to have a contribution from the small grains which provide the short wavelength extinction but have little effect in the visible.

The contradiction in Figure 1 can be resolved if the number of large grains per H-atom, $x_1$, is less in the H II regions than in the general diffuse IS medium. $x_1\sigma_v$ in the H II regions would then be less than the value shown in Figure 1. A depletion of at least a factor 3 is required. Such a decrease has been suggested by many authors (e.g. O'Dell and Hubbard, 1965; Schiffer and Mathis, 1974; Mezger, this conference).

REFERENCES

Churchwell, E., Mezger, P.G., Huchtmeier, W. 1974, Astron. Astrophys., 32, 283.

Comte, G., Monnet, G. 1974, Astron. Astrophys., 33, 161.

Huchtmeier, W. Batchelor, R.A. 1973, Nature, 243, 155.

Jenkins, E.B., Savage, B.D. 1974, Ap.J., 187, 243.

Mezger, P.G., Smith, L.F. Churchwell, E. 1974, Astron. Astrophys., 32, 269.

Mezger, P.G., Wilson, T.L., Gardner, F.F., Milne, D.F. 1970, Astrophys. Letters, 6, 35.

O'Dell, C.R., Hubbard, W.B. 1965, Ap.J., 142, 591.

Pauls, T., Mezger, P.G., Churchwell, E. 1974, Astron. Astrophys., 34, 327.

Rogerson, J.B., York, D.G., Drake, J.F., Jenkins, E.B., Morton, D.C., Spitzer, L. 1973, Ap.J., 181, L110.

Savage, B.D., Jenkins, E.B. 1972, Ap.J., 172, 491.

Schiffer, F.H.III, Mathis, J.S. 1974, Ap.J., 597, 82.

Searle, L. 1971, Ap.J., 168, 327.

Smith, E. 1975, Preprint

Witt, A.N., Lillie, C.F. 1973, Astron. Astrophys., 25, 397.

OPTICAL STUDIES OF GALACTIC NEBULAE AND OF THEIR EXCITING STARS : THE
ROLE OF DUST ON THE RADIUS OF THE IONIZED HYDROGEN SPHERE.

M.C. LORTET

Département d'Astrophysique Fondamentale
Observatoire de Meudon
Meudon, France.

ABSTRACT

From new observational data, Georgelin, Lortet-Zuckermann and Monnet
(1975) and Chopinet and Lortet-Zuckermann (1975) have rediscussed the
use of the Zanstra method in order to derive the rate of stellar ultraviolet photons from hot stars and have studied the relation between
the excitation of the nebular spectrum and the spectral type of the
exciting star. I wish to show here that dust grains mixed within the
ionized gas cannot be an efficient competitor with hydrogen atoms for
the absorption of stellar Lyman photons, and thus that this mechanism
cannot affect significantly the effective temperature scale derived by
Georgelin, Lortet-Zuckermann and Monnet.

We start from empirical results obtained by Georgelin, Lortet-Zuckermann and Monnet (1975) for the ratio of the Lyman photons luminosity
to the visible luminosity $N_L/\pi F_V$, the corresponding effective temperature
(for luminosity class V stars), and the excitation parameter u (Table 1 ;
the visual absolute magnitude used is also indicated).

Table 1. Parameters for hot stars (from Georgelin, Lortet-Zuckermann and
Monnet, 1975).

| Sp. Type | $N_L/\pi F_V$ | $T_{eff} \times 10^{-3}$ | $u$ (pc $cm^{-2}$) | $M_V$ |
|---|---|---|---|---|
| O5 | (12.14) | (43.4) | (86) | -5.5 |
| O6 | 11.92 | 39.4 | 66 | -5.2 |
| O7 | 11.72 | 36.8 | 55 | -5.1 |
| O8 | 11.54 | 35.0 | 45 | -4.9 |
| O9 | 11.32 | 33.0 | 33 | -4.4 |
| O9.5 | 11.17 | 32.1 | 26 | -4.0 |
| B0 | 10.95 | 30.9 | 20 | -3.7 |
| B0.5 | (10.64) | 29.5 | 14 | -3.4 |

We mean to study how the results should be corrected, if we assume that
the observed regions contain dust grains mixed within the ionized gas,
and able of absorbing a fraction 1 - f of the stellar Lyman photons. In
this case, only a fraction f of the stellar photons has been measured by
the observed quantities, and we should use a corrected ratio $(N_L/\pi F_V)$
higher by $\Delta \log N_L/\pi F_V$ = colog f.

The fraction f has been shown by Mathis (1971), Petrosian, Silk and
Field (1972) to depend only on the true ultraviolet absorption $\tau$ by dust
grains, on the length $r_o$ of the dusty Strömgren sphere. It is interesting to notice a very simple approximation for the fraction f ($\tau$),
namely :

$$colog_{10} f = \tau/3 \tag{1}$$

in the range $0 < \tau < 4$.

Now, the true optical depth for a stellar photon going through the
dusty nebula of radius $r_o$ is

$$\tau = (1 + s) \frac{n_d}{n_H} \, \sigma_{uv} \; x \; n_H \, r_o \qquad (2)$$

with the notations of Mezger, Smith and Churchwell (1974) ($n_H$ is the density of hydrogen in the ionized gas, $n_e \simeq n_H$). The radius $r_o$ is related to the empirical excitation parameter scale by the classical relation

$$n_H \, r_o = n_H^{1/3} \, u \; x \; 3 \; x \; 10^{18} \qquad (3)$$

(u in parsecs cm$^{-2}$). The factor s takes into account a possible increase of the effective path length of a photon by scattering on grains. Numerically we put

$$\tau = 2.7 \; x \; 10^{-2} \; x \; D^{-1} \, u \qquad (4)$$

where the parameter D involves several parameters pertaining to grains and also to the nebular density, so that :

$$(1 + s) \frac{n_d}{n_H} \, \sigma_{uv} \left( \frac{n_H}{10^3} \right)^{1/3} = 9 \; x \; 10^{-22} \, D^{-1} \qquad (5)$$

Table 2 gives the fraction of photons absorbed by the gas and the corrected effective temperature for three values of D. The relation between $N_L / \pi F_V$ and the effective temperature is taken from the best available model atmospheres, following the discussion by Panagia (1973).

Figure 1 illustrates the results and shows how they compare with Conti's effective temperature scale (1973). It has been emphasized by Georgelin et al (1975) that the difference between their scale and Conti's one should be interpreted very carefully, because the error bars on each scale are large ; they also stress that Conti's effective temperature is expecte d to be an upper limit, because of the large uncertainties in the presently available model atmospheres.

From Fig. 1, we see that it is likely that dust grains absorb less than 0.3 or 0.4 of the total Lyman photons flux, because otherwise the effective temperature scale would be shifted towards too high temperatures, mainly for stars hotter than about 08.

We should await further empirical data on the nebulae and more elaborated stellar model atmospheres in order to obtain information on the amount of dust mixed within the ionized gas, and to decide whether dust grains do affect the ionization structure of optical nebulae, as was suggested by Mezger, Smith and Churchwell (1974).

Table 2. Fraction of the Lyman photons absorbed by the gas and corrected effective temperature, $T_{eff}$ x $10^{-3}$.

| Sp. Type | D | | | | | |
|---|---|---|---|---|---|---|
| | 1 | | 3 | | 10 | |
| | f | T | f | T | f | T |
| O5 | 0.16 | 70 ? | 0.56 | 49.6 | 0.84 | 45 |
| O6 | 0.25 | 54 | 0.64 | 43 | 0.88 | 40.4 |
| O7 | 0.32 | 45 | 0.69 | 38.8 | 0.90 | 37.3 |
| O8 | 0.38 | 40 | 0.74 | 36 | 0.92 | 35.0 |
| O9 | 0.50 | 35.6 | 0.80 | 33.6 | 0.94 | 33.2 |
| O9.5 | 0.59 | 33.6 | 0.84 | 32.5 | 0.95 | 32.2 |
| B0 | 0.66 | 32.0 | 0.88 | 31.2 | 0.96 | 31.0 |
| B0.5 | 0.75 | 30.0 | 0.91 | 30.1 | 0.97 | 29.6 |

REFERENCES

Chopinet, M., and Lortet-Zuckermann, M.C. 1975, Low dispersion spectra of small H II regions and planetary nebulae (Paper VIII), submitted to Astr. and Ap.
Conti, P.S. 1973, Ap.J., 179, 181.
Georgelin, Y.M., Lortet-Zuckermann, M.C., and Monnet, G. 1975, The rate and fate of stellar ultraviolet photons (Paper VII), submitted to Astr. and Ap.
Mathis, J.S. 1971, Ap.J., 167, 261.
Mezger, P.G., Smith, L.E., and Churchwell, E. 1974, Astr. and Ap., 32, 269.
Panagia, N. 1973, Astr. J., 78, 929.
Panagia, N. 1974, Ap.J., 192, 221.
Petrosian, V., Silk, J., and Field, G.B. 1972, Ap.J. (Letters), 177, L69.

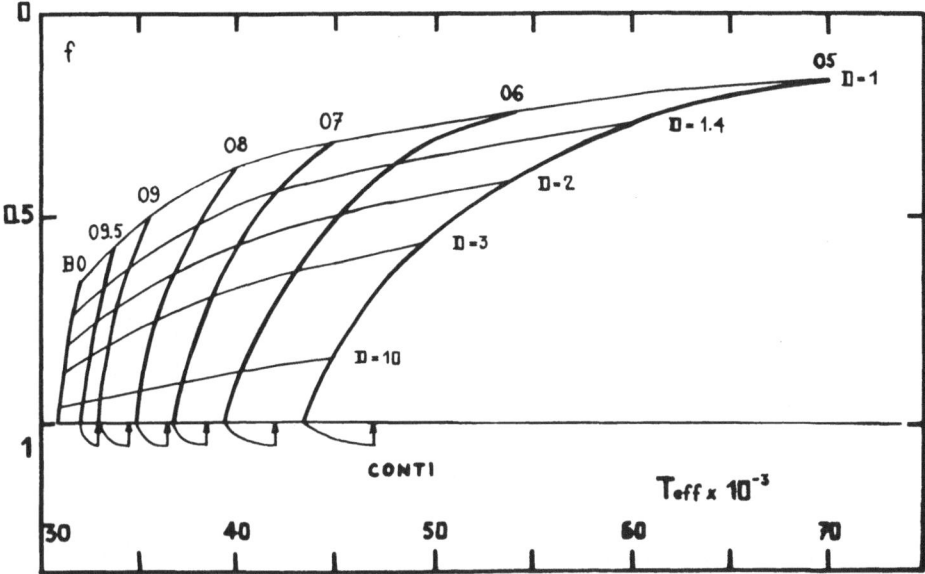

Fig. 1  Corrected effective temperature, as a function of the fraction f
of stellar Lyman photons absorbed by the gas. The lines for the
same spectral type (thicker lines) are obtained by varying the
parameter D. Conti's effective temperature is indicated by an
arrow.

# MOLECULAR MASERS IN HII REGIONS

B. F. BURKE

Department of Physics and Research Laboratory of Electronics
Massachusetts Institute of Technology
Cambridge, Massachusetts

## I.  The Principal Question

A recent summary (Burke, 1974) of problems associated
with cosmic masers was given in the form of six questions,
none of which could be answered in a convincing way.  The
six questions, and a summary of the answers given at that
time, were:

1) Where are their geometrical properties?  The OH and
$H_2O$ masers occur as clusters of compact emitters, each compact emitting condensation exhibiting a narrow line whose
width corresponds to a doppler shift of 1 km/s or less.
Each condensation appears to be $10^{13} - 10^{14}$ cm in size, and
the entire cluster is $10^{16} - 10^{17}$ cm in extent with a total
velocity spread of $\sim$20 km/s.  The intensities of $H_2O$ sources
change dramatically in a time scale of months, although velocity changes are small and possibly negligible, implying

small acceleration OH sources vary only slightly.

2) What is the pump mechanism?  Several possible models
have been suggested, and more than one may be relevant.  Observational tests so far have not given much guidance.

3) Are the masers saturated or unsaturated?  The intense
ones are probably saturated, but the evidence is suggestive,
not conclusive.

4) How are the masers excited?  The combined solutions
for homogeneous, uniformly pumped masers show that the apparent size of the maser is not necessarily the true size.
The simple models show that a spherical, homogeneous model
develops an unsaturated core, which is the apparent object,
surrounded by a saturated envelope that is pure amplifier,
strengthening the quasi-spherical waves from the core without being visible themselves.

5) What is the role of magnetic fields?  No clear-cut
Zeeman patterns had been identified although the strong circular polarization of the OH sources, and the linear polarization observed in $H_2O$ sources could be taken for evidence
of milligauss fields.

6) What use are the masers?  They are transient phenomena probably having a lifetime of the order of 300 years.
The suspicion is very strong that they are related to star
formation, and their great intensity therefore holds the
hope of studying some aspects of star formation processes
throughout the galaxy.

II.  Maser Statistics

The six questions that have been summarized are, for the
most part, not yet settled.  There is not yet general agreement on the pumping mechanism, or on the proof of saturated

conditions. A recent extensive study by Lo (1974), however, of occurrence of $H_2O$ masers in HII regions verifies that there is approximately a 10% probability of finding an $H_2O$ maser complex associated with a compact HII region. A total of 63 compact HII regions, most of them provided by F. Israel and H. Habing from Westerbork SRT observations, were examined using the Haystack Radio Telescope with a maser receiver, and seven $H_2O$ maser complexes were found.

The positional accuracy was better than 30 arc-seconds, enabling a positional comparison to be made between the $H_2O$ masers, the OH masers, radio continuum condensation, and infrared sources. In the seven cases examined, three $H_2O$ masers (S255, S269, and G45.1+0.1) were clearly not coincident with the radio continuum condensation; two (S235 and G45.5+0.1) had some degree of positional correlation; and for one (UA27-IR) the relevant continuum data were not available.

A recent extensive VLBI analysis of $H_2O$ maser sources has been completed by the MIT-SAO-NRL consortium (Moran et al., 1975), and maps of twelve $H_2O$ maser complexes associated with HII regions (plus three associated with stars) have been developed. The relevant data are shown in Table 1.

Table 1

Properties of $H_2O$ Maser Complexes

| Source | Distance | Velocity Range | Peak Flux Density | No. of Features | Resolved Features | Region Size | Largest Feature |
|---|---|---|---|---|---|---|---|
| | Kpc | Km/s | Jy | | | $10^{16}$cm | $10^{13}$cm |
| W3(C) | 2.8 | -41/-33 | 500 | 6 | NO | 13 | <8 |
| W3(OH) | 2.8 | -55/-46 | 3100 | 7 | NO | 6 | <8 |
| Orion A | 0.5 | -2/18 | 13000 | 9 | YES | 25 | ~2 |
| S255 | 1.5 | 5/16 | 600 | 3 | NO | 10 | <5 |
| S269 | 2.5 | 16/21 | 50 | 2 | YES | 22 | ~10 |
| W31 | 2/18 | -4/2 | 450 | 4 | NO | 0.6/5 | <6/50 |
| M17 | 2.1 | 7/20 | 900 | 4 | NO | 2 | <6 |
| W49(N) | 14 | -8/13 | 40000 | 9 | NO | 15 | <40 |
| ON1 | ~3 | 8/22 | 900 | 3 | NO | 9 | <9 |
| ON2 | 5.5 | -6/10 | 250 | 6 | YES | 4 | ~25 |
| W 75(S) | 3 | -7/4 | 100 | 5 | YES | 31 | ~15 |
| NGC7538(S) | 4.4 | -62/-52 | 200 | 5 | NO | 2 | <13 |

Inspection of the table reveals that the sizes of the complexes range from 2 - 31 x $10^{16}$ cm, and that individual masers are typical, of apparent size 0.5 - 13 x $10^{13}$ cm. Thus the typical apparent size range can still be taken as $10^{12}$-$10^{13}$ cm, with overall dimensions for a complex of $10^{16}$-$10^{17}$ cm. The separation between features is still compatible with the Goldreich-Keely-Litvak (1972, 1973) "hot spot" model. Calculations of more realistic models have not, so far, been published.

The energy requirements of the masers remain a difficult puzzle. The super-energetic maser in W49 ($L_{maser}$ = 6 x $10^{33}$ ergs/sec) would require enormous pump energy, and even the more modest examples such as S235 ($L_{maser}$ = 4 x $10^{29}$ ergs/sec) still make strong demands on any direct photon-pumping model. The evidence, therefore, favors collisional pumping models such as those of deJong (1972).

III.  Demonstration of Magnetic Fields in OH Masers

The discovery that the star V1057 Cyg was associated with a 1720 MHz OH maser of extraordinary intensity (Lo and Bechis, 1973) encouraged a new search for Zeeman patterns since the spectrum consisted of a close doublet exhibiting opposite circular polarizations. An interferometric measurement by Lo and Bechis (1974) using the Owens Valley Observatory Interferometer showed that the lines were coming from sources that were within 1" of the same position, but since there are examples of maser complexes completely contained within 1", the proof was not convincing. Several authors have exhibited other examples, but all suffer from the same failing.

A VLBI program was undertaken by the MIT-SAO-NRL group to settle the question more convincingly. By the time of the experiment, the V1057 Cyg source had faded below measurability, but the W3OH 1720 MHz OH maser was examined, and a similar pair of oppositely polarized lines was demonstrated

to share the same position to within 0".001.  In addition,
the similar variation of fringe phase across both lines
showed that they had similar sub-structure.  Lo et al. (1975)
therefore concluded that the lines were indeed a Zeeman doublet, with a magnetic field of 1.3 x $10^{-3}$ gauss.

## IV.  An Evolutionary Maser Scheme

In Section II it was noted that Lo found poor correlation between $H_2O$ source positions and the radio continuum
condensations.  In addition, the OH and $H_2O$ positions are
usually not coincident, although OH masers usually agree
very well with the positions of the continuum condensations,
and with positions of infrared sources.  Generally, the $H_2O$
masers are in the same general region as the OH, continuum,
and IR sources, separated by a few arc minutes at most.  Although they are not the same objects, they surely share some
common features, and there is general agreement that the
process of star formation is in some way related to the occurrence of masers.

The coincidences, and lack of coincidences, led Lo
(1974) to suggest an evolutionary scheme for all the phenomena.  In every known case so far, there is an infrared
source in the vicinity of, but not coincident with, the $H_2O$
maser complexes.  It seems likely that at a very early stage
in star formation, the dense regions necessary for $H_2O$ maser
processes are provided by the copious infrared energy, not
directly but through the heating effects.  The dynamical
time scale for $H_2O$ masers seems to be short - the characteristic time determined by variations is of the order of $10^7$
sec, while the entire complexes have a dynamical time scale
of a few hundred years.  Eventually, the star breaks from
its dust cocoon, and the direct radiation destroys the $H_2O$,
setting up a plasma in which OH is found in sufficient
quantity to provide the raw maser material, in the midst of
a dense HII condensation.  The OH pump source, whether IR or

UV, is not yet clear. The time scale observed for individual OH masers seems longer (many years, as a rule), but eventually the UV radiation field destroys the OH as well. The HII region develops into a compact HII region, and the maser phase of star formation is finished. The occurrence of maser source complexes is consistent with the observed rate of star formation, and if the scheme is a realistic description of the events, there would seem to be eventually a good chance of describing the sites of active star formation throughout the galaxy through the occurrence of maser sources.

## V. Other Masers

The OH and $H_2O$ maser sources, through VLBI observations, are the best-established examples of interstellar molecular masers, but two new classes have been raised as possibilities. The work described by P. Thaddeus at this symposium has given strong reasons to suspect that the vibrationally incited SiO rotational lines arise from inverted populations.

In addition, the 1.25 cm lines of $CH_3OH$ studied by Barrett et al. (1974) have shown evidence of time-variation that strongly suggests small size and high brightness temperature. Similar results have also been obtained by Hills (1975). VLBI observations of these methyl alcohol transitions (Moran & Barrett, 1975) have so far given negative results, but negative VLBI results can often be caused by experimental difficulties. Final conclusions about angular sizes of the sources must wait for further experiments.

## References

Barrett, A. H., Ho, P. T., Martin, R. N. 1974, in press, Ap. J. (Letters).

Burke, B. F. 1974, "Maser Sources in HII Regions", F. J. Kerr and S. C. Simonson, III (eds.), Galactic Radio Astronomy, 267-273, IAU Symposium #60.

deJong, T. 1972, Thesis, University of Leiden.

Goldreich, P. and Keely, D. A. 1972, Ap. J., 174, 517.

Hills, R., Pankonin, V. and Landecker, T.L. 1975, Astron.&Astrophys., 39, 149.

Litvak, M. M. 1973, Ap. J., 182, 711.

Lo, K. Y. 1974, Ph.D. Thesis, Massachusetts Institute of Technology.

Lo, K. Y., Burke, B. F., and Haschick, A. D. 1975, in preparation.

Lo, K. Y., and Bechis, K. P. 1973, Ap. J. (Letters), 185,L71.

Lo, K. Y., and Bechis, K. P. 1974, Ap. J. (Letters), 190, L125.

Moran and Barrett, private communication.

Moran, J., et al. 1975, in preparation.

# MASERS ASSOCIATED WITH GALACTIC H II REGIONS

V.V. BURDJUZHA, T.V. RUZMAIKINA and D.A. VARSHALOVICH

Academy of Sciences of the USSR
Institute for Space Research, Moscow

Academy of Sciences of the USSR
O. Schmidt Institute of Physics of the Earth, Moscow

Academy of Sciences of the USSR
A.F. Jioff Physical-Technical Institute, Leningrad

## ABSTRACT

A large number of maser sources of OH and $H_2O$ are now known to be associated with H II regions. In general, they are the most powerful maser sources discovered in our Galaxy. This report deals with three problems: (a) a pumping mechanism for OH and $H_2O$ masers; (2) formation of maser condensations; (3) mechanisms for the polarization of maser emission. We then summarize the complete picture for OH and $H_2O$ maser emission.

## (1) Pumping Mechanism

Litvak (1969) noted that transitions at $\lambda \sim 53$ μ and $\lambda \sim 34$ μ can under certain conditions contribute to the pumping mechanism, as can transitions at $\lambda \sim 120$ μ and $\lambda \sim 79$ μ, $[^2\Pi_{3/2}, J = 3/2 \rightarrow {}^2\Pi_{3/2}$ $J = 5/2$, and ${}^2\Pi_{3/2}$ $J = 3/2 \rightarrow {}^2\Pi_{1/2}$ $J = 1/2]$. Multiple scattering of the radiation transfers the population from the F = 2 to F = 1 level or from the F = 1 to F = 2 level depending on which transition predominates. In fact, as Burdjuzha and Varshalovich (1973) and Burdjuzha (1974a) have shown, the transitions ${}^2\Pi_{3/2}, J = 3/2 \rightarrow {}^2\Pi_{1/2}, J = 3/2$ ($\lambda \sim 53$ μ) and ${}^2\Pi_{3/2}, J = 3/2 \rightarrow {}^2\Pi_{1/2}, J = 5/2$ ($\lambda \sim 34$ μ) can result in overpopulation of the upper levels, F = $2^+$, $1^+$ of the F = $2^+ \rightarrow 2^-$ (1667 MHz) transition and F = $1^+ \rightarrow 1^-$ (1665 MHz) transition, and these are the lines that have the maximum intensity in sources associated with H II regions. As Harper and Low (1971) have shown, H II regions in which the powerful maser sources of OH and $H_2O$ are found, are also powerful sources of IR-emission with $\lambda \sim 50 - 100$ μ.

This model of pumping by quanta with $\lambda \sim 53$ μ is based on the fact that in passing through a region containing OH molecules the IR spectrum is distorted, so that on interaction with the OH molecules an effective transfer of population between levels of various parities take place. This leads to a population inversion of the upper levels of the Λ-doublet for the ground state ${}^2\Pi_{3/2}, J = 3/2$, $F^\pi = 1^+$, $2^+$ relative to the lower ones, $F^\pi = 1^-$, $2^-$.

The model of pumping by quanta with $\lambda \sim 34$ μ is similar: at certain optical depths the distorted IR-emission still allows the 34 μ (F = $2^+ \rightarrow$ F = $2^-$) transition, which after an IR-cascade provides the transfer of population from the inverted level F = $2^+$ to the inverted level F = $1^+$, making the line F = $1^+ \rightarrow 1^-$ at 1665 MHz a maximum. For these pumping models to work, it is necessary to have the kinetic temperature in maser sources lower ($\lesssim 40$ K) than the kinetic temperature of the surrounding medium ($\sim 100$ K). That is, it is necessary to have structure within the cloud medium. Since the probability of OH-molecule inter-actions with far IR emission ($\lambda \sim 53$ μ and 34 μ) is largest in the H II region environment, these pumping mechanisms are apparently basic at least for densities $\leq 10^6$ cm$^{-3}$.

The near IR-emission flux is a pump for the $H_2O$ masers (Litvak 1972), Goldreich and Kwan 1974). This flux is radiated by dust particles mixed with ionized gas (Wynn-Williams et al. 1972). Photons with $\lambda \sim 6.3$ $\mu$, corresponding to the transition from the ground vibrational level, $v = 0$, to the first excited one, $v = 1$, are absorbed by the $H_2O$ molecule more intensively from the $J = 5_{23}$ level. Thus the $5_{23}$ level turns out to be inverted relative to the $J = 6_{16}$ level, leading to maser action $J \rightarrow J = 6_{16} \rightarrow 5_{23}$.

## (2) Formation of Maser Condensations

The theory that OH maser sources are protostars has received much attention. This idea was first proposed by Shklovski in 1966 and then developed by Mezger and Robinson (1968). Recently, observational data obtained by Sullivan (1971) and Habing et al. (1972) has somewhat changed the idea of the protostar nature of maser sources. Evidently maser sources are gas clouds surrounding compact H II regions. Strelnitsky and Sunyaev (1972), Burdjuzha (1974b) and Strelnitsky et al. (1972) have proposed new ideas on the protoplanetary nature of maser sources. Some properties of maser sources point to their cloud structure: high dispersion of radial velocities, independence of the intensity variation for structures with a small velocity separation, variations of polarization (change of sign of circular polarization in several months). An estimate of the mass of maser sub-sources ($\sim 10^{24}$ - $10^{28}$ g) can be obtained from the observed luminosity, the apparent size, the probable rate of pumping, and the assumption of the normal element abundances. The density limits suggest that the maser sub-sources cannot be gravitationally bound even under the most favourable conditions (for example, at temperatures $\sim 10$ K, during a gravitational instability that may have occurred before a star ignited). That is, the maser sub-sources cannot have been formed as a result of gravitational instability.

Maser clouds are evidently developed during the fragmentation of the medium as a result of thermal instabilities (Burdjuzha and Ruzmaikina 1974). These instabilities can grow behind the shock wave expanding ahead of the ionization front that bounds a compact H II region, if its velocity is sufficiently small (a D-type front according to Kahn). Thermal instability in a cooling medium results in the formation of clouds which are preserved against destruction by an outer pressure which develops if $(\frac{\partial L}{\partial T})_P - \frac{L}{T} < 0$, where L is a specific function of cooling in units of erg g$^{-1}$ sec$^{-1}$.

The criterion for thermal instability under the stationary conditions $(\frac{\partial L}{\partial T}) < 0$ was obtained by Field (1965). The minimum sizes, $\ell_{cr}$, of clouds that can be formed due to thermal instability is determined by the thermal conductivity of the medium. The critical sizes and masses of such clouds, $\ell_{cr} \sim 10^{13} - 10^{16}$ cm and $M_{cr} \sim 10^{21} - 10^{27}$ g, agree with the observed sizes and masses for maser sources. The clouds in which the kinetic temperature and density are respectively lower and higher than the ambient values, produce conditions for the operation of a maser with pumping by emission. We believe that the difference observed in the spatial locations of OH and $H_2O$ maser sources that has been indicated by Hills et al. (1972) and Burke et al. (1970) can be explained as a consequence of this picture. The clouds located between the shock wave and ionization fronts behave as OH maser sources (temperature in clouds $T_{kin} < 100$ K). Within the ionization front nearer the compact H II region (the transition zone between H I and H II regions is very small for high densities), the clouds are heated up to temperatures $T_{kin} \sim$ 500 - 1000 K and conditions are generated in them for $H_2O$ maser action. Since OH maser sources are typically located at distances of 0.1 - 1 pc from the compact H II regions (Habing et al. 1972), and since the size of the compact H II regions is also 0.1 - 1 pc (Wynn-Williams 1971), $H_2O$ maser sources are evidently immersed in the H II regions or in the transition region. The radial velocity range observed for OH maser sources is up to 15 km sec$^{-1}$ which is in agreement with the fact that the observed radial velocity interval cannot be higher than the shock wave velocity, $U_s$, of 20 km sec$^{-1}$. As a rule, the $H_2O$ maser sources have the higher range of velocities. We believe that this phenomenon is explained by one of the mechanisms for cloud acceleration suggested by

Strelnitsky and Sunyaev (1972).

It is natural to assume that the compact H II region is  the source of excitation for OH and $H_2O$ masers. It is possible that an older H II region could also play a part in the case when the OH and $H_2O$ sources are seen in projection. The degree of anisotropy for OH sources is $\Omega/4\pi \sim 10^{-4} - 10^{-5}$ (anisotropy is necessary due to the energy and illumination considerations (Burdjuzha 1974b); for $H_2O$ maser sources, $\Omega/4\pi \sim 10^{-2} - 10^{-3}$ (anisotropy is also necessary for the same reasons but it is probably determined by geometry (Sullivan 1971). If any spatial configuration of the OH clouds is formed after the passage of the ionization front, an elongated structure of the clouds is more probable. This defines the anisotropy for maser radio emission of $H_2O$. The lifetime for OH maser sources is at least equal to the time which the ionization front takes to traverse the distance that the shock wave passed during the lifetime of the compact H II region ($10^4 - 10^5$ years), i.e. $t_i \sim 3 \times 10^3 - 10^4$ years. If this picture is true, OH and $H_2O$ maser sources are formed in the regions where D-type ionization fronts can exist. Since maser emission is not typical of all the compact H II regions, it is probable that the criterion for the presence of masers is the possibility for D-type fronts to exist.

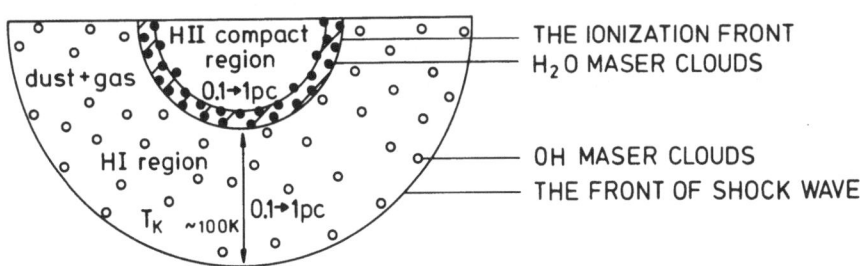

Fig. 1   Schematic diagram of the model for H2O and OH masers near
         H II regions.

## (3) Mechanisms for the Polarization of Maser Emission

Let us briefly discuss the polarization peculiarities of galactic masers. The radio emission of cosmic maser sources of OH associated with H II regions consists of narrow, intense, emission lines that usually have strong, circular polarization (Palmer and Zuckerman 1967). The character of this polarization does not correspond to the usual Zeeman pattern and the spectrum lacks the appropriate symmetry of the $\sigma_-$, $\pi$ and $\sigma_+$ components. More than ten papers have been devoted to the problem of OH maser polarization. Recently Goldreich, Keely and Kwan (1973) suggested that the polarization of cosmic maser emission occurs as a result of Zeeman pattern enhancement in which the linear polarization is smeared by the Faraday effect, or by non-resonant molecular scattering. However, as the authors noted, the lack of Zeeman $\sigma_-$ and $\sigma_+$ components was still not accounted for. Goldreich et al. also consider polarization peculiarities of $H_2O$ masers, such as the presence of linearly polarized components in some sources. Litvak also noted that linear polarization has a Zeeman origin. So the linearly polarized components both in OH and $H_2O$ maser sources have been explained. Zuckerman et al. (1972) also considered the Zeeman origin of galactic maser polarization. Why, then, is it still necessary to analyze the Zeeman picture? The fact that the circular polarization of maser emission occurs only in the presence of the magnetic field $\vec{H}$ as a result of the Zeeman effect is consistent with the data observed, although no obvious Zeeman pattern has been observed in any ground state spectra (in the $^2\Pi_{3/2}$ J = 5/2 spectra of excited OH, Zeeman components were fitted by Zuckerman et al. 1972). Nevertheless circular polarization is observed only for the lines that correspond to the transitions with a high gyromagnetic "g"-factor. For example, for the transitions between sublevels of the hyperfine structure of the $\Lambda$-doublets of OH in the J = 3/2 ground state, where $g_{F=1} = 1.17$, $g_{F=2} = 0.702$, and J = 5/2 state, where $g_{F=2} = 0.567$, $g_{F=3} = 0.405$. For lines associated with energy levels with low gyromagnetic "g"-factor (g << 1), such as transitions between the hyperfine sublevels of the OH rotational state $^2\Pi_{1/2}$, J = 1/2 or transitions J → J = $6_{16}$ → $5_{23}$ in $H_2O$, circular polarization has never been observed. Varshalovich and Burdjuzha (1975) have shown that the lack of one Zeeman component ($\sigma_-$ or $\sigma_+$) can be

related to the redistribution of population between the appropriate
sublevels. This new mechanism is thus added to those previously
suggested for suppression of the circularly polarized modes. These were:
(i) the Cook and Shklovsky mechanism,  where polarization is related to
the motion of the medium (acceleration or rotation) and (ii) the mechanism
of mode suppression due to Heer and Settles (1967), which although
criticized, is still important, as noted by Slysh (1973).

As a result of resonant scattering by the far IR emission of OH
molecules in the outer layers of a molecular cloud, the initial IR
spectrum will be distorted in passing through the cloud, i.e. it will
be cut by IR absorption lines, and the resonant frequency will be
shifted from the center to the wing of a line because of the velocity
gradient with increasing optical depth. With the steep slopes in the
spectrum, the IR emission intensities appropriate to the various Zeeman
components will differ, i.e. for the pumping mechanism of Burdjuzha and
Varshalovich (1973) and Burdjuzha (1974a), the IR pump intensity varies
sufficiently across the Zeeman sublevels to explain the circular polarization. The velocity gradient is required to be about $0.1 - 1$ km sec$^{-1}$
per a.u. It should be noted that the polarization pumping mechanism
suggested by Varshalovich and Burdjuzha (1075), together with that of
Zeeman mode suppression of Cook and Shklovsky, can provide pumping of
the 1665 MHz radio line. The magnetic field in maser condensations and
in their environment is the interstellar field. Its high value  of
$\sim 10^{-2} - 10^{-3}$ Gauss (three to four orders of magnitude greater than the
normal interstellar field) is explained by the fact that the maser zone
is located within the shell of a compact H II region in which the
densities $10^4 - 10^8$ cm$^{-3}$ differ significantly from the interstellar ones
(Zuckerman and Evans 1974). We believe that each polarization feature is
a separate maser-emissive condensation. The distributions of the
relative intensities $I_{\sigma+}$ and $I_{\sigma-}$ for components of the ground $J = 3/2$
and excited $J = 5/2$ states of OH are different. For the ground state it
is impossible to fit the distribution of $I_{\sigma+}$ and $I_{\sigma-}$ components to the
form of a regular Zeeman pattern, but such a pattern may be observed in
the excited state. The differences can be explained by three effects:
1) The optical depth in the lines of the $J = 5/2$ level are $\sim 2 - 3$
times lower than those for the lines of the $J = 3/2$ level; 2) the $g_F$-

factors for the J = 5/2 state are lower than those for J = 3/2, so
that the dips in the IR spectrum have less effect on the J = 5/2 state;
3) for the polarization pumping by the wings of the IR absorption lines
in the J = 5/2 state, the distribution of population between the Zeeman
sublevels is determined only by the influence of the IR transition,
whereas transitions through the states $^2\Pi_{1/2}$ J = 1/2, $^2\Pi_{1/2}$ J = 5/2
influence the population of the Zeeman components of the J = 3/2 state;
these transitions disturb the regularity of the simple picture of
polarization.

The spatial grouping of left-hand and right-hand circular polarizations such as that seen in W3 (Cooper et al. 1971), is caused by ir-
regularity in the medium. That is, we assume that the W3 maser condensations -41.2R, -41.7R and -42.2R are located at a region where the
velocity gradient or the magnetic field gradient is opposite to that
in the region containing condensations -44.3L. -45.0L, -45.4L, -45.6L.
If the velocity gradient is absent in the medium or if it is small, the
Zeeman pattern will not be distorted even for the ground state and the
components of the left-hand and right-hand circular polarizations will
be observed. Some of the circularly polarized maser components should
not dominate, that is, the distribution function for $\sigma_+$ and $\sigma_-$ components must be of a random character. The polarization cycle (the redistribution of the population of the magnetic sublevels by quanta of
the IR-line $\lambda \sim 120~\mu$) links the collisions that populate the upper
signal levels. Such a mechanism of polarization therefore acts independently of the details of the pumping mechanism. Due to the conservation of angular momentum and because the directed OH maser emission is
almost completely circularly polarized, IR-emission scattered by the
same molecules of OH must also be circularly polarized to a high degree.

## SUMMARY

To summarize this complete picture: OH and $H_2O$ maser sources are
clouds surrounding compact H II regions. The density near the compact
region is rather high, i.e. it is possible to have a D-type shock wave
front. OH and $H_2O$ clouds are formed as a result of the thermal instability under the non-stationary conditions in the cooling medium

after the passage of the shock wave. $H_2O$ maser clouds are located at the transition zone between H I and H II regions or at the H II region, i.e. the ionization front has passed through them. The flux of far IR-emission generated by the shell of the compact H II region is probably the pump for OH masers. But for $H_2O$ masers pumping is performed by the flux of near IR-emission generated by heated dust particles mixed with ionized gas. The masers emit directionally (anisotropy extent $\Omega/4\pi \sim$ $10^{-2} - 10^{-4}$), and their number is rather great. The lifetime for OH masers is at least $10^4$ years. Quanta with $\lambda \sim 120\ \mu$ are absorbed by OH molecules in the medium with a small velocity gradient and cause the repopulation of the Zeeman sublevels. Hence, the population of the magnetic sublevels appropriate to one of the Zeeman triplet modes of the ground state dominates, and an observer sees maser emission of only one polarization from a given condensation. If there are no depolariza-tion factors in the medium, or they are small, linear polarization of the OH and $H_2O$ masers takes place. Each condensation is responsible for one detail in a maser spectrum. As there are about $10^4 - 10^5$ such con-densations, and since they emit directionally, the total picture repre-sents the spectrum formed by the single circularly polarized structures.


## REFERENCES

Burdjuzha, V.V., Varshalovich, D.A. 1973, Astron. Zh. 50, 481.
Burdjuzha, V.V. 1974a, preprint ISR Academy of Sciences, USSR No. 199, Astrophys. Lett. 15, 189, 1973.
Burdjuzha, V.V. 1974b, Astron. Zh. 51, 26.
Burdjuhha, V.V., Ruzmaikina, T.V. 1974, Astron. Zh. 51, 346.
Burke, B.F., Papa, D.C., Papadopoulos, G.D., Schwartz, P.R., Knowles, S.H., Sullivan, W.T., Meeks, M.L., Moran, J.H. 1970, Astrophys. J. 160, L63.
Cook, A.H. 1967, Nature 211, 503.
Cooper, A.J., Davies, R.D., Booth, R.S. 1971, Monthly Notices Roy. Astron. Soc. 152, 383.
Field, G.B. 1965, Astrophys. J. 142, 531.
Goldreich, P., Keeley, D.A., Kwan. J.Y. 1973, Astrophys. J. 179, 111.
Goldreich, P., Kwan, J.Y. 1974, Astrophys. J. 191, 93.
Habing, H.J., Israel, F.P., de Jong, T. 1972, Astron. & Astrophys. 17, 329.
Harper, D.A., Low, F. 1971, Astrophys. J. 165, L9.
Heer, V.V., Settles, R.м. 1967, J. Mol. Spectrosc. 23, 448.
Hills, R., Janssen, M.A., Thornton, D.D., Welch, W.J. 1972, Astrophys. J. 175, L69.
Litvak, M.M. 1969, Astrophys. J. 156, 471.

Litvak, M.M. 1972, Atoms and Molecules in Astrophysics, eds. T.R. Carson
    and M.J. Roberts, New York: Academic 201-276,
Mezger, P.G., Robinson, B.J. 1968, Nature 22, 1107.
Palmer, P., Zuckerman, B. 1967, Astrophys. J. 148, 727.
Shklovsky, I.S. 1966, Astron. Zircul. 372.
Shklovsky, I.S. 1969, Astron. Zh. 46, 3.
Slysh, V.I. 1973, Astrophys. Lett. 14, 213.
Strelnitsky, V.S., Sunyaev, R.A. 1972, Astron. Zh. 49, 704.
Sterlnitsky, V.S., Sunyaev, R.A., Varshalovich, D.A. 1972, Comm. Astro-
    phys. and Space Phys. 5, 155.
Sullivan, W.T. 1971, Astrophys. J. 166, 321.
Varshalovich, D.A., Burdjuzha, V.V. 1975, Astron. Zh. (in press).
Wynn-Williams, C.G. 1971, Monthly Notices Roy. Astron. Soc. 151, 397.
Wynn-Williams, C.G., Becklin, E.E., Neugebauer, G.N. 1972, Monthly
    Notices Roy. Astron. Soc., 160, 1.
Zuckerman, B., Yen, J.L., Gottlieb, C.A., Palmer, P. 1972, Astrophys.
    J. 177, 59.
Zuckerman, B., Evans, N.J. 1974, Astrophys. J. 192, L149.

# ON THE NATURE OF THE SiO MASER IN THE KLEINMANN-LOW NEBULA

H.J. HABING, F.M. OLNON, P.J. BEDIJN, T. DE JONG

Sterrewacht, Leiden. The Netherlands.

## ABSTRACT

Recently several emission lines from rotational transitions in vibrationally excited levels of SiO have been detected. The first source
was found in the Kleinmann-Low (KL) nebula in Orion. Most likely the radiation source is an astrophysical maser, perhaps associated with the
Becklin Neugebauer (BN) object. Following the detection of the SiO maser
in Orion, the same radiation has been found from a number of late type
stars associated with $OH/H_2O$ line emission. However, no SiO masers have
been found at the positions of OH masers in    H II regions. This is
somewhat puzzling since the KL nebula and the BN object are generally
considered to be very young objects.

It is known from the work of Larson        that during the last stages of the formation of massive stars the central, embryonal star is already on the main sequence, while still accreting matter at a high rate.
It can be shown that the high gas density in the envelope pushes the
Strömgren sphere of the star onto the stellar surface. We suggest that in

the neutral gas envelope close to the star the conditions are favour-able for SiO maser formation. More specifically we propose that the BN object is the source of maser emission observed in the direction of the KL nebula.

We have constructed a model of the BN object to explain its near-infrared spectrum using the methods of Larson (1969, Monthly Not. R.A.S. 145, 297) but adding foreground extinction. The parameters of the best fit are a luminosity of the central star of $L = 5000 L_\odot$, a mass inflow rate in the cocoon of $\dot{M} = 2 \times 10^{-5} M_\odot$ and an extinction in the KL nebula of $A_v = 30^m$. The physical conditions at the point in the cocoon where we expect maser emission are as follows: $n(H_2) \approx 10^8$ cm$^{-3}$, the infall velo-city $u \approx 20$ km sec$^{-1}$, the kinetic temperature of the gas is $\approx 500$ K, the temperature of the radiation field $\approx 2000$ K diluted by a factor of about $10^{-2}$ to $10^{-3}$. These conditions are very similar to those in the expanding envelopes of late-type stars, for which detailed SiO maser models have recently been published by Kwan and Scoville (1974, Ap. J. Lett. 194, L97). Thus the SiO maser lines in the KL nebula could very well be produced in the cocoon around the BN star.

In general we expect that infrared sources associated with H II re-gions with spectra peaking at near-infrared wavelengths are good candi-dates for SiO maser emission.

(A more extensive account of this work has been submitted to Astro-nomy and Astrophysics).

OBSERVED PARAMETERS OF O TYPE STARS

PETER S. CONTI

Joint Institute for Laboratory Astrophysics
University of Colorado and National Bureau of Standards
Boulder, Colorado  80302

ABSTRACT

A review is made of the determinations of the effective temperature, luminosity and masses of O type stars.  This also involves the
status of the comparison of the observations with non-LTE plane-parallel
models, which can then be used to derive other parameters about the
stars.  For most O type stars, there seems to be good understanding of
the temperature scale and overall comparison to models.  For a significant minority, the Of stars and supergiants, there are still uncertainties in the models due to the necessity of including spherical geometry
effects and stellar wind dynamics.

## I.  INTRODUCTION

It is my intention in this paper to review the various determinations of the three most important physical parameters of O type stars, namely the effective temperature, $T_{eff}$; the luminosity, L; and the mass, M.  Two other parameters, the radius, R, and surface gravity, g, can be derived from the first named ones by the well-known relations $R^2 \sim LT^{-4}$ and $g \sim MR^{-2}$.  Although it is possible in principle to determine R and g directly for certain kinds of eclipsing binaries, in practice this cannot yet be done with sufficient accuracy for the stars to be discussed here.

Once physical parameters have been specified then atmospheric and interior stellar models can be constructed.  The comparison with real stars comes about by knowledge of their appropriate observed parameters. The <u>models enable one to predict other observables</u> of stars, for example the flux below the Lyman continuum, $L_c$, which are of particular importance to the various fields of astrophysics.  The problem of determining, say, $L_c$, is two fold:  one must have correctly specified the physical parameters for the stars and one must have physically meaningful stellar models.  The discussion of these points is the intent of this paper.  It should be stated at the outset that so far all work with models has assumed a solar composition for the material making up the stellar opacity. In particular, the He/H ratio by number is 0.1.  Also, all stellar models have neglected rotation.

## II.  NOMENCLATURE

Inasmuch as this symposium is concerned with HII regions, I have restricted my discussion to O type stars since they are the primary ionizing sources.  B type stars have essentially no effect on discrete HII regions although they may contribute slightly to the overall ionization balance of the interstellar medium.  O stars are those objects in which HeII absorption lines are seen, primarily the $\lambda4541$ and $\lambda4686$ lines visible in the blue region of the spectrum.  The subdivision of O types proceeds primarily by the ratio of the absorption lines of HeI/HeII, in particular, $\lambda\lambda4471/4541$, a point to which I will return shortly.

Some O stars have emission lines. The most typical emission lines observed are listed in Table 1.

### Table 1
Typical emission lines observed in O stars

| Wavelength | Ion | Mechanism |
| --- | --- | --- |
| $\lambda 4634, 40$ | NIII | Intrinsic |
| $\lambda 5696$ | CIII | Intrinsic? |
| $\lambda 4686$ | HeII | Extrinsic |
| $\lambda 6562$ | Hα | Extrinsic |

A small fraction of O stars have additional emission lines present but this need not concern us here. Traditionally astronomers constructing model atmospheres for stars use a plane-parallel assumption: that is, the line formation can be considered to be in a slab of material whose thickness, $r$, is small compared to the radius of the star. For a few stars, it is clear that this assumption will not hold although for most O types, it works well. It should be kept in mind that a line formation region can be plane parallel for some lines, but not necessarily for all lines. If the line formation is not plane parallel, it is said to be "extended," and $r \gtrsim R$.

The underlying mechanism causing a line to be in emission in a stellar atmosphere is an overpopulation of its upper level. The physical reason behind this is understood in at least a few cases for O stars. Mihalas, Hummer, and Conti (1972) have found that the NIII lines at $\lambda\lambda 4634,40$ can come into emission by a recombination into their upper level from dielectronic transitions in NIV. Due to the transition probabilities involving these levels and others, an emission is the most probable exit from the excited states. These authors find that the NIII lines can be in emission even if the line formation is plane parallel although it is also true that an extended envelope will strengthen the lines. We call the emission mechanism for the NIII lines "intrinsic" since it is independent of the geometry of the stellar atmosphere. The physical mechanism underlying the $\lambda 5686$ CIII line is not as well understood but it is thought to be similar to that for the NIII lines.

By contrast, the $\lambda4686$ HeII and H$\alpha$ lines can only be in emission if the line(s) are formed in an extended region of a stellar atmosphere (Mihalas 1974). We refer to these emission lines as "extrinsic" since they depend on the geometry of the stellar atmosphere. We call stars with emission in the $\lambda4686$ line Of stars, those with $\lambda4686$ weak or absent O(f) stars, and those with $\lambda4686$ more or less normal O((f)) stars (Walborn 1972; Conti and Leep 1974). (In all three cases, NIII emission is observed.) The Of and possibly the O(f) stars have, in this context, evidence for extended envelopes, and it may be necessary to use different physical characteristics to describe the appropriate models.

Supergiant stars (OI types) are those objects that are very luminous for their spectral type. It is likely they too have extended geometry for some lines and we shall consider them in a similar set as the Of stars. We shall refer to the two types collectively as If stars.

Of stars have definite evidence of outflow of material from the star itself as shown by the widths of the emission lines and the appearance, in some cases, of P Cygni profiles. The wind velocities range up to 1500 km sec$^{-1}$, as seen by observations in the visible portion of the spectrum, Hutchings (1968). Some OI stars, without evidence for emission in the visible region of the spectrum, have winds when observed in the rocket or satellite UV region (Morton 1967). In these cases too, the wind velocities range up to 1500 km sec$^{-1}$, but the absence of an effect in the visible region must mean that the density-velocity relation can be different for different stars. An important implication of this result is that a wind can be present without the necessity for extended geometry for all lines.

### III. CLASSIFICATION AND TEMPERATURE SCALE

The classification of O type stars has recently undergone a very detailed investigation by Walborn (1971) and by Conti and Alschuler (1971). Both authors use as the primary classification criterion the ratio of $\lambda4471$ HeI to $\lambda4541$ HeII, the former from eye estimates and standard stars, the latter from measures of coude plates. The two independent methods agree well according to Conti and Leep (1974). The physical basis of this classification, which is intended to arrange the

O stars into a temperature sequence, is reasonably well in hand.  Auer
and Mihalas (1972) have constructed plane-parallel non-LTE models for O
stars and made predictions of line strengths for various lines of HeI,
HeII and H.  Figure 1 shows the predicted run of the line ratio 4471/4541
for various effective temperatures and (log) surface gravity (Conti
1973b), expressed as a log of the equivalent width W.  Auer and Mihalas
(1972) pointed out that gravities $\leqslant 10^{3.3}$ would make the stars unstable to
continuum radiation pressure.  The line ratio depends on both $T_{eff}$ and g,
and in practice one usually specifies the gravity independently.  The
measured line ratio then gives an estimate of the temperature.  The gravity for luminosity type V, or main sequence stars, is taken to be $10^4$,
whereas that for giants and supergiants is presumably lower.  These gravities are no more certain than a factor 2, which introduces at least a
5% uncertainty in $T_{eff}$, and is possibly systematic.

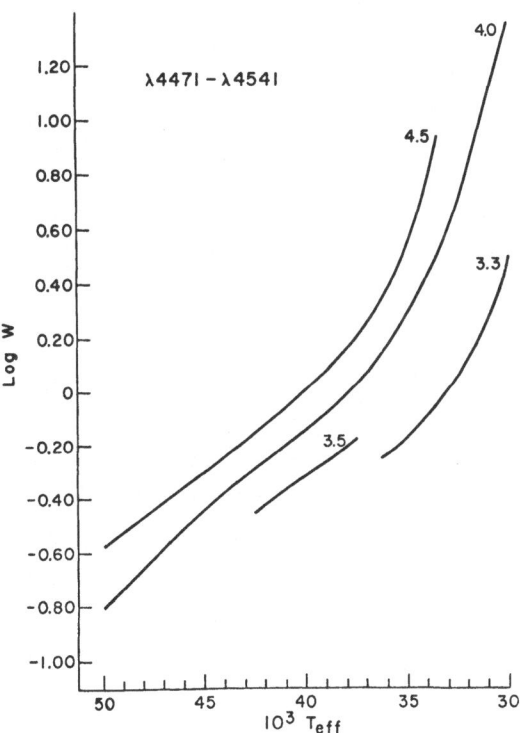

Fig. 1.  Predicted log W for λ4471 HeI minus λ4541 HeII, for different
(log) gravities, from the non-LTE plane-parallel models of Auer and
Mihalas (1972).  The temperature scale is based on this ratio and the observed mean values for the spectral types.  (From Conti 1973b-copyright,
University of Chicago Press, used with permission.)

It is merely a convenience to assign the classification scale by the HeI/HeII line ratio, even though this tends to "quantize" the assignment of temperature to a particular star. The accuracy of estimation of temperatures by this method is one spectral subclass, or about 5%, with random error. In addition to the line ratios, we also have a large number of W measures for HeI, HeII, and H lines for many stars. These have been compared with the Auer-Mihalas models elsewhere (Conti 1973a, 1974). For our purposes here, we will discuss only a limited sample of O stars, namely those having well established L (§IV; Conti and Burnichon 1975).

Figure 2 shows the line strength prediction (log W) for $\lambda4471$ HeI from the Auer-Mihalas models and the observed measures. Figure 3 shows the line strength predictions for $\lambda4541$ HeII and the observed measures. For both lines, the agreement between the observed and predicted line strengths for the O stars is gratifying. This statement also holds true for other HeI and HeII lines (Conti 1973a; 1974), aside from $\lambda4686$ HeII. The gravity for the O stars is $\sim10^4$ whereas that for the Of and supergiant stars appears to be lower, $\sim10^{3.5}$. The general absence of emission lines that require an extended atmosphere, and the good agreement between the observed and predicted line strengths, means that the Auer-Mihalas models are good estimates of the real O stars. One has confidence then, that other predictions of these models, e.g. $L_c$, can be adopted with some justification.

On the other hand, we have evidence of extended geometry in the Of and supergiant stars from the presence of certain emission lines, and we note that the gravities are nearer the limit for stability against continuum radiation pressure. If we arbitrarily adopt gravities of $10^{3.3}$, and then use Figure 1 to estimate temperatures for Of and OI stars, we find a scale about 10% cooler than that for the main sequence (Conti 1973b; Panagia 1973). However, one should be cautious in accepting this interpretation because the effect of extended geometry on the HeI/HeII lines is unknown.

There are two other possible methods to obtain the temperature scale for O stars, and we shall discuss each of them. Both give results consistent with the HeI/HeII scale, but neither can answer the question of the temperature scale for the Of and OI stars satisfactorily.

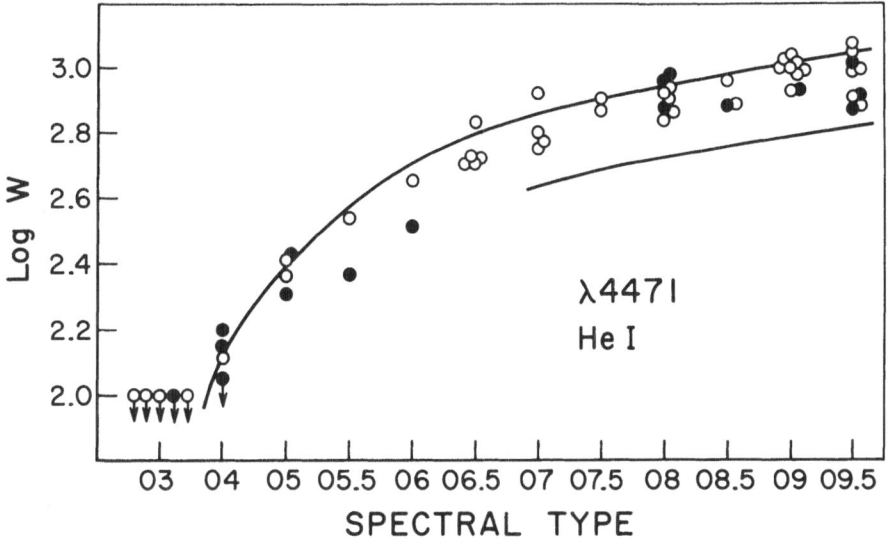

Fig. 2. Observed and predicted log W for λ4471 HeI. The solid lines are from the non-LTE plane-parallel models of Auer and Mihalas (1972), the upper being log g = 4.0, the lower, log g = 3.3-3.5. The open circles are the O stars, the filled circles, the Of and OI stars, with well determined luminosities (§IV). There is good agreement between theory and observation for this line for all stars.

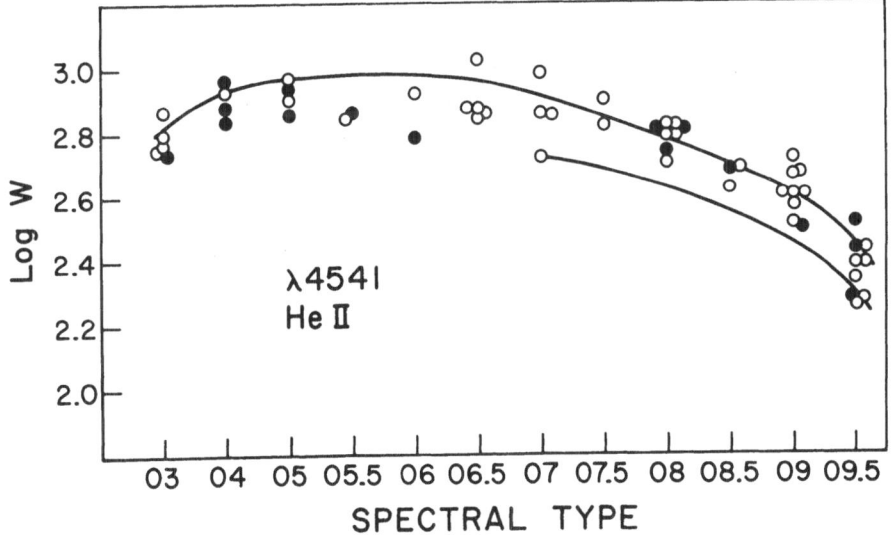

Fig. 3. Same as Figure 2 for the line λ4541 HeII. There is good agreement between theory and observation for this line for all stars.

Morton (1969), following work by previous investigators, derived a
Zanstra temperature scale for O stars by measuring parameters in the HII
region excited by and surrounding certain stars and equating this to a
$L_c$ measure. He then used the $L_c$ and LTE models to derive temperatures.
The non-LTE models of Auer and Mihalas were subsequently constructed and
superseded previous ones, but this does not make much difference in the
present calculation. Figure 4 shows a Zanstra temperature-spectral type
diagram for most of the stars in Morton's (1969) paper, with spectral
types derived by Conti (1973b). Morton's own temperature scale was an
average at each spectral type. The two dashed lines are the HeI/HeII
temperature scale discussed above. The Zanstra temperatures are consistent with this scale although there are some stars that appear to be
cooler, as if their gravities were near $10^{3.3}$. None of the stars plotted
in Figure 4 is an Of or OI star, however.

At least part of the reason for the scatter in this diagram is the
presence of dust in the HII regions. In this case some of the $L_c$ photons
are absorbed by the dust and thereby are not available to be accounted
for by the emission measures. The temperature of the star is thereby
underestimated. (This is also true of nebulae which are density
bounded.) On the other hand, if the HII region is complex, it is possible that other stars contribute to the ionization, and the stellar temperature is overestimated. A more detailed discussion of a Zanstra mechanism $T_{eff}$ scale is contained elsewhere in this symposium (Lortet-
Zuckermann 1975).

Morrison (1975) has derived a temperature scale for O stars by very
careful measures of the Stromgren photometric indicies uvbyβ. She compares her results for 85 stars with the colors predicted by Mihalas
(1972) from the non-LTE models. For stars cooler than 40000°K (type O6)
on the main sequence, the temperature scale agrees with that of Conti
(1973b). For Of and OI stars of latter type there is a tendency for the
photometric indicies to indicate a temperature scale not as cool as indicated by Conti (1973b). For all types earlier than O6, there is a
tendency for the O stars to have indicies indicating temperatures appreciably cooler than indicated by Conti (1973b). In both cases, Morrison
(1975) suggests that effects of spherical geometry are important. For
the later Of and OI stars, she is able to show, using extended models

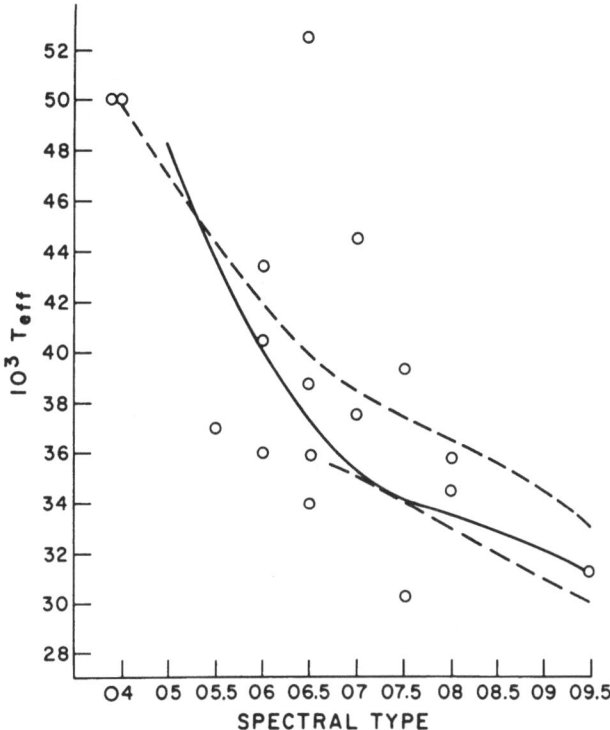

Fig. 4. Comparison of the HeI/HeII temperature scale with a Zanstra mechanism scale. The open circles are placed at ordinate positions corresponding to their Zanstra temperatures (Morton 1969). The thin line is
Morton's temperature scale. The upper dashed line corresponds to the
main-sequence temperature scale of Conti (1973b); the lower line, one
~10% cooler at each spectral type, corresponding to a gravity of ~$10^{3\cdot3}$.
This line is probably a lower limit to the temperature of Of and OI stars
(see text). There is reasonably good agreement between the ionization
temperature scale and the Zanstra scale, although the scatter is a little
large.

constructed by Mihalas and Hummer (1974), that the effects of spherical

geometry on the photometric indicies go in a direction consistent with

the observed colors. For the earlier type stars, the comparison is not

consistent with what is expected from spherical geometry alone. This

should make us a little cautious about a definitive determination of the

temperature scale for the hottest 0 type stars.

The problem with the extended geometry is basically that physically

realistic self-consistent models have not yet been constructed. Mihalas

and Hummer (1974) have constructed arbitrarily extended models to study

effects on the continuum, but have not included dynamical effects.

Castor, Abbott, and Klein (1975) have considered dynamical models but have not yet included full non-LTE calculations. Both groups, however, are now attempting to include all of the known physics and to construct models with both dynamical effects and geometrical extension. An improved solution to the problem of the temperature scale for Of and OI stars will await these models.

In the meantime we can make some comments on the likely range of the results. An extended atmosphere does not normally affect the $L_c$ since many lines, and the continuum, are formed in a more or less plane-parallel layer. However, there might be major blanketing and backwarming effects in the lines in the Lyman continuum. This is because the extended geometry will tend to "desaturate" some lines that would otherwise be at full strength in a plane-parallel case. One can thereby imagine greater blanketing and backwarming effects in an extended star than in one without this geometry. It is possible that a somewhat cooler star than expected could then mimic the visual appearance of a non-extended hotter one. The blanketing effect is most important in the Lyman continuum where much of the stellar energy emerges, whereas the backwarming occurs in the visible region. There probably are limits on how much of an effect there can be since the total blanketing in the region $\lambda > 912$ Å, which is observable, is small. The maximum expected effect, if the blanketing in the unobservable region $\lambda < 912$ Å is similar to that $> 912$ Å is about 10%. That is, a star with extended geometry might be 10% cooler than expected due to increased backwarming into the visible region. However, detailed line calculations are necessary before it can be asserted that this estimate is realistic.

One other effect should be mentioned. It might turn out that line blocking at the HeI and HeII ionization edges in the region $\lambda < 912$ Å might produce additional non-LTE effects in the HeI/HeII ratio. This might affect the temperature scale in either direction, but the best guess at the present is that this is unimportant for stars with plane-parallel or extended geometry.

## IV. LUMINOSITY AND MASS

It is best to estimate luminosities of O stars from membership in clusters with distances established by methods independent of the O stars themselves. This calibration has been made by Conti and Alschuler (1971) and by Walborn (1971). The two methods agree reasonably well although in most cases the same stars are being considered. Conti and Alschuler picked a number of clusters in which B stars had been used to find the distances either by UBV photometry and spectral types, or by Hβ photometry alone. The calibration of O star magnitudes then depends on the B star distances, which are thought to be well known.

From the $M_v$ it is then necessary to apply a bolometric correction to obtain L. These are taken from Morton's (1969) discussion, which although it uses LTE models, is not substantially different from the newer non-LTE work. Geometrical extension probably has no effect on the determination of the bolometric correction. The kinetic energy contained in an O star wind, although of the order of a solar luminosity or more, is also a negligible contribution to the bolometric correction.

Figure 5 shows an HR diagram for O stars discussed by Conti and Burnichon (1975) which is based on the above references. Theoretical mass evolution tracks were obtained from Stothers (1972). It is important to note that from about type O5 and cooler, there is good agreement between the observed lower limit in luminosity for O stars and the theoretical ZAMS. This gives us confidence that we have correctly specified the $T_{eff}$ and L for real stars. It will also be noted that most Of stars, and of course OI stars, are evolved. There are an appreciable number of stars with $M \geq 60_\odot$ even up to $120_\odot$. The "canonical" limit to main sequence stellar masses was near 60 or so, based on linear calculations of stability against pulsation (Swarzchild and Harm 1959) or limits on infall accretion during the early stages of star formation (Larson and Starrfield 1971). As Conti and Burnichon (1975) point out, however, theoretical calculations involving non-linear pulsation might show a different mass limit against pulsational instability.

It is recognized that several stars in Figure 5 seem to fall below the theoretical ZAMS. Conti and Burnichon (1975) suggest this may be due to a still incorrect choice of effective temperature for these very hot stars, or problems with the stellar interior models. It should be noted that the theoretical evolution tracks given by Stothers (1972) are

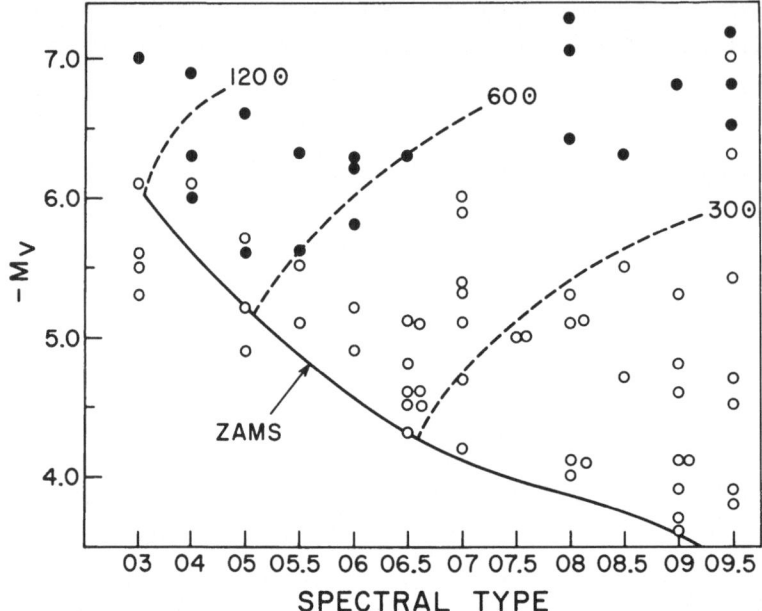

Fig. 5. HR diagram of O type stars with "well determined" luminosities. The open circles are O stars, the filled circles, Of or OI stars. The lines are from theoretical evolutionary tracks of Stothers (1972), as discussed by Conti and Burnichon (1975).

without mass loss. This effect is probably very important in O stars (Castor, Abbott and Klein 1975) and modifications to these tracks may be necessary.

Better determinations of the masses of O stars could come, in principle, from observations of double line and/or eclipsing binary stars. Data on O type systems are scanty, but a perusal of the Batten (1967) catalogue indicates that the masses derived from the few binaries with well established orbits are consistent with the results shown in Figure 5. Further work on O type binaries will be very important in this connection.

## V.  CONCLUSIONS

A summary of the observed parameters derived for the O type stars is contained in Table 2. These numbers represent the best present estimates of the quantities. The $M_V$ come from the calibration of Conti and Alschuler (1971) with slight modifications due to the inclusion of the

Table 2

Observed parameters of O type stars

| Spectral Type | $M_V$* | | $T_{eff}$ | | $\log L/L_\odot$ | | | $M/M_\odot$ | $\log R/R_\odot$ | | $\log g$ |
|---|---|---|---|---|---|---|---|---|---|---|---|
| | ZAMS | V | V | If | ZAMS | V | If | ZAMS | V | If | V |
| O3 | -5.7 | -6.1[†] | 55000[†] | 52000[†] | 6.08 | 6.24[†] | 6.40 | 120[†] | 1.16 | 1.28 | 4.2 |
| O4 | -5.5 | -5.8[†] | 50000 | 47500[†] | 5.86 | 5.98[†] | 6.28 | 90[†] | 1.13 | 1.30 | 4.1 |
| O5 | -5.3 | -5.5 | 47000 | 44500[†] | 5.70 | 5.78 | 6.20 | 60[†] | 1.07 | 1.32 | 4.1 |
| O5.5 | -5.1 | -5.3 | 44500 | 42500[†] | 5.56 | 5.64 | 6.12 | 45 | 1.04 | 1.32 | 4.0 |
| O6 | -4.8 | -5.1 | 42000 | 40000[†] | 5.34 | 5.46 | 6.04 | 37 | 1.01 | 1.34 | 4.0 |
| O6.5 | -4.5 | -4.9 | 40000 | 38000 | 5.16 | 5.32 | 5.96 | 30 | 0.98 | 1.34 | 3.95 |
| O7 | -4.2 | -4.7 | 38500 | 36500 | 4.98 | 5.18 | 5.92 | 28 | 0.94 | 1.36 | 4.0 |
| O7.5 | -4.1 | -4.6 | 37500 | 35500 | 4.90 | 5.10 | 5.88 | 25 | 0.92 | 1.36 | 4.0 |
| O8 | -3.9 | -4.5 | 36500 | 34500 | 4.80 | 5.04 | 5.86 | 23 | 0.92 | 1.37 | 3.95 |
| O8.5 | -3.8 | -4.4 | 35500 | 33500 | 4.72 | 4.96 | 5.84 | 21 | 0.90 | 1.39 | 3.95 |
| O9 | -3.7 | -4.3 | 34500 | 33000 | 4.66 | 4.90 | 5.82 | 19 | 0.89 | 1.39 | 3.95 |
| O9.5 | -3.6 | -4.2 | 33000 | 31500 | 4.58 | 4.82 | 5.72 | 18 | 0.89 | 1.38 | 3.9 |

*$M_V$ for all If types taken to be $-6^m.7$.

[†]Very uncertain.

O3 stars discussed by Conti and Burnichon (1975). $M_v$ for all If stars is taken to be $-6^m_{.}7$ with an uncertainty of $\pm 0^m_{.}5$. The temperature scale is that of Conti (1973b), with that for the supergiants taken to be ~5% cooler than the main sequence, as discussed in §III. The masses are very roughly interpolated from Figure 5, and are only <u>estimated for stars on the ZAMS</u>. One should not make estimates of masses of <u>evolved</u> stars from Figure 5 or this table since the evolutionary tracks have been computed without mass loss, which is undoubtedly important.

It should be noted that the gravities derived from M and R, in Table 2, are more or less consistent with those implied by Figures 2 and 3. However, the accuracy is not better than a factor two. The L/M ratio for the most luminous main-sequence stars, $\sim 10^4$, is consistent with the upper limit for accretion in massive stars discussed by Kahn (1974). (This would not be true if M was appreciably less than $100_\theta$.) The careful reader will notice that there are no entries for luminosity type III stars. This is because their values cannot be determined independently of the data for V and If stars and is rather uncertain for the latter types in any case. Type III can be interpolated half way between the V and If columns but it should be realized that this will be a guesstimate.

The entries of Table 2 are similar to those derived by Panagia (1973) from nearly identical data. There are slight differences in the values for the hottest stars and in the temperature scale for the If types. For other astrophysically important parameters for O type stars, Panagia's paper provides a good discussion and additional tables. The parameters for the main-sequence stars appear to be well in hand but there is still some basic uncertainty about the temperature scale for the If type stars.

This research has been partly supported under National Science Foundation grant MPS72-05062 A02.

## REFERENCES

Auer, L. H., and Mihalas, D.  1972, Ap. J. Suppl., 24, 193.
Batten, A. H.  1967, Pub. Dom. Ap. Obs., 13, 119.
Castor, J. I., Abbott, D. C., and Klein, R. I.  1975, Ap. J., 195, 157.
Conti, P. S.  1973a, Ap. J., 179, 161.
_____.  1973b, ibid., 179, 181.
_____.  1974, ibid., 187, 539.
Conti, P. S., and Alschuler, W.  1971, Ap. J., 170, 325.
Conti, P. S., and Burnichon, M.-L.  1975, Astron. and Ap. (in press).
Conti, P. S., and Leep, E. M.  1974, Ap. J., 193, 113.
Hutchings, J.  1968, M.N.R.A.S., 141, 219.
Kahn, F.  1974, Astron. and Ap., 37, 149.
Larson, R. B., and Starrfield, S.  1971, Astron. and Ap., 13, 190.
Lortet-Zuckermann, M.-C.  1975, this volume.
Mihalas, D.  1972, Ap. J., 176, 139.
_____.  1974, A. J., 79, 1111.
Mihalas, D., Hummer, D. G., and Conti, P. S.  1972, Ap. J. Letters, 175, L99.
Mihalas, D., and Hummer, D. G.  1974, Ap. J. Letters, 189, L39.
Morrison, N.  1975, Ap. J. (in press).
Morton, D.  1967, Ap. J., 147, 1017.
_____.  1969, Ap. J., 158, 629.
Panagia, N.  1973, A. J., 78, 929.
Stothers, R.  1972, Ap. J., 175, 431.
Swarzchild, M., and Härm, R.  1959, Ap. J., 129, 637.
Walborn, N.  1971, Ap. J. Suppl., 23, 257.
_____.  1972, A. J., 77, 312.

# RECENT OPTICAL OBSERVATIONS OF SOME EVOLVED HII REGIONS

J. MEABURN

Dept. of Astronomy, The University,
Manchester, England.

## ABSTRACT

A variety of recent optical observations of the HII regions in the Magellanic Clouds, M42, M8, M16 and M17 are compared with observations of molecular complexes and thermal radio emission. It appears that all these HII regions are composed of a mass of ionization fronts entering into the densest regions of inhomogeneous masses of HI, dust and molecules. Many of the motions appear simply to be flows from these ionization fronts. However, some motions are not easily explained in this way.

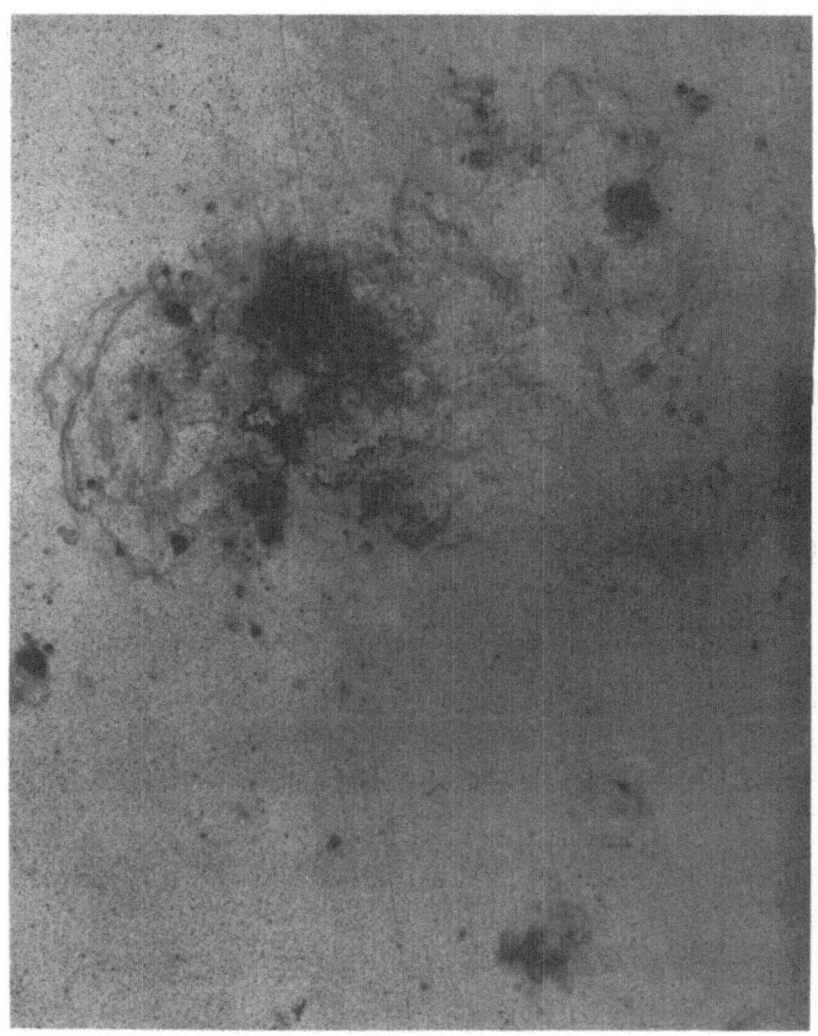

Fig. 1 A section of the 5 hour exposure of the LMC through the 100Å
        Hα    filter in the focal plane of the SRC 48-inch Schmidt camera.

(Davies, Elliott and Meaburn, 1975)

## 1) INTRODUCTION

It is illuminating to compare a variety of recent optical observations of the extensive HII regions M42, M17, M16 and M8 particularly with the large scale distribution of molecules and the thermal radio emission. In several cases this process permits models to be constructed of both the overall and the detailed structure of these 'evolved' HII regions that bear some semblance to reality.

The most relevant observation of each of these HII regions will now be considered in this way.

## 2) THE HII REGIONS IN THE MAGELLANIC CLOUDS

Before looking at the detailed structure of particular HII regions in the Galaxy it is instructive to examine the large scale structure and distribution of these in the Magellanic Clouds.

The complexes of HII regions and supernova remnants have now been photographed in both the LMC and SMC with the SRC 48-inch Schmidt camera through a 100Å bandwidth interference filter centred on Hα and [NII] (Davies, Elliott and Meaburn, 1975). This filter, of my own design, is of a mosaic construction, 15 inches square and has high index, high order spacing layers and a square profile. When placed in the convergent beam in front of the photographic plate the variation in transmission of Hα and [NII] both over the $6^\circ$ field and within the f/2.5 convergent cone is less than 5 percent. It permits exposures of up to 5 hrs. to be obtained. In Figure 1 part of a plate of the LMC obtained with a 5 hour exposure is presented. The Doradus nebular complex is shown to be surrounded by many fainter filaments which are possibly ionization fronts moving into the huge neutral masses. Ionized material of this nature covers the whole of the LMC and SMC.

## 3) THE STRUCTURE OF M42

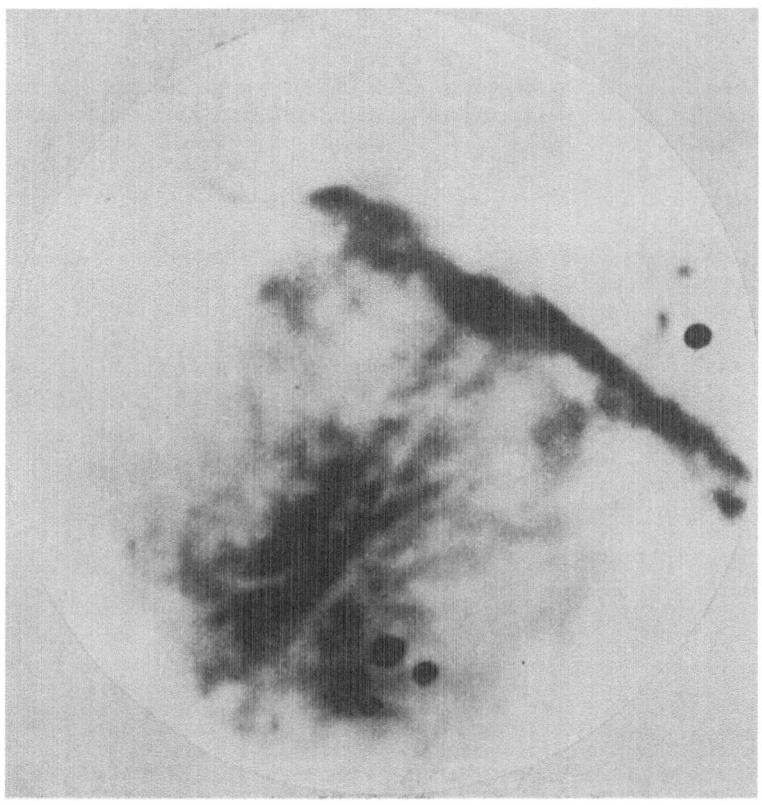

Fig. 2  A 5 min.  [NII]  photograph of M42 at the f/15 focus of the 98-
inch Isaac Newton Telescope (after Elliott and Meaburn, 1974a).

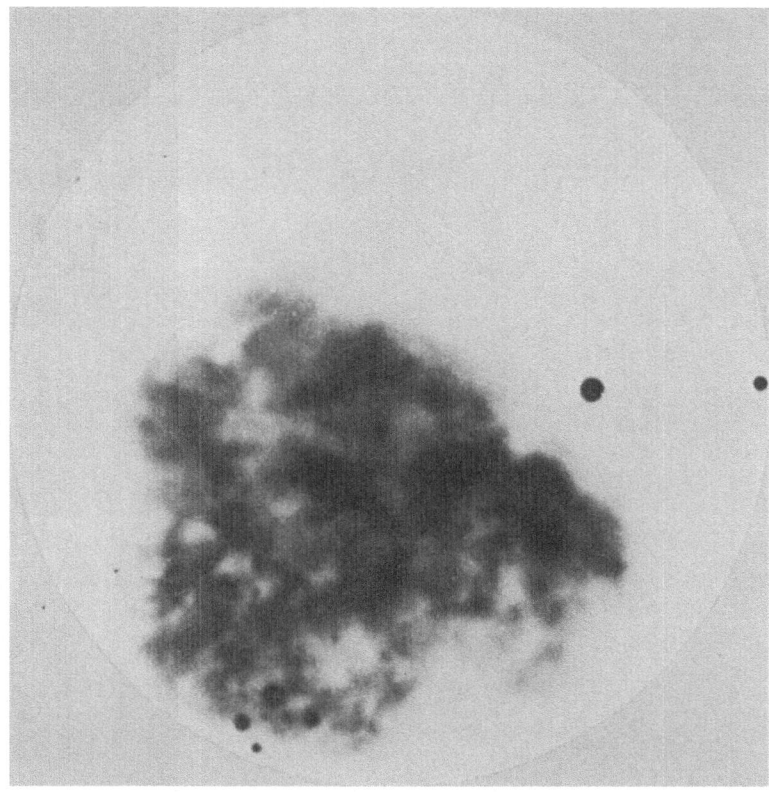

Fig. 3 A 2.5 min. exposure of M42. Similar to Figure 2 but at [OIII] (after Elliott and Meaburn, 1974a).

Some of the most recent optical observations of M42 have already been reviewed in detail (Meaburn, 1974 and Dopita, Dyson and Meaburn, 1974). Here only the principal conclusions from these will be considered.

Osterbrock and Flather (1959), Danks and Meaburn (1971), Caplan (1972), Elliott and Meaburn (1973) have all demonstrated from measurements of the [OII] and [SII] line ratios that the local electron density, $Ne_{local}$, varies from around 500 e/cm$^3$ in the outer regions of M42 to around 15000 e/cm$^3$ near the Trapezium group of stars. Elliott and Meaburn (1973) conclusively demonstrated that $Ne_{local}$ reaches its maximum value $0.5$ arc min from this group of exciting stars. Moreover, it was demonstrated that the ionized material only occupied about 0.05 of the volume of the nebula. This clumpy distribution of the ionized material was indicated directly by Elliott and Meaburn (1973) for large variations, on a 3 arc sec scale, of the [OII] ratio were observed in the core of M42. These demonstrated that density variations of < 1000 to >15000 e/cm$^3$ occured over distances of this order.

The small scale variations in the ionization structure in the core of M42 are indicated in Figures 2 and 3. Here image tube photographs (Meaburn, 1974) at the f/15 focus of the 98-inch Isaac Newton Telescope (Elliott and Meaburn, 1974) in the light of [NII] and [OIII] respectively are shown. The [NII] line originates in regions with no ionizing radiation harder than 29.6 eV. and [OIII] in regions with none less than 35.1 eV.

It can be seen that there are many extensive maxima of the brightness ratio, B [NII] / B [OIII] . Münch and Taylor (1974) have since demonstrated the presence of many filaments in the core of M42 in the light of OI 8446Å. These coincide exactly with the maxima of B [NII] / B [OIII] shown here. The OI line originates from lower energy regions containing photons less than 13 eV. They indicate that similar filamentary structure was not observed in their [OI] 6300

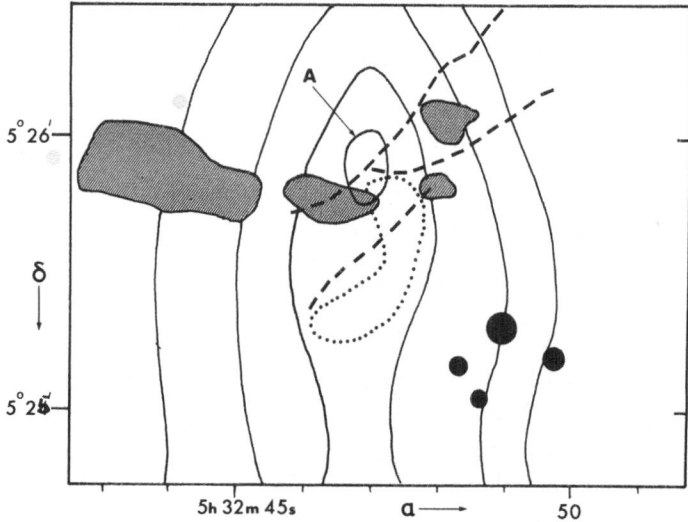

Fig. 4 A small scale view of some of the observed phenomena near the
Trapezium stars in M42. The dashed line indicates the ionization
fronts, the solid lines are some of the contours surrounding peak,
A, of the HCHO emission. The shaded areas are where the
[OIII] line is split and the dotted region encloses the region
where the local Ne has its maximum values (after Elliott and
Meaburn, 1974b).

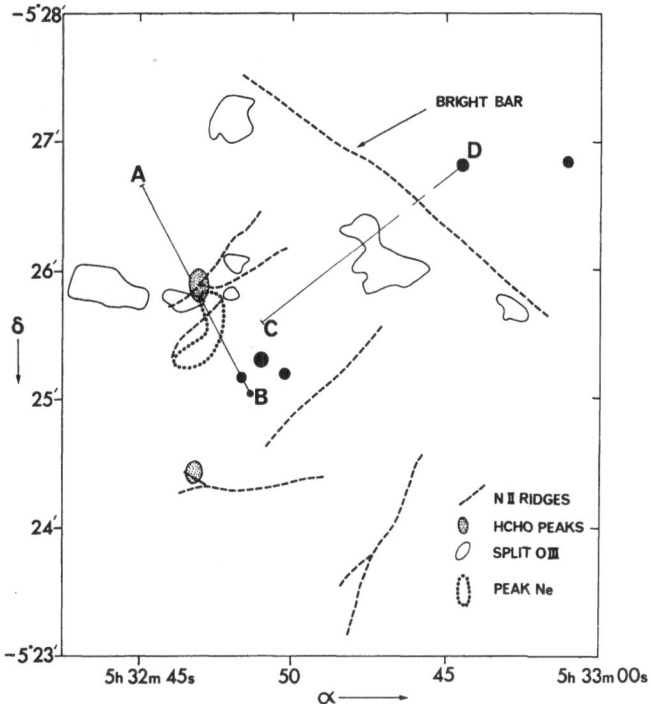

Fig. 5 A medium scale view of the phenomena in M42. The major
ionization fronts [NII] ridges are shown dashed. The
splitting and HCHO emission are shown to be closely related to
these fronts (after Elliott and Meaburn, 1974a).

photograph. The likely explanation is, however, that the [OI] 6300 image of the filaments was swamped by the brighter image in the light of the scattered continuum. This is a less important effect at 8446Å where the continuum is diminished.

These ridges of maximum B [NII] / B [OIII] and the filaments of OI emission are most certainly very close to ionization fronts which are moving into neutral material. Only in these regions, shielded from the hard ionizing radiation can such low energy ions exist near the Trapezium stars. This emphasises that in the very core of M42 there is a large quantity of neutral material.

These ionization fronts are shown in Figures 4 and 5 compared to several other of the phenomena observed in the core of M42. In particular, it is notable that the most prominent fronts which are nearest the Trapezium stars appear closely associated with one of the maximum of HCHO emission (Thaddeus et al., 1971) and the maximum of local electron density (Elliott and Meaburn, 1973). Moreover, the regions where the [OIII] line is split (Wilson et al., 1959) are also closely associated with these fronts. This is shown in detail in Figure 4 against the contours of HCHO (1 arc min resolution). In Figure 5 it is also demonstrated that [OIII] splitting is also clearly associated with the larger fronts.

In Figure 6 several similar associations on a much larger scale and with an 8 arc sec resolution are illustrated (Dopita, Isobe and Meaburn, 1974). Here ridges where B Hα / B [NII] have minimum values are shown compared to regions where the scattered continuum are highest and regions where the [NII] line is split (Dopita, Dyson and Meaburn, 1974; Dopita, Gibbons, Meaburn and Taylor, 1971; Deharveng, 1973). It can be seen that all these phenomena appear to be closely related. The low energy regions where B Hα / B [NII] have minimum values again most certainly indicate the presence of major ionization fronts.

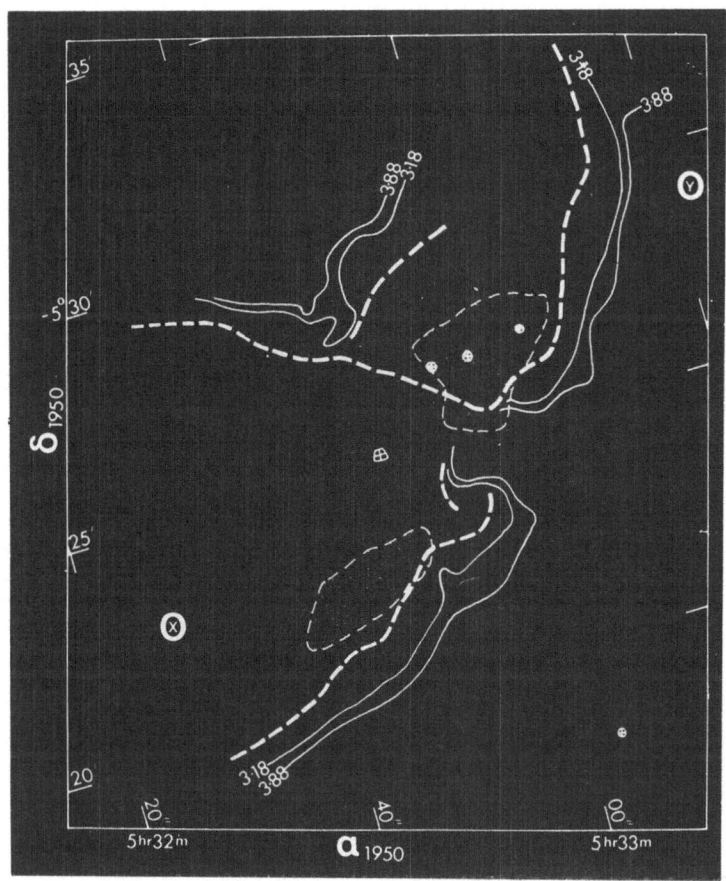

Fig. 6 A large scale view of the observed optical phenomena in M42. The ionization fronts are shown as heavily dashed lines, whereas the regions where the [NII] line is split are dotted areas enclosed by lightly dashed lines. Two contours of the continuum light of the Trapezium stars scattered by the concentrations of dust are also shown (after Dopita, Isobe and Meaburn, 1974).

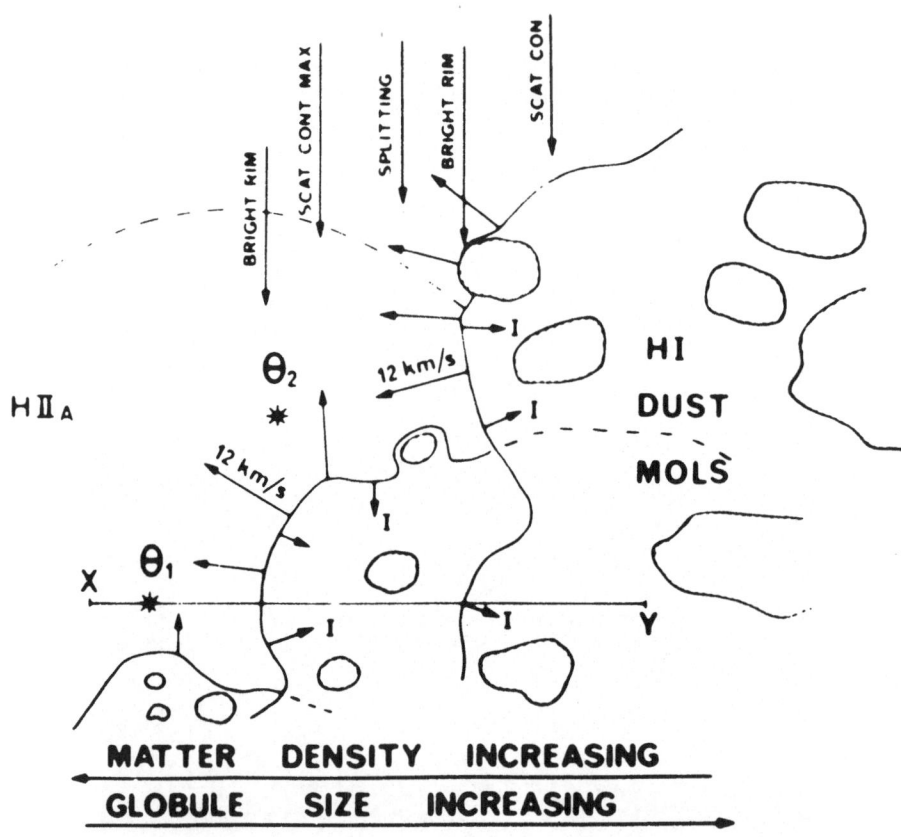

Fig. 7 A schematic model of M42 that attempts to explain all the
phenomena shown in Figures 5, 6 and 7. The line splitting is due
to seeing two separate flows from ionization fronts along one line
of sight.

A very schematic model of M42 that attempts to explain all these phenomena is shown in Figure 7. In this it is simply proposed that the line splitting is caused by seeing separate flows along one line of sight of ionized material streaming from separate ionization fronts. In this model the scattered continuum must be that of the starlight scattered by the dust in the neutral material behind the fronts. It would be very interesting to examine its properties more closely in the future for it originates from the regions as yet unaffected by the hard ionizing radiation from the stars.

The neutral material must be very inhomogeneous for the ionized material is observed to be in a very condensed state and partially ionized globules can be observed in it.

The close proximity to the Trapezium stars ($\sim$ 0.1 pc) of the HCHO molecules that are observed in emission as proposed in this model is further substantiated by the measurements of the OI and [OI] radial velocities of 28 km/s (heliocentric) by Münch and Taylor (1974). These low energy lines are emitted by the neutral material closest to an ionization front and consequently have a negligible flow velocity with respect to this front and the neutral material. This velocity of 28 km/ sec is then very close to that of 26 km/s for the HCHO molecules and 29 km/s for the Trapezium stars. The radial velocities from OI and [OI] can be obtained to a high degree of accuracy for these lines are observed at zero radial velocity in the airglow spectrum and their rest wavelengths measured.

Kaler (1969) claimed to have detected a systematic change of the heliocentric radial velocity with ionization potential of the ion being observed in the core of M42. He suggested lines from high energy ions were being emitted by material approaching the observer faster than those from low energy ions. However, Fehrenbach (1974) has suggested from his own observations that this may not be the case for all positions and is investigating this suggestion more thoroughly.

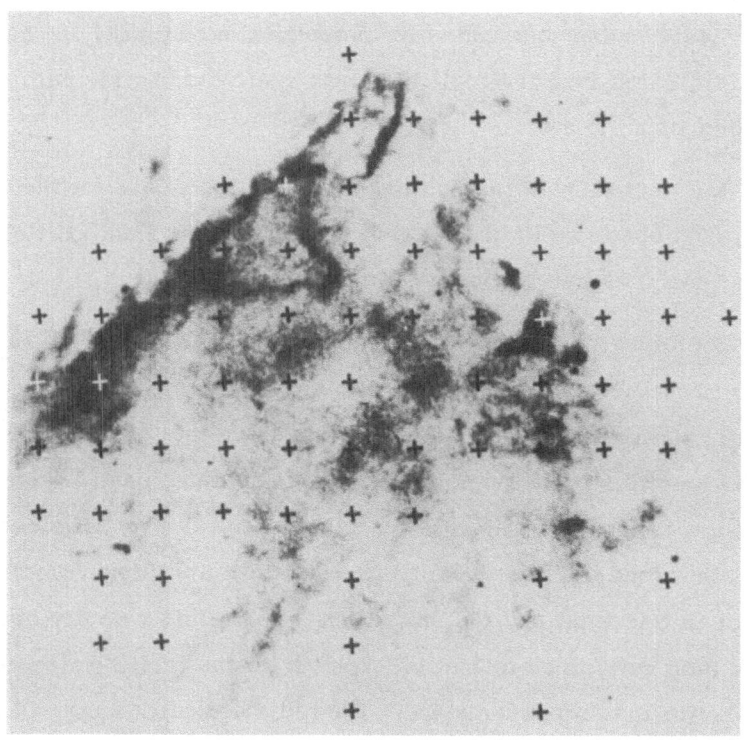

Fig. 8 An [NII] photograph of M17 (after Smith, 1972).

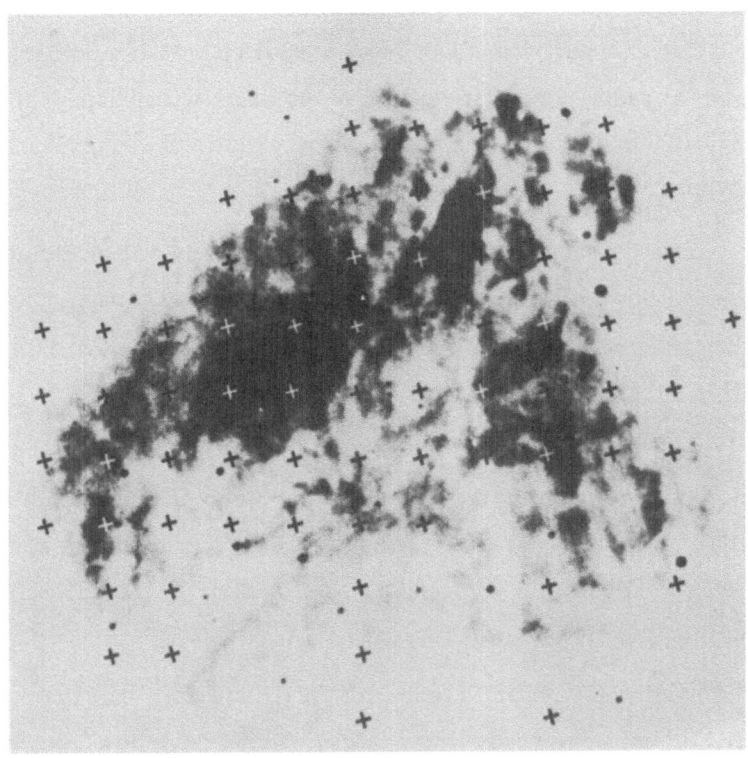

Fig. 9 An [OIII] photograph of M17 (after Smith, 1972).

However, Kaler's results are very consistent with this model for the high energy lines are emitted furthest from the ionization fronts where the flow velocities are greatest.

Scheglov (1968), Lee (1969), Meaburn (1971), Dopita, Gibbons, Meaburn and Taylor (1973) have all suggested that faint (one percent) components of the emission lines are present which indicate a large volume though small amount of the material of M42 is approaching the observer at radial velocities of up to 100 km/s with respect to the bulk of M42. If this is the case, it cannot be explained simply by flow from ionization fronts.

No really satisfactory explanation exists as yet of these phenomena but radiation pressure on dust may play a part.

4) M17, M16, M8

It can be seen from the [NII] and [OIII] photographs of Smith, 1972 shown in Figures 8 and 9 that many enhancements of B [NII] / B [OIII] similar to those in M42 occur also in M17. The nebula is then again being formed by a complex of ionization fronts eating into massive neutral regions.

Scheglov (1968) and Meaburn (1971) indicated that the motions in this HII region were particularly large. These have now been surveyed very thoroughly (Elliott and Meaburn, 1975) at 95 positions (some are shown on Figures 8 and 9 as crosses) with a Fabry-Perot monochromator (Meaburn, 1972) on the 74-inch Radcliffe telescope but now with a wavelength resolution of $0.13\overset{\circ}{A}$ and angular resolution of 28 arc sec.

Many large regions where the [OIII] line had up to four components were discovered. Some samples of these [OIII] profiles are shown in Figures 10 and 11. The radial velocity zero is arbitrary.

Two most curious phenomena are noticed in these results. When the [OIII] line becomes double the components split to both negative and positive radial velocities wrt. to the mean radial velocity of the

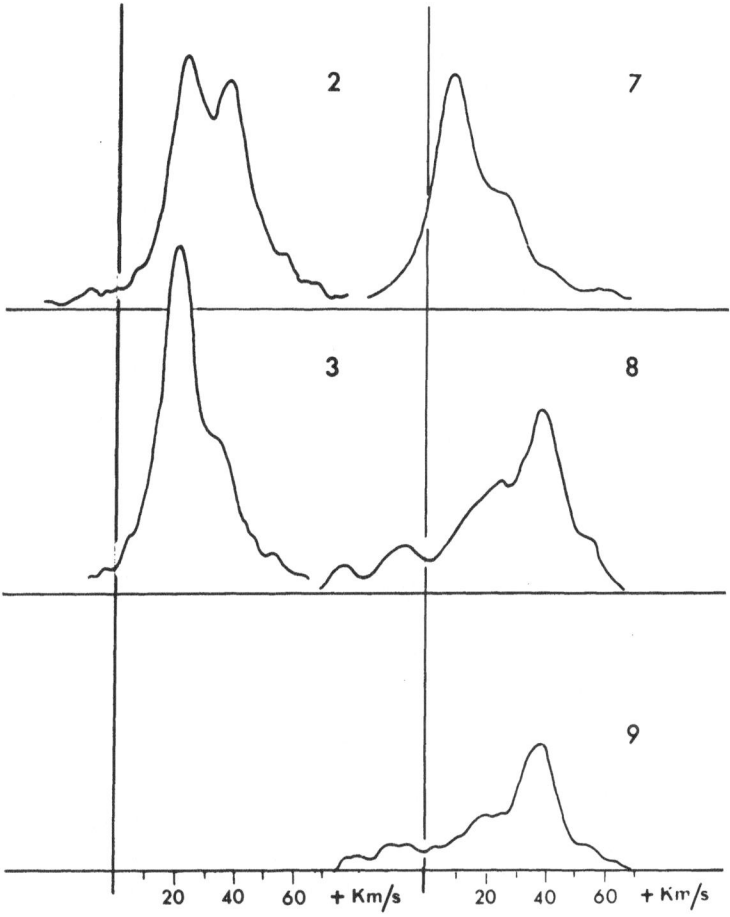

Fig. 10  Fabry-Perot profiles of the  [OIII]  line from the positions
marked as crosses at the extreme bottom left in Figures 8 and 9
(after Elliott and Meaburn, 1975a).

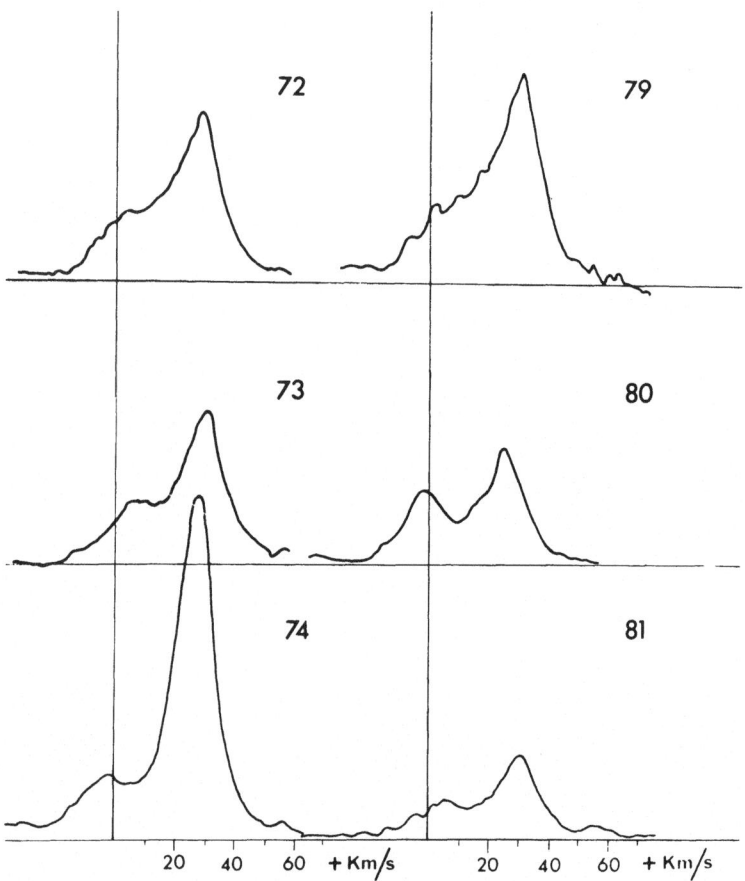

Fig. 11  As for Figure 10 but for the positions at the extreme right of
Figures 8 and 9.

line for the nebula though negative components are the largest (up to 50 km/s from the mean). Also, all this splitting occurs over the large neutral regions where [OIII] is the faintest.

The regions where the [OIII] line has more than one component are shown compared to many of the other phenomena observed in this nebula in Figure 12.

A schematic model is again presented in Figure 13 that attempts to explain these observations simply with flows from ionization fronts.

It is presumed that the exciting stars are obscured and are in the densest region near the maximum of the thermal radio emission, and that again the splitting is due to seeing many separate flows along each line of sight. However, these flow velocities would have to be much higher in M17 than M42 in this simple explanation. An alternative model could include a supernova remnant embedded in the ionized material.

There are many examples of supernova remnants being closely associated with HII regions. The Rosette and NGC 2264 have one between them, so have M20 and M8. Likewise, IC 410 is closely associated with a strong non-thermal radio ridge and IC 443 is expanding into an HII region. In many cases high velocities of the ionized material are observed in the regions where collisions are occuring (Meaburn, 1968 and 1971). However, in M17 non-thermal radio emission has not yet been discovered.

Presumably M17 is produced by a complex of ionization fronts eating into huge neutral clouds containing C O mapped by Lada, Dickinson and Penfield (1974). This is substantiated by the radial velocity measurements of Courtes et al. (1968) of H$\alpha$ for M17 of 25.1 km/s ($V_{LSR}$). This is very near the $\sim$ 20 km/s observed for CO.

Elliott and Meaburn (1975 b and c) have shown the [OIII] line split also over the large neutral intrusions in M16 (Figure 14) and M8. Likewise, Smith's (1972) photographs show a similar complex of fronts

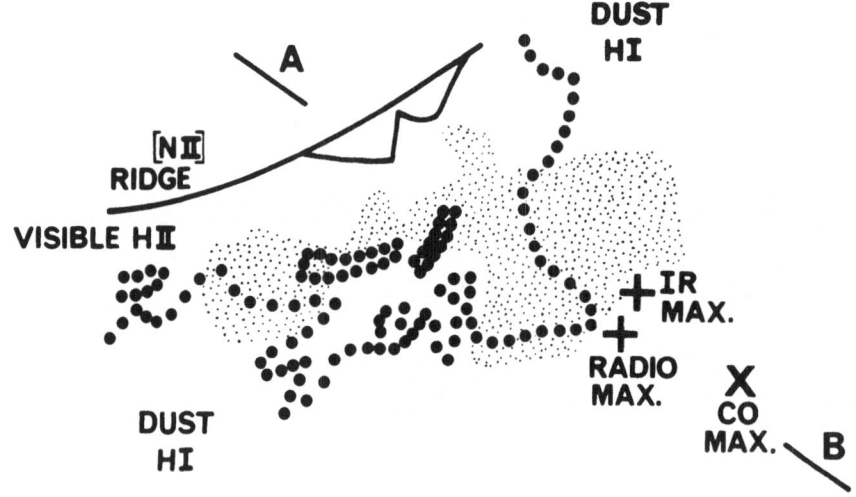

Fig. 12  Many of the observed optical and radio phenomena of M17 are
shown.  The shaded area is where the  [OIII]  line is observed
to be split (Elliott and Meaburn, 1975).  The heavily dotted
lines enclose the brightest part of the unobscured HII region
visible at optical wavelengths.  The maximum of the extensive
regions of 2 cm. thermal radio emission (Schraml and Mezger,
1969), the 21 $\mu$  peak from hot dust (Lemke and Low, 1972)
and the CO emission (Lada et al., 1974) are shown as crosses.

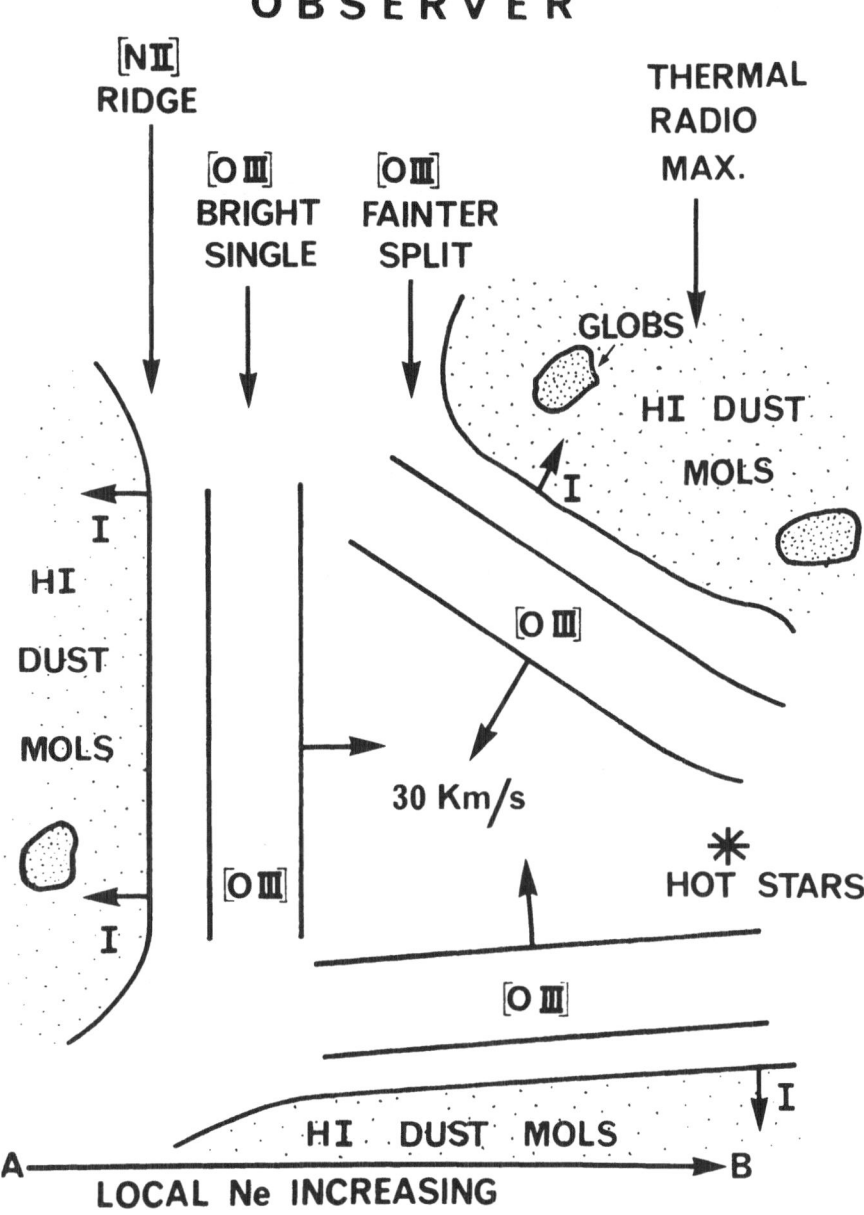

**OBSERVER**

[NII] RIDGE

[OIII] BRIGHT SINGLE

[OIII] FAINTER SPLIT

THERMAL RADIO MAX.

GLOBS

HI DUST MOLS

I

HI DUST MOLS

[OIII]

30 Km/s

[OIII]

HOT STARS

[OIII]

I

HI DUST MOLS

A — LOCAL Ne INCREASING — B

Fig. 13  A model through A B in Figure 12 of M17 to explain the phenomena in Figure 12 is presented. The motions here are simply explained by flows from ionization fronts, I.

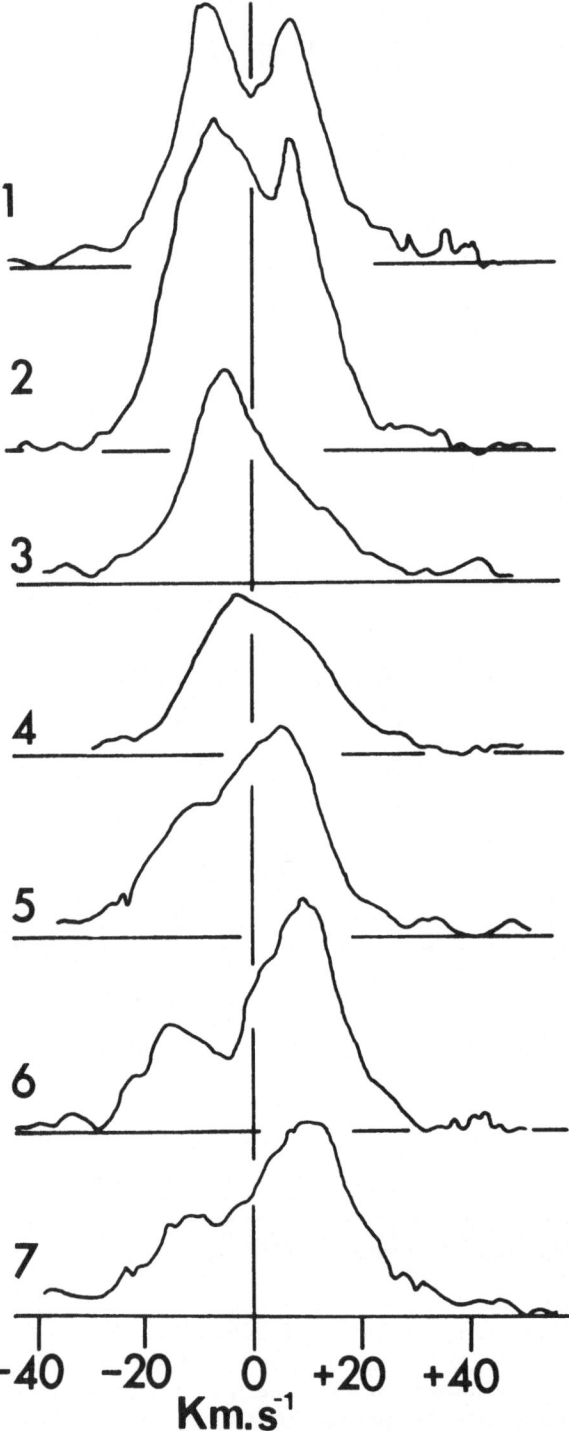

Fig. 14  A line of  [OIII]    Fabry-Perot profiles across the neutral
intrusions in M16 (after Elliott and Meaburn, 1975b).

in      M16, M8 and M20.  In particular, the 'Hourglass' in M8 appears to be a set of fronts very analogous to the 4 arc min core of M42.  Also from measurements of the  $[OII]$  and  $[SII]$  ratios (Meaburn, 1969 and Bohuski, 1974) the ionized material in M8 is shown to have $Ne_{local}$ of 300 e $/cm^3$ in the outer regions rising to 5000 in the central 'Hourglass'.  It is also in a similar condensed state to M42.

Similar models to those for M42 and M17 can be constructed for these nebulae to explain these phenomena.

## REFERENCES

Courtes, G., Georgelin, Y., Georgelin, Y., Monnet, G. and
      Pourcelot, A.  1968, Interstellar Ionized Hydrogen (Benjamin Inc.),
      571, (1968).
Bohuski, T. J. 1973, Astrophys. J., 184, 93.
Caplan, J. G. 1972, Astron. Astrophys., 18, 408.
Danks, A. C. and Meaburn, J. 1971, Astrophys. Space Sci., 11, 398.
Davies, R. D., Elliott, K. H. and Meaburn, J. 1975 (in prep.).
Deharveng, L. 1973, Astron. Astrophys., 29, 341.
Dopita, M. A., Dyson, J. E. and Meaburn, J. 1974, Astrophys. Space
      Sci., 28, 61.
Dopita, M. A., Gibbons, A. H., Meaburn, J. and Taylor, K.T.1973,
      Astrophys. Letts., 13, 55.
Dopita, M. A., Isobe, S. and Meaburn, J. 1975, Astrophys. Space Sci.
      (in press).
Elliott, K. H. and Meaburn, J. 1973, Astron. Astrophys., 27, 367.
Elliott, K. H. and Meaburn, J. 1974, Astron. Astrophys., 34, 473.
Elliott, K. H. and Meaburn, J. 1974, Astrophys. Space Sci., 28, 351.
Elliott, K. H. and Meaburn, J. 1975, Astrophys. Space Sci., (in press).
Elliott, K. H. and Meaburn, J. 1975, Mon. Not. Roy. Astron. Soc.,
      (in press).
Elliott, K. H. and Meaburn, J. 1975, (in prep.).
Fehrenbach, C. 1974, (private comm.).
Kaler, J. B. 1969, Astrophys. J., 148, 925.
Lada, C., Dickinson, D. F. and Penfield, H. 1974, Ap. J. Letts.,
      189, L 35.
Meaburn, J. 1968, Astrophys. Space Sci., 2, 115.
Meaburn, J. 1969, Astrophys. Space Sci., 3, 600.
Meaburn, J. 1971, Astrophys. Space Sci., 13, 110.
Meaburn, J. 1972, Astron. Astrophys., 17, 106.
Meaburn, J. 1974, Appl. Optics Dec.
Meaburn, J. 1974, Proceedings of the Second Regional Meeting IAU,
      Trieste, Italy.
Münch, G. and Taylor, K. T. 1974, Ap. J. Letts. 192, L 93.

Osterbrock, D. E. and Flather, E. M. 1959, Astrophys. J., <u>129</u>, 235.

Scheglov, P. V. 1968, I.A.U. Symp. No. <u>34</u>, Planetary Nebulae, 270.

Thaddeus, P., Wilson, R. W., Kutner, M., Penzias, A. A. and
    Jefferts, K. B. 1971, Astrophys. J. Letts., <u>168</u>, 59.

Wilson, O. C., Münch, G., Flather, E. M. and Coffeen, M. F. 1959,
    Astrophys. J. Suppl. <u>4</u>, 199.

Lemke, D. and Low, F. J. 1972, Ap. J., (Letters), <u>177</u>, L53.

Schraml, J. and Mezger, P. G. 1969, Ap. J., <u>156</u>, 269.

# EVOLVED H II REGIONS

E. CHURCHWELL

Max-Planck-Institut für Radioastronomie,
Bonn, FRG

## ABSTRACT

A probable evolutionary sequence of H II regions based on six
distinct types of observed objects is suggested. Two examples which may
deviate from this idealized sequence, are discussed. Even though a size-
mean density relation for H II regions can be used as a rough indication of whether a nebula is very young or evolved, it is argued that
such a relation is not likely to be useful for the quantitative assignment of ages to H II regions. Evolved H II regions appear to fit into
one of four structural types: rings, core-halos, smooth structures, and
irregular or filamentary structures. Examples of each type are given
with their derived physical parameters. The energy balance in these
nebulae is considered. The mass of ionized gas in evolved H II regions
is in general too large to trace the nebula back to single compact H II
regions. Finally, the morphological type of the Galaxy is considered
from its H II region content.

## I. INTRODUCTION

Most of the emphasis at this symposium has been on the earliest stages in the development of O-stars and compact H II regions. In contrast, I will consider some aspects of the later evolutionary phases of H II regions. The basis for my discussion of "evolved" H II regions is laid in Section II where a probable evolutionary sequence of star formation and subsequent nebula development is suggested. The question of whether it is feasible to calibrate a mean density size relation in order to assign ages to H II regions, is considered. In Section III, the physical properties and energy balance of several "evolved" H II regions which have been studied at radio wavelengths, are discussed. A consideration of the morphological type and luminosity class of the Galaxy based on its H II region content is undertaken in Section IV.

## II. EVOLUTION OF H II REGIONS

One might well ask whether enough is known about the evolution of H II regions to merit a discussion of "evolved" H II regions as a separate group of objects. As a framework upon which to base further discussion, we must first try to define what we mean by evolved H II regions and to see whether, in fact, observational evidence supports our definition. Probably the most comprehensive review on the evolution of H II regions is that of Mathews and O'Dell (1969); although their main emphasis was on the theoretical aspects of evolution. In Trieste, I (Churchwell, 1974) discussed six observed types of objects which suggest an evolutionary sequence. These are listed in abbreviated form in Table 1; the order is believed to be chronological.

It is doubtful whether all H II regions go through the last stage; but if Torres-Peimbert et al. (1974) are right, then a significant fraction must end up as naked O-stars. This depends critically on whether H II regions ultimately become density-bounded at some point in their evolution (see discussion by Churchwell, 1974).

T A B L E   1

Probable Sequence of Evolutionary Stages of O-Star Formation
and Subsequent Nebula Development

| Evolutionary Stage | Type of Observed Object | Observational Probes | Examples | Comments |
|---|---|---|---|---|
| 1. Proto-stellar clouds | Cool, dense molecular clouds | mm-wave lines of molecules such as HCN, HNCO, and $H_2CO$ | 2-mm-$H_2CO$ cloud centered on KL nebula in Orion | Thaddeus et al. (1971) find for 2 mm $H_2CO$ cloud $N_{H_2} \sim 2\times10^5$ cm$^{-3}$, $M \sim 200\ M_\odot$, $T_K \lesssim 70$ K, $D^2 \sim 0.6$ pc. |
| 2. Pre-main-sequence O-stars surrounded by an optically thick shell | Some compact 20µ IR emission sources | IR continuum emission $H_2O$ masers | IRS 5 in W3 | No observable radio continuum emission. Size $\lesssim 0.05$ pc. Strong emission at 20µ but not observable at 2µ (Wynn-Williams et al., 1972). Not possible to distinguish between evolved super-giant and pre-main-sequence O-star surrounded by a dust-shell. Position of IRS 5 suggestive of latter. |
| 3. Earliest detectable H II region | Compact H II regions | radio continuum emission IR continuum emission usually $H_2O$ and/or OH masers | W3 (OH), ON-1, compact components in DR21 and NGC 7538 | Radio flux densities infer a range in spectral types between 08 for W3 (OH) to B0-B1 for ON-1. Usually not observable at optical wavelengths. |
| 4. Expanded H II regions | Dense intermediate-sized H II regions | the whole range of radio, optical, and IR techniques including spectroscopy | Orion A, M17, RCW 38 | $10^2 - 10^3$ larger and less dense by $< 10^{-2}$ than compact objects such as W3 (OH). These objects have high surface brightness, at radio and optical wavelengths. |
| 5. Very expanded H II regions | Large, diffuse H II regions | same as 4 | IC 434, NGC 7822, NGC 7000 - IC 5070 η Carina | $10 - 10^2$ larger and less dense by $< 10^{-2}$ than Orion A. Low surface brightness and large extent make radio studies very difficult. Easily observable at optical wavelengths. |
| 6. Naked O-stars | O-stars with no observable associated H II region | Typical stellar observational methods | ζ Pup. | Torres-Peimbert et al. (1974) find that $\sim 50$ % of a sample of 285 O-stars have no observable (E < 50 pc cm$^{-6}$) associated H II region. |

The sequence given in Table 1 is certainly oversimplified and one can easily think of deviant cases. Let us consider two examples.

1) In Zuckerman's (1973) model of the Orion A nebula, the Trapezium stars are located on the outer edge of a large dense molecular cloud which lies behind the nebula. This in itself is a rather curious situation, since one would normally expect the stars to have formed near the center of the dense molecular cloud. In addition, the Trapezium is moving toward the center of the molecular cloud (due to gravitational forces according to Zuckerman 1973). As the Trapezium stars sink further into the denser regions of the neutral gas, it is probable that the H II region will contract as it evolves.

2) The opposite situation could also occur in which an O-star, during the process of contraction onto the main-sequence, is accelerated (perhaps by a close encounter with another member of the developing star cluster) and thereby rapidly moves out of the cloud in which it was formed. In this case it would be possible for a very large nebula to form around such a star even though the star itself is quite young.

Attempts to calibrate a size-mean density relation in order to quantitatively assign ages to H II regions have failed. This is because both the size and mean density of an H II region depend on numerous other parameters as well as age. For example, nebular expansion may be retarded by surrounding neutral gas at different rates in different galactic locations. The ionized gas may be continuously replenished by embedded neutral globules and surrounding neutral gas, thereby maintaining a high electron density throughout the lifetime of the ionizing star. In addition, stellar winds, the effects of dust on both evolution and ionization structure, multiple ionization sources, and possible variable initial densities and masses out of which H II regions are formed could all have important influences on the size and mean density of a nebula. The upper limit on the age of an H II region is the main-sequence lifetime of its ionizing star; an approximate lower limit is the sound crossing time of an evolved nebula (the _initial_ Strömgren sphere may be formed much faster than the sound crossing time).

In spite of the uncertainties mentioned above I will assume that, in a statistical sense, large ($D_{H\ II} \geq 5$ pc which implies a sound crossing time of $\sim 2 \times 10^5$ yr), low density ( $N_e$ <100 cm$^{-3}$) H II regions are "evolved". I will therefore confine the rest of this discussion to objects in categories 4 and 5 in Table 1.

## III. PHYSICAL PROPERTIES, ENERGY BALANCE AND CONSIDERATIONS ON THE MASSES OF EVOLVED H II REGIONS

### a) Physical Properties

The apparent brightness distribution of evolved H II regions can be classified into four structural types: rings, core-halos, smooth structures and irregular or filamentory structures. The physical properties of several evolved H II regions which have been studied at radio wavelengths, are given in Table 2; examples of all four structural types are included (column 1). The physical parameters are those given in the references unless otherwise noted. The symbols have their usual meaning: $N_{erms}$ is the rms electron density; $M_{H\ II}/M_o$ is the mass of ionized gas in solar units; and E is the emission measure. $N_c$ (column 9) is the Lyman continuum photon flux required to ionize the nebula. It was calculated using the relation (see Mezger 1973; p. 166)

$$\left[\frac{N_c}{s^{-1}}\right] \geq 5.0457 \times 10^{46} \left[\frac{T_e}{K}\right]^{-0.8} \left[\frac{U}{pc\ cm^{-2}}\right]^3$$

where the electron temperature was assumed to be $10^4$ K, except in IC434 where $T_e$ = 3000 K was adopted. For most sources listed in Table 2, $N_c$ is a minimum value since a substantial fraction of the total flux density probably resides in the outer low brightness regions which are either below the detection limit of present radio telescopes or cannot be unambiguously separated from the background radiation. To relate $N_c$ to the stellar UV flux, one must know if the nebula is ionization-bounded and what fraction of the UV flux is absorbed by dust. $M^*/M_o$ (Column

T A B L E   2

Examples of giant H II regions in the Galaxy and the LMC

| Structure | Nebula | Component | Angular Diameter (arc min) | Physical Diameter (pc) | $N_{e_{RMS}}$ (cm$^{-3}$) | $M_{HII}/M_o$ | E (pc cm$^{-6}$) | $N_c$ (s$^{-1}$) | $M^*/M_o$ | Refs. | Notes |
|---|---|---|---|---|---|---|---|---|---|---|---|
| Ring | NGC 7822 (W1) | Ring | ~120 | ~30 | <20 | >1611 | <$10^4$ | >$1.2 \times 10^{49}$ | ≥540 | 1 | 1 |
| | NGC 2237-46 (Rosette) | Ring | ~120 | ~50 | ~16 | 11000 | ~$3 \times 10^4$ | ~$1.5 \times 10^{49}$ | ≥670 | 2 | 2 |
| | | Hole | ~18 | ~7 | ~2 | | | | | | |
| Core-Halo | S206 (NGC 1491) | Core | ~3 | ~2.6 | 155 | 105 | $9 \times 10^4$ | >$1.9 \times 10^{49}$ | >850 | 3 | 3 |
| | | Halo | >20 | >17 | <14 | >2800 | <$5 \times 10^3$ | | | | |
| | S209 | Core | 2.9 | 4.6 | 140 | 490 | $1 \times 10^5$ | >$4.9 \times 10^{49}$ | >2200 | | |
| | | Halo | >20 | >32 | <8 | >9600 | <$3 \times 10^3$ | | | | |
| Smooth | NGC 7000 - IC 5070 (North America -Pelican Nebula) | | ~185 | 54 | 9 | 18000 | 4350 | ~$3.1 \times 10^{49}$ | ≥1390 | 4 | 4 |
| | IC 434 | | ~130 | 16 | 6 | 266 | ~580 | $1.9 \times 10^{48}$ | 85 | 5 | 5 |
| | IC 1318 b | | ~25 | 10 | ~38 | ~1830 | $2.3 \times 10^4$ | ~$2.3 \times 10^{49}$ | ≥1030 | 6,7 | 6 |
| Filamen-tary | NGC 2070 (30 Doradus) | | >3 | >50 | <70 | >$3 \times 10^5$ | ~$4.0 \times 10^5$ | >$8 \times 10^{51}$ | >$4 \times 10^5$ | 8,9 | 7 |

REFERENCES:

1. Churchwell and Felli (1970)
2. Menon (1962)
3. Walmsley et al. (1975)
4. Wendker (1968)
5. Caswell and Goss (1974)
6. Wendker (1970)
7. Baars and Wendker (1973)
8. McGee et al. (1972)
9. Huchtmeier and Churchwell (1974)

## NOTES TO TABLE 2

1. I will use NGC 7822 to refer to the whole ionized complex ($2^o$ ring structure plus the radio condensations G118.6+4.8 and G118.1+5.0). The flux density in the complete ring has not been measured, thus only limits can be given for the physical parameters.

2. The E-value was determined from the size and density given in columns 4 and 5.

3. The diameters quoted for S206 and S209 are the geometric mean half-power widths. The total extent of the halos has not been accurately measured, but they are known to have a diameter of at least one degree (i.e. $\sim$ 87 pc for S209 and $\sim$ 52 pc for S206). Thus only limits can be placed on the physical parameters of these nebulae.

4. Embedded in the smooth component are about 18 local maxima. Since ref. 4 did not calculate a value for U, I have estimated it by using the flux density of the smooth component (430 Jy at 1.414 GHz), a distance of 1 kpc and an electron temperature of 8000 K (which was assumed for the derivation of the other parameters).

5. The E-value was derived from the product $N_{e_{rms}}^2 d$ where the values were taken from columns 4 and 5. Other values are as given in ref. 5. The lower limit for $N_c$ was derived assuming a $T_e$ value of 3000 K. Caswell and Goss (1974) derive a value $T_e \leq 3500$ K.

6. IC 1318 b also has some local maxima which account for $\leq$ 20 % of the total flux density. The parameters $N_e$, $M_{HII}/M_o$, E and U were derived by assuming a spherical model with uniform density, a distance of 1.5 kpc (ref. 6), an electron temperature of $10^4$ K, and 80 % of the flux density given by ref. 5 at 11 cm.

7. The source seen at radio wavelengths is only the small brightest core of 30 Dor. Shapley and Paraskevopoulos (1937) give a total diameter of 25' arc or 400 pc. Thus the parameters derived from the radio continuum data are only limits.

11) is a crude estimate of the total mass contained in the stellar cluster to which the ionizing star(s) belong. This has been calculated using $<N_c>/<M> = 2.23 \times 10^{46}$ photons $s^{-1}$ $M_o^{-1}$ derived by Mezger et al. (1974) for the range of spectral types 04 - K5 (the Salpeter (1955) "original luminosity function" is assumed). The truncation at K5 stars certainly causes an underestimate of the total stellar mass. In addition, the $M^*/M_o$ values are underestimates because $N_c$ is underestimated.

With the exception of 30 Dor all the nebulae listed in Table 2 can be ionized by a single 05 or cooler star. The rough estimates of total stellar mass contained in the stellar cluster to which the ionizing star(s) belong (excluding 30 Dor) seem to agree with the range of values given by Blaauw (1964) for open clusters in the solar neighborhood.

### b) Energy Balance in Evolved H II Regions

In the following I will briefly summarize the presently available information on the energy balance in the nebulae listed in Table 2 whose ionizing stars are known. 30 Doradus will be discussed for other reasons. The stellar Lyman continuum fluxes are all taken from Churchwell and Walmsley (1973).

#### i) NGC 7822

Churchwell and Felli (1970) found that the combined stellar radiation from the CEPH IV association could not account for the ionization of the combined radio components G 118.6+4.8 and G 118.1+5.0. However, the new Lyman continuum photon fluxes derived by Churchwell and Walmsley (1973) increase the stellar flux of ionizing photons by a factor of $\sim 2.4$. Thus the situation is reversed; now the stellar photon flux is slightly greater than required to ionize the brightest two components in the ring. Ionization of the low density parts of the ring may account for the rest of the stellar UV photons.

#### ii) Rosette Nebula

The stars in the cluster NGC 2244, believed to be responsible for the ionization of the Rosette nebula, have been listed by Osterbrock and

Stockhausen (1960). The combined $N_c^*$ value of these stars (an O5, O6, three O8, and an O9 V-star) is $\sim 9.8 \times 10^{49}$ photons $s^{-1}$. This is about a factor of six greater than that required to ionize the nebula. On the other hand, it is likely that the flux density (and $N_c$ in direct proportion) would be significantly increased if the outer regions were mapped with a higher temperature resolution than that used to the present time. Also it is likely that some of the stellar UV photons are absorbed by grains and degraded into IR radiation. The most detailed studies of this nebula at radio wavelengths are those of Menon (1962) and Bottinelli and Gouguenheim (1964). Davies and Tovmassian (1963) have reported a low density H I shell around the nebula, although the physical association of the H I with the nebula is uncertain, if real, it would imply that the nebula is ionization-bounded.

### iii) S206

The O6-star BD $50^{\circ}$ 886 has been proposed as the ionization source for S206 by Georgelin et al. (1973). This star can easily account for the ionization of S206 including the extended halo. The synthesis map of Israel at $\lambda$ 6 cm (private communication) shows a point source just to the SE of the peak of S206 but it contains only a small fraction of the total flux density and therefore probably is not an important additional energy source. Observations of OH, $H_2CO$ and H I line absorption profiles reveal no clearly identifiable neutral gas associated with either S206 or S209 (Walmsley et al. 1975).

### iv) IC 434

The ionization source of this nebula is $\sigma$ Ori (Caswell and Goss 1974) which is an O9.5 V star. The $N_c^*$ value for this star is $\sim 2.2 \times 10^{48}$ photons $s^{-1}$ which can more than account for the ionization of the nebula. Caswell and Goss (1974) point out that the $N_c^*$ values given by Churchwell and Walmsley (1973) for early B-stars decrease much more rapidly than is consistent with observations of nebulae ionized by early B-stars.

Perhaps the most interesting aspect of this nebula is the surprisingly low electron temperature ($T_e < 3500$ K) derived by Caswell and Goss

(1974) from a combination of Hα and radio continuum data. If the Hα intensity scale is seriously in error, the inferred $T_e$ value could be significantly higher. On the other hand, if the low $T_e$ value is real, it would indicate that not all physical processes have been included in the thermal balance for low density H II regions.

### v) 30 Doradus

The stellar UV flux required to account for the measured radio free-free emission is $\geq 8 \times 10^{51}$ photons s$^{-1}$. This is the equivalent of $\sim 80$ O4-stars. No cluster in the spiral arms of our Galaxy contains so many hot stars. The central cluster in 30 Dor contains at least 9 WN 6 - 7 stars, one identified O8 star, and a central WR-like emission object. If the star cluster responsible for the ionization of 30 Dor has the same luminosity function as those in our Galaxy, then one would predict a stellar mass $\gtrsim 4 \times 10^5$ M$_o$. This is 1 - 2 orders of magnitude greater than any open cluster in our Galaxy. Clearly, the ionization of 30 Dor requires either a different luminosity function (in the sense that a larger percentage of very hot massive O-stars are formed relative to cooler lower mass stars) or the ionizing star cluster is much more massive than any in the solar neighborhood. Huchtmeier and I (1974) have argued in favor of the latter.

### c) Mass of Ionized Gas

The ionized gas masses associated with the diffuse nebulae listed in Table 2 are surprisingly large (a factor of two - ten greater than that of the associated star cluster). An H II region will have roughly attained its maximum ionized mass when it reaches pressure equilibrium with its surroundings. At this point the maximum ionized gas mass is related to the "initial" ionized mass by

$$M_{max} \sim (2\, T_{H\,II}/T_{H\,I})\, M_i \sim (200 - 2000)\, M_i$$

where $T_{H\,II}$ is the kinetic temperature of the ionized gas and $T_{H\,I}$ is that of the surrounding neutral gas. Obviously the nebulae listed in Table 2 are too massive to be traced backwards in time to single compact

H II regions such as W3 (OH) whose mass is $\sim 0.1$ $M_o$. The discrepancy would be even worse if the electron temperature in evolved H II regions is as low as 3000 K. Clearly this rough extrapolation does not include the possibility of ionization by several sources which is known to be the case for NGC 7822 and the Rosette. In the two cases where we believe only a single star ionizes each nebula, IC 434 and S206, the predicted initial mass of IC 434 is approximately compatible with that of a compact H II region, but that of S206 is too high by about an order of magnitude. The general inability to extrapolate the masses back to those of compact H II regions may imply that the ionized gas masses have been overestimated (perhaps due to unresolved clumpiness), or the extrapolation method is too crude, or our concept of the later evolutionary phases of H II regions is inadequate.


IV. <u>GALACTIC TYPE FROM H II REGIONS</u>

To what morphological type does our Galaxy belong, based on its H II region content? Since the pioneer work of Sersic (1960) it has been known that the diameters of the few largest H II regions in spiral and irregular galaxies are a function of galactic type. From a comparison of the optical diameters of 143 H II regions in our Galaxy with those of 369 regions in M33, Georgelin (1971) concluded that our Galaxy is of type $S_c$-. Figure 1 shows the histograms on which Georgelin based his analysis.

The five largest H II regions in our Galaxy according to Georgelin (1971) are:

$$
\begin{array}{ll}
\text{RCW 102, 104, 106} & d = 216 \text{ pc} \\
\eta \text{ Carina (RCW 53)} & = 175 \text{ pc} \\
\text{G } 328\overset{o}{.}5 -1\overset{o}{.}0 & = 175 \text{ pc} \\
\text{RCW 113} & = 164 \text{ pc} \\
\text{Orion loop} & = 149 \text{ pc}
\end{array}
$$

Since the work of Georgelin, Sandage and Tammann (1974) have shown that the diameter of the largest H II regions is a function not only of the

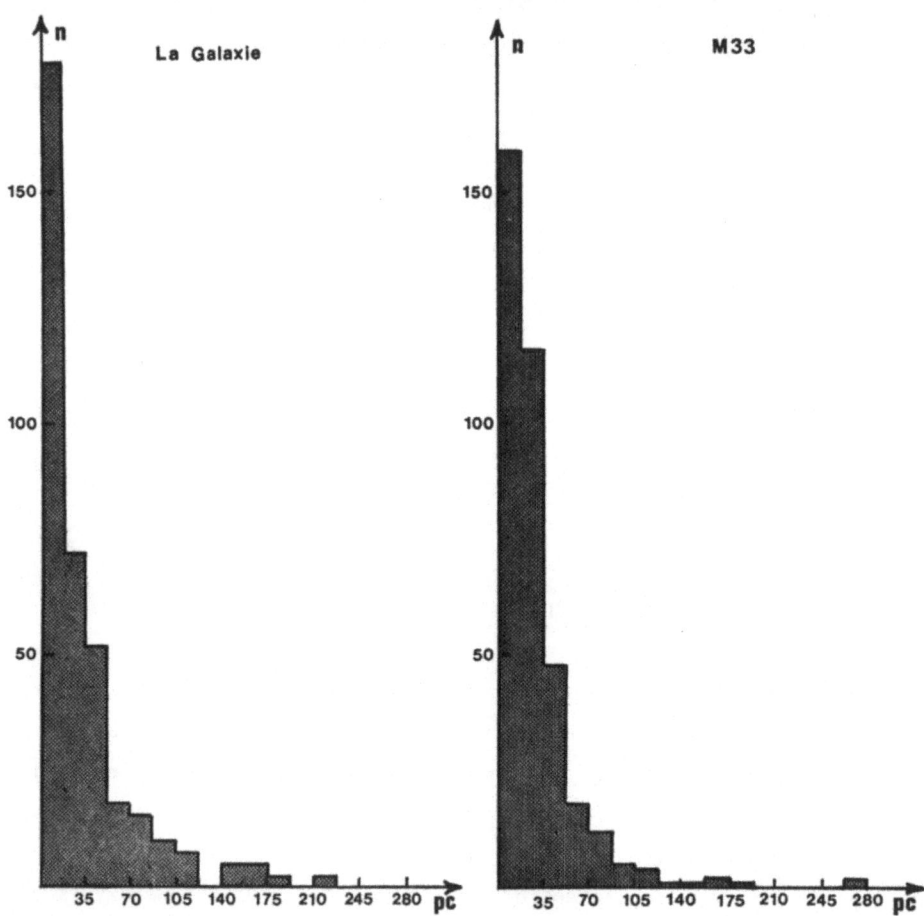

Fig. 1   Histograms of H II region diameters in the Galaxy and in M33 as
derived by Georgelin (1971).

morphological type but also of the luminosity class of a galaxy. Figure
2 shows the mean H II diameter - galaxy type (morphological type and
luminosity class) relation derived by Sandage and Tammann (1974) from
17 galaxies in the local, M81 and M101 groups.

The diameter of the largest and the three largest H II regions in our
Galaxy in conjunction with these calibration curves predict type S IV
for our Galaxy. From this plot alone, it could also be an Ir IV but from
other evidence, we believe our Galaxy to be a spiral. The distinction
between $S_b$, $S_b+$, $S_c-$ and $S_c$ cannot be made from the data of Sandage and
Tammann (1974) since they did not include $S_a$ or $S_b$ types in their cali-

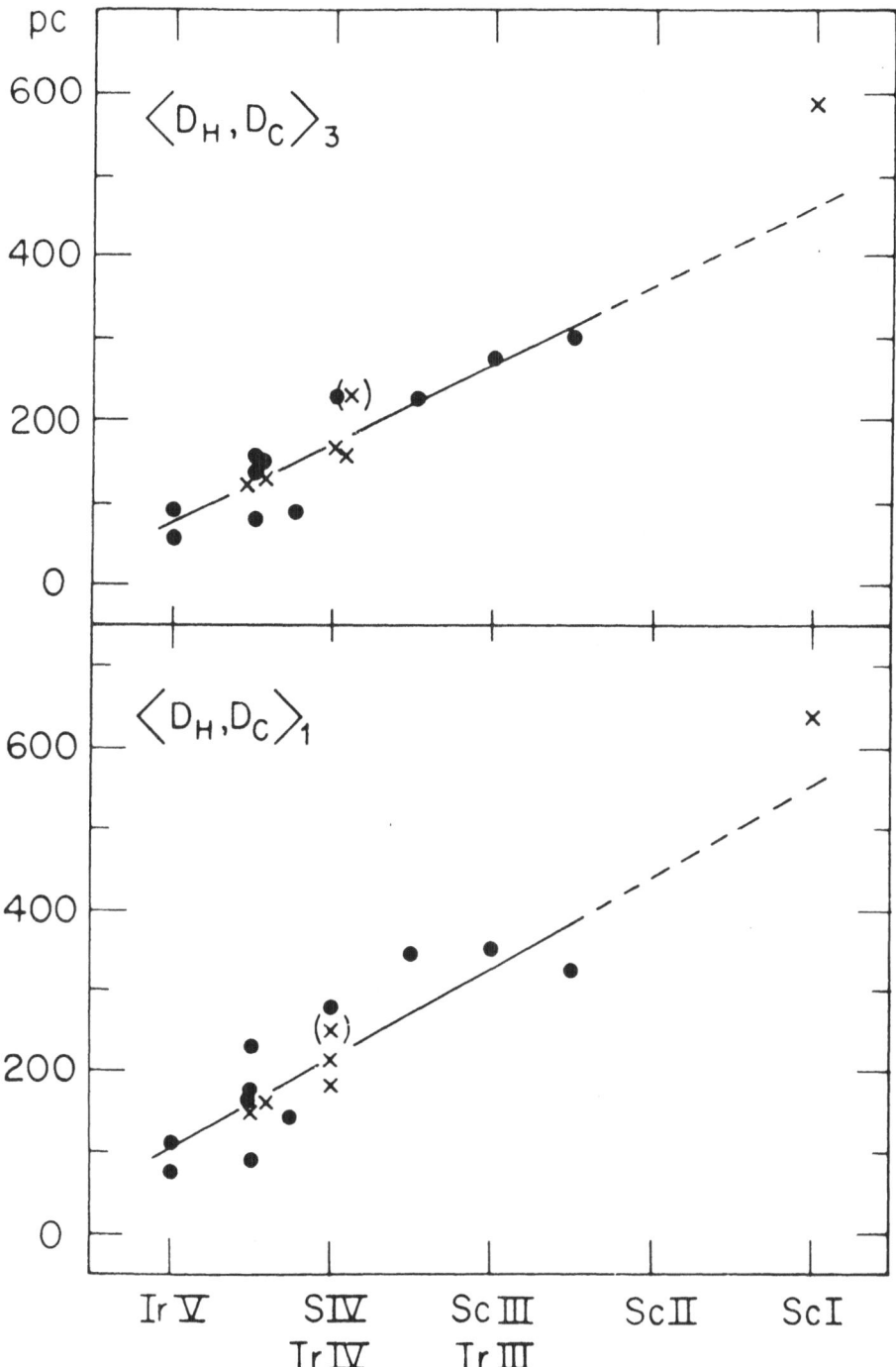

Fig. 2  Calibration curves of Sandage and Tammann (1974) of the mean
halo-core diameters of the three largest H II regions (above)
and the largest H II region (below) against galaxy type and
luminosity class. The line is the best fit to the measured data
(dots and crosses).

bration sequence. The luminosity class derived in this way does not appear to be correct for our Galaxy. It predicts that the luminosity of our Galaxy is less than that of M33, which is a factor of 10 less massive. It is, of course, possible that our Galaxy has a peculiar mass-to-luminosity ratio or its mass has been overestimated or the largest H II regions have not yet been identified. However, if our Galaxy is an $S_b$ or intermediate between $S_b$ and $S_c$ (as indicated by Georgelin) then it is probably not valid to use the calibration curves of Sandage and Tammann. An important extension to their work would be to include $S_b$ types in this calibration sequence.


## V. CONCLUDING REMARKS

In this review I have scratched the surface of only a few of the interesting problems with evolved H II regions still to be solved. Many problems have not even been mentioned, such as embedded condensations, mass loss from the ionizing star, turbulence, flows and ionization structure in H II regions. One of the more interesting projects still to be done is the optical determination of abundances, particularly of O and N, in enough evolved nebulae with a significantly large range in galactic radius to determine whether an abundance gradient occurs in our Galaxy. Also clarification of the possible low electron temperature in IC 434 by several independent techniques is certainly desirable. At radio wavelengths it would be interesting to have more high sensitivity maps which can unambiguously delineate the outer boundaries of several large H II regions whose ionizing stars are known. Low frequency radio recombination line observations in several evolved H II regions would provide an important check on $T_e$ and the kinematics in these objects.


## REFERENCES

Baars, J.W.M., Wendker, H.J. 1973, I.A.U. Symposium No. 60, Maroochydore, Australia
Blaauw, A. 1964, Ann. Rev. Astr. & Astrophys. 2, 213
Bottinelli, L., Gouguenheim, L. 1964, Ann. Rev. Astr. & Astrophys. 27, 685

Caswell, J.L., Goss, W.M. 1974, Astr. & Astrophys., 32, 209

Churchwell, E. 1974, paper presented at the 2nd European Regional Meeting in Astronomy, Trieste, Italy

Churchwell, E., Felli, M. 1970, Astr. & Astrophys. 4, 309

Churchwell, E., Walmsley, C.M. 1973, Astr. & Astrophys., 23, 117

Davies, R.D., Tovmassian, H.M. 1963, M.N.R.A.S. 127, 61

Feast, M.W. 1961, M.N.R.A.S., 122, 1

Georgelin, Y.P. 1971, Astr. & Astrophys. 11, 414

Georgelin, Y.M., Georgelin, Y.P., Roux, S. 1973, Astr. & Astrophys. 25, 337

Huchtmeier, W.K., Churchwell, E. 1974, Astr. & Astrophys. 35, 417

Mathews, W.G., O'Dell, C.R. 1969, Ann. Rev. Astr. & Astrophys. 7, 67

McGee, R.X., Brooks, J.W., Batchelor, R.A. 1972, Austr. J. Phys. 25, 581

Menon, T.K. 1962, Ap. J. 135, 394

Mezger, P.G. 1973, in "Interstellar Matter", Proc. 2nd Adv. Course, Swiss Soc. of Astr. and Astrophys., Publ. Geneva Observatory, Geneva

Mezger, P.G., Smith, L.F., Churchwell, E. 1974, Astr. & Astrophys. 32, 269

Osterbrock, D.E., Stockhausen, R.E. 1960, Ap. J. 131, 310

Salpeter, E.E. 1955, Ap. J. 121, 161

Sandage, A., Tammann, G.A. 1974, Ap. J. 190, 525

Sersic, J.L. 1960, Zeit. f. Astrophys., 50, 168

Shapley, H., Paraskevopoulos, J.S. 1937, Ap. J. 86, 340

Thaddeus, P., Wilson, R.W., Kutner, M., Penzias, A.A., Jefferts, K.B. 1971, Ap. J. 168, L59

Torres-Peimbert, S., Lazcano-Araujo, A., Peimbert, M. 1974, Ap. J. 191, 401

Walmsley, C.M., Churchwell, E., Kazès, I., Le Squéren, A.M. 1975, Astr. & Astrophys., in press

Wendker, H.J. 1968, Zeit. f. Astrophys. 68, 368

Wendker, H.J. 1970, Astr. & Astrophys. 4, 378

Wynn-Williams, C.G., Becklin, E.E., Neugebauer, G. 1972, M.N.R.A.S. 160, 1

Zuckerman, B. 1973, Ap. J. 183, 863

RADIO CONTINUUM SPECTRA

M. J. SEATON

Department of Physics & Astronomy,
University College London,
Gower Street, London WC1E 6BT
England

ABSTRACT


    Radio continuum flux spectra $S_\nu$ depend on electron temperatures
$T_e$ and on angular areas $\omega(E)$ which enclose contours of constant
emission measure E.  Different methods of analyzing observations are
discussed.

In 1958 a Symposium on Radio Astronomy was held in Paris, organized jointly by the International Astronomical Union and the International Scientific Radio Union (Bracewell 1959). In wide-ranging reviews of the entire subject, the problems of thermal emission from ionized gas were barely mentioned. At that time the theory of thermal emission was understood in principle and there were more interesting unsolved problems to discuss. It must also be recalled that further work on the interpretation of thermal emission required advances in the absolute calibration of radio telescopes and in obtaining improved angular resolution.

An important paper on the interpretation of radio continuum observations of H II regions was published by Wade in 1958. The basic mechanism is

$$X^+ + e \; \rightleftarrows \; X^+ + e + h\nu \tag{1}$$

where $X^+$ is $H^+$ or $He^+$ and the electron density is $N_e = N(H^+) + N(He^+)$. The absorption coefficient for (1) is given by an expression of the form

$$\kappa_\nu = N^2_e \, \alpha \, \nu^{-2} \, (b + \tfrac{3}{2} \ln T_e - \ln \nu )/T^{3/2}_e \tag{2}$$

where $T_e$ is the electron temperature. Due to the factor $\nu^{-2}$ in (2), $\kappa_\nu$ becomes large for $\nu$ small and an equation of radiative transfer must therefore be solved:

$$\frac{d \, I_\nu}{d \, \ell} \; = - \; \kappa_\nu I_\nu + j_\nu \tag{3}$$

where $j_\nu$ is the emissivity. The source function $(j_\nu / \kappa_\nu)$ depends only on the velocity distribution of the free electrons and since this is Maxwellian we have

$$(j_\nu / \kappa_\nu) = B_\nu \, (T_e) \tag{4}$$

where, at radio frequencies,

$$B_\nu \, (T) \; = \frac{2\nu^2}{c^2} \; kT \, . \tag{5}$$

Some variations of $T_e$ occur within H II regions but these variations are not large and their effects are of minor importance. In the present review I assume $T_e$ to be constant. The solution of (3) for the emergent intensity is then

$$I_y = B_y (T_e) (1 - \exp( - \tau_y )) \tag{6}$$

where

$$\tau_y = \int \kappa_y d\ell , \tag{7}$$

the integral being along the line of sight. The observed intensity is often expressed in terms of a brightness temperature, $T_b(y)$:
$I_y = B_y (T_b(y))$. From (5) and (6),

$$T_b(y) = T_e(1 - \exp(- \tau_y )). \tag{8}$$

From (2) and (7),

$$\tau_y = E_x \chi (y, T_e) \tag{9}$$

where

$$E = \int N_e^2 d\ell \tag{10}$$

and

$$\chi = a y^{-2}(b + \tfrac{3}{2} \ln T_e - \ln y )/T_e^{3/2} . \tag{11}$$

For a given line of sight the intensity $I_y$ depends on two parameters, the emission measure $E$ and the electron temperature $T_e$. In principle one can deduce these parameters from observations of $I_y$ as a function of $y$, but in practice one does not have good angular resolution for all frequencies which are of interest. One therefore considers the total flux, $S_y$, from a source.

A "source" can be defined as an object contained within some prescribed angular area $\Omega$. Background emission is subtracted and the intensity $I_y$ from the source therefore goes to zero on the boundary of the area $\Omega$. The total flux from the source is

$$S_y = \int_0^\Omega I_y \, d\omega . \tag{12}$$

Using (6) and (9), we have

$$S_y = B_y \int_0^\Omega \left\{ 1 - \exp\left[ -\chi E(\omega )\right] \right\} \, d\omega . \tag{13}$$

The brightness contours may be of a quite irregular shape, but one can obtain the integrated area enclosed by the contours at any given brightness level. The function $\omega (E)$ is defined to be such that all emission with $E \geqslant E_o$ comes from an integrated angular area $\omega (E_o)$. The function $E(\omega )$ in (13) is the inverse of the function $\omega (E)$.

In (13) we may replace the upper limit $\Omega$ of the integral by $\infty$, since by definition of "a source", $E(\omega) \to 0$ as $\omega \to \Omega$.

Integrating by parts,

$$S_y = B_y \times \int_0^\infty \omega(E) \, e^{-xE} \, dE . \qquad (14)$$

We have here replaced the frequency variable $y$ by the variable $x$, defined by (11). In place of the flux, $S_y$, we may work with the variable

$$y(x) = S_y \, / (B_y x) . \qquad (15)$$

We then have

$$y(x) = \int_0^\infty \omega(E) e^{-xE} \, dE . \qquad (16)$$

Two methods of analysis have been considered.

Method (i) was first used by Wade (1958). Suppose that, at one frequency $y_0$, measurements have been made of angular areas $A(T_b(y_0))$ enclosed by contours of brightness $T_b(y_0)$. Assuming a value for $T_e$, for each value of $T_b(y_0)$ we can use (8) and (9) to obtain E and hence $\omega(E) = A(T_b(y_0))$. Evaluation of (16) then gives $y(x)$ for all values of $x$ and hence $S_y$ for all values of $y$. The assumed value of $T_e$ can then be adjusted to give a best fit to the observed flux spectrum. In many of the earlier attempts to use this method, low values of $T_e$ were obtained (see, for example, the review by Mezger, 1968). This can be explained as a consequence of incomplete angular resolution. This point is best appreciated by considering a simple limiting case. If the frequency is so low that $(x E) \gg 1$ for all lines of sight through the source, (13) gives

$$S_y = B_y \Omega = \frac{2 y^2 k}{c^2} \times T_e \Omega . \qquad (17)$$

If the total angular area $\Omega$ has been over-estimated, the deduced electron temperature will be under-estimated. Recent continuum observations, with greatly improved angular resolution (Harris 1973; Martin 1973), give higher values of $T_e$ which are in better agreement with values deduced from other observations.

<u>Method (ii)</u> has been discussed by Salem and Seaton (1974)[*]. We see
that (16) gives $y(x)$ as the Laplace transform of $\omega(E)$. The inverse
transform is

$$\omega(E) = \frac{1}{2\pi i} \int_C y(x)\, e^{xE}\, dx \tag{18}$$

where $C$ is a suitable integration contour in the complex plane. The
procedure is to assume $T_e$, obtain $y(x)$ from the observed flux spectrum,
fit $y(x)$ to an analytic function, and evaluate $\omega(E)$ and hence
$A(T_b(\nu_o))$ at frequencies $\nu_o$ for which high resolution observations
have been made. As in method (i), the assumed value of $T_e$ can be
adjusted so as to get the best agreement with observations. In some
cases one cannot get good agreement for any value of $T_e$. In such
cases, method (ii) provides a measure of the extent to which angular
resolution is incomplete. This basic idea is not new (it has been used
earlier by Ryle and Downes 1967 ; Mezger, Schraml and Terzian 1967;
and Hjellming, Andrews and Sejnowski 1969): the work of Salem and Seaton
is an advance in that it provides a more complete mathematical theory.

In a recent review (Seaton, 1974) I have discussed the general
problem of determining $T_e$ in nebulae. A few years ago rather low
values ($T_e \lesssim 5000$ K) were often given in the literature, but higher
values (in the range 8000 K $\leqslant$ T $\leqslant$ 10000 K) are indicated by more
recent work, using various techniques. From the standpoint of the
thermal balance, $T_e$ is determined by the energy distribution of the
ionizing radiation and by the abundances of the heavy elements which
produce most of the cooling. It is insensitive to $N_e$ so long as $N_e$ is
not too large or too small[**].

---

[*] The normalization of $x$ and $y(x)$ used by Salem and Seaton differs
from that used in the present review; the former is more convenient in
practice, but the latter provides a simpler general presentation.

[**] For $N_e > 10^4 \text{cm}^{-3}$, collisional de-excitation reduces the cooling
rates and $T_e$ increases. For rather extreme cases of low densities
and very dilute radiation the gas is only partially ionized, cooling
is more effective and $T_e$ decreases.

Accurate values of $T_e$ are required for good determinations of abundances from optical observations. For this type of work it is best to use values of $T_e$ deduced from temperature-sensitive intensity ratios of forbidden lines.

So far as radio continuum flux spectra are concerned, I think that the best procedure may often be to assume $T_e$ to be known, and to use the observations to deduce information about angular areas enclosed by emission measure contours. This technique may be particularly useful for distant sources with small angular areas.

The availability of computer programs for the interpretation of radio continuum observations, and radio line observations (Brocklehurst and Seaton 1972), is discussed in an accompanying paper by Michael Salem.

REFERENCES

Bracewell, R.N. 1969, "Paris Symposium on Radio Astronomy", Stanford Univ. Press.
Brocklehurst, M., and Seaton, M.J. 1972, Mon.Not.R.astr.Soc., 157,179.
Harris, S. 1973, Mon.Not.R.astr.Soc., 162, 5P.
Hjellming, R.M., Andrews, M.H., and Sejnowski, T.J. 1969, Ap.J., 157, 573.
Martin, A.H.M. 1973, Mon.Not.R.astr.Soc., 163,141.
Mezger, P.G. 1968, "Interstellar ionized hydrogen" (ed.Y.Terzian), p.459, Benjamin, New York.
Mezger, P.G., Schraml, J., and Terzian, Y. 1967, Ap.J., 150, 807.
Ryle, M., and Downes, D. 1967, Ap.J., 148, L.17.
Salem, M., and Seaton, M.J. 1974, Mon.Not.R.astr.Soc., 167, 493.
Seaton, M.J. 1974, Quart.J., R.astr.Soc., 15, 370.
Wade, C.M. 1958, Aust.J.Phys., 11, 388.

# SOME RECENT CALCULATIONS AND COMPUTER PROGRAMS

M. SALEM *

Department of Physics & Astronomy,
University College London

## ABSTRACT

The construction of a unique spherical model with filling
factor unity of a thermal radio source is briefly described.  The function $\Omega(E)$ must previously be obtained from radio flux observations as
described by Seaton at the Symposium.  Information on the availability
of computer programs for determination of $\Omega(E)$ and construction of the
spherical model is given.  Information is also given on computer programs for the computation of the hydrogen and helium line and continuum
spectrum of $H^+$ regions, and for computation of the departure coefficients from thermodynamic equilibrium, $b_n$, at low electron temperatures.

An approximation formula for the computation of the absorption

---

\* Present address:  Cavendish Laboratory, Cambridge.

coefficient of hydrogenic ions at all electron temperatures and radio frequencies of current astrophysical interest is also given. It is expected that this formula will be of most use at electron temperatures below 1000 K, for which the standard approximation becomes inaccurate.

## 1. Spherical model construction

A function $\Omega(E)$ can be determined from the radio frequency spectrum of a thermal source, where $\Omega$ is the total angular area within which the emission measure is not less than E (Seaton 1975).

Let us assume the source to be spherically symmetric. The electron temperature, $T_e$, is assumed to be approximately constant, but the electron density, $N_e$, is a function of the distance from the centre. For any given point, let $\rho$ be the distance to the centre, $r$ be the perpendicular distance from the line of sight through the centre, and s be the projection of $\rho$ along the line of sight (Fig.1). R is the total radius of the object, and $2\sigma$ is the total length of the line of sight within the object.

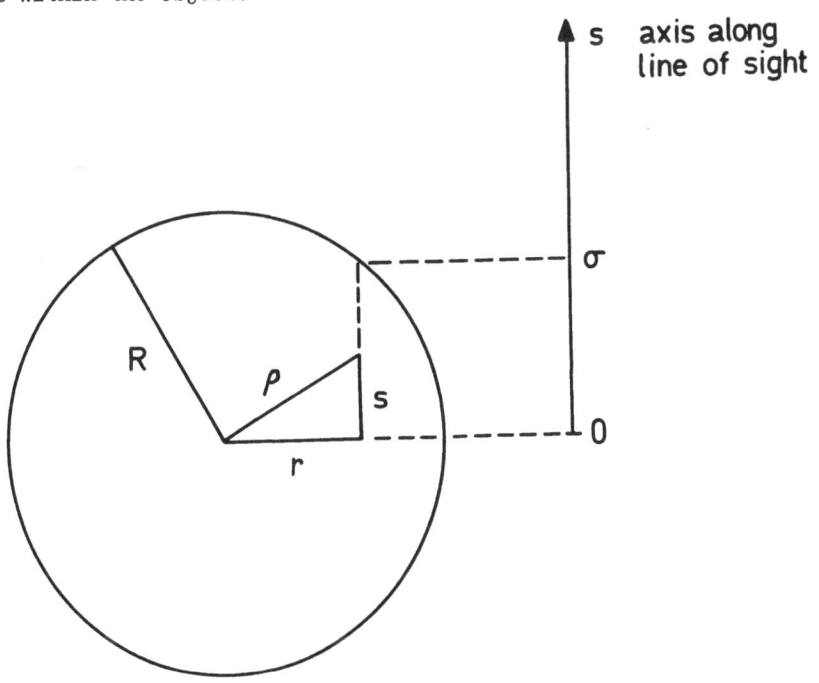

For such a geometry, the contours of constant emission measure are clearly concentric circles, and we have

$$\Omega = \frac{\pi r^2}{z^2},$$ (1)

where z is the distance of the object. We also have

$$\rho^2 = r^2 + s^2.$$ (2)

For any value of $\Omega$, and hence, through (1), of r, the emission measure is defined by

$$E(r) = \int_{-\sigma}^{\sigma} N_e^2(\rho)\,ds.$$ (3)

Using (2) we obtain

$$E(r) = \int_{r^2}^{R^2} \left[ N_e^2(\rho^2)/(\rho^2-r^2)^{\frac{1}{2}} \right] d(\rho^2)$$ (4)

This is an integral equation which may be solved for $N_e^2(\rho^2)$; details are given by Salem (1974). The solution is

$$zN_e^2(\rho^2) = \pi^{-\frac{1}{2}} \int_0^{E\rho} \left[ \Omega(E) - \Omega(E_\rho) \right]^{-\frac{1}{2}} dE,$$ (5)

where $E_\rho$ is defined by

$$\rho^2 = \Omega(E_\rho)z^2/\pi.$$

Equation (5) can be integrated numerically.

A similar procedure can be carried out for different assumed geometries. It should be emphasized that, given the assumptions of a known and constant electron temperature, and a given geometry, the model is uniquely determined by the radio continuum spectrum.

2. Computer programs for continuum spectrum analysis, model construction, and line spectrum computation

A computer program, ILTHII (Inverse Laplace Transform - HII region), has been written and documented for the calculation of $\Omega(E)$ from the continuum spectrum, the construction of the spherical model, and related calculations (Salem 1975a). The theory is given by Salem & Seaton (1974) and Salem (1974).

Program RCMBLN (ReCoMBination LiNe), which computes the intensities and profiles of hydrogen and helium radio recombination lines, is also available (Brocklehurst & Salem 1975a). The theory is described by Brocklehurst & Seaton (1972). The input for this program is a spherical or plane-parallel isothermal model; in particular, models constructed by ILTHII are suitable. Line profiles are computed taking pressure broadening into account. The departure coefficients from thermodynamic equilibrium, $b_n$ , are required for this calculation; for convenience a separate program, BN, CN SELECT, which selects the desired values from a very large table has been provided (Salem 1975b).

3. Computer programs for low temperature $b_n$ factors

      Computer programs have been written for the computation of the $b_n$ coefficients at low electron temperatures (Brocklehurst & Salem 1975b), and for the interpolation of desired values from tables (Salem 1975c). These programs, although operational, are not yet fully documented; further information can be obtained from the present author. The theory of the method is described by Brocklehurst (1973). The accuracy of the $b_n$ coefficients is essentially determined by that of the cross-sections of Banks, Percival & Richards (1973) (about 20%).

4. The absorption coefficients for hydrogenic ions

      Oster (1961) calculated the absorption coefficient for hydrogenic ions, using classical (non-quantum) methods. The formula obtained (Oster's equations 121-123) involved complicated integrals; he consequently derived an analytical approximation suitable for electron temperatures higher than a few hundreds of K.

      More recently studies of the interstellar medium have required knowledge of the absorption coefficient for lower electron temperature. Quantum calculations have been performed by Gayet (1970) and Oster (1970); in both cases, numerical results were published in graphical form.

      It appears not to have been appreciated that the exact classical formula (Oster 123) gives results in very good agreement with quantum calculations if it is evaluated numerically. Furthermore, the form of the classical formula is very suitable for analytical approximation or one-dimensional table interpolation.

A useful approximation is given below. A more detailed discussion, a table suitable for interpolation, and other approximate expressions will be published elsewhere.

The absorption coefficient, $K_\nu$ , is given in $pc^{-1}$, the frequency, $\nu$ , in GHz, the electron temperature, $T_e$, in K, and the electron and ion densities in $cm^{-3}$. We define $t = T_e/100$. The absorption coefficient is given (exactly) by:

$$K_\nu = 4.6460 \times 10^{-3} \, N_e N_i \; \nu^{-7/3} t^{-3/2} \times$$

$$\left[ \exp(4.7800 \times 10^{-4} \nu /t)-1) \right] \times I(Mo), \tag{6}$$

where

$$Mo = 0.04497 \; \nu^{2/3}/t. \tag{7}$$

The function $I(Mo)$ is given approximately by

$$\log_{10} I(Mo) \; \simeq \; \begin{cases} -1.124945v + 0.378765 & (-5 \leqslant v \leqslant -2.6) \\ -1.232644v + 0.098747 & (-2.6 < v \leqslant -0.25) \\ -1.084191v + 0.135860 \, , & (-0.25 < v \leqslant +1.4) \end{cases} \tag{8}$$

where $v \equiv \log_{10} Mo$.

Approximation (8) is accurate to within $\pm 7$ per cent. For $1.4 < v \leqslant 3.0$, the error is still less than 30 per cent. The range of values of v covers all conditions currently of interest.

The values of $I(Mo)$ calculated numerically agree well with the results of Gayet (1970).

REFERENCES

Banks, D., Percival, I.C. & Richards, D., 1973, Astrophys.Lett.14, 161.
Brocklehurst, M., 1973, Astrophys.Lett. 14, 81.
Brocklehurst, M. & Seaton, M.J., 1972, Mon.Not.R.astr.Soc. 157,179.
Brocklehurst, M. & Salem, M., 1975a, Computer Physics Communications,
                                                    in press.
Brocklehurst, M.& Salem, M., 1975b, in preparation.
Gayet, R., 1970, Astron.& Astrophys. 9, 312.
Oster, L., 1961, Rev.Mod.Phys. 33, 525.
Oster, L., 1970, Astron.& Astrophys. 9, 318.
Salem, M. & Seaton, M.J., 1974, Mon.Not.R.astr.Soc. 167, 493.
Salem, M., 1974, Mon.Not.R.astr.Soc., 167, 511.
Salem, M., 1975a, Computer Physics Communications, in press.
Salem, M., 1975b, in preparation.
Seaton, M.J., 1975, this Symposium.

# SURVEYS OF THE GALACTIC PLANE AT 5 GHz.

W.J. Altenhoff

Max-Planck-Institut für Radioastronomie

Bonn, Germany

## ABSTRACT

A new survey of the galactic plane between $l=359^{\circ}$ and $55^{\circ}$, $b=\pm2^{\circ}$, has been made at 5 GHz with the 100-m telescope, with a half-power beamwidth of 2.6'. Preliminary contour maps show more than 1000 peaks. The strong increase in the number of apparent sources is mainly due to the resolution of the well known source complexes into separate components.

All of the sources with antenna temperatures greater than 1 K in the continuum survey are being observed in the $H110\alpha$ and $H_2CO$ lines. The two lines are observed simultaneously. The purpose of these surveys is to identify which sources are thermal and to derive their kinematic distances.

The participants in these continuum and line surveys are W.J. Altenhoff, J. Bieging, D. Downes, T.A. Pauls, J. Schraml, T.L. Wilson and J.E. Wink.

# THE GALACTIC DISTRIBUTION OF MOLECULES (A CO SURVEY)[†]

N. Z. SCOVILLE[††]

Owens Valley Radio Observatory
California Institute of Technology

## ABSTRACT

The $J = 1 \rightarrow 0$ emission of CO has been surveyed in the galactic plane between $\ell = -10°$ and $+90°$ with a 1' beam sampling every degree. Molecular clouds emitting in the CO line are plentiful over the inner region of the galaxy. Their greatest number occurs in the galactic nucleus and at a radius of 5.5 kpc - a distribution similar to radio H II regions but very different from that previously derived for atomic hydrogen. The total mass in molecular clouds is found to be $1 \rightarrow 3 \ 10^9$ $M_\odot$, and each typically has $10^5 \ M_\odot$, an average density of 500 $cm^{-3}$, and a temperature of 7°K. These results suggest that most of the interstellar medium in the interior of the galaxy is molecular $H_2$.

[†]I describe the results of a joint investigation with P. M. Solomon, SUNY Stony Brook, which will also be published in the Astrophysical Journal 1975.

[††]Present address:  Physics and Astronomy Department, University of Massachusetts, Amherst, Massachusetts  01002.

## I.  INTRODUCTION

Prerequisite for understanding galactic evolution is a knowledge of the large scale distribution and total mass of dense interstellar clouds.  Within these, the birth of stellar associations and clusters is currently taking place on a time scale short compared to the age of the galaxy from a mixture of processed and primeval gas.  These clouds are the active regions of the Milky Way.  They are grossly underplayed in measurements of the 21-cm atomic hydrogen (HI) emission since most of the gas there will be molecular ($H_2$) and the little HI which remains will generally be cool.  Molecular hydrogen has no permanent dipole moment and therefore its pure rotational or vibrational transitions are weak infrared quadrapole lines.  Though this radiation may be barely detectable in the very densest clouds with high temperature cores, it will never be possible to survey the galaxy for cold molecular clouds by this technique.  Observations of the ultraviolet resonance lines of $H_2$ at $\lambda \sim 1000$ Å have been obtained from numerous clouds in front of nearby O and B stars, with about 0.1 - 1.2 mag. of visual extinction and provide the only direct evidence for the ubiquity of molecular hydrogen (Spitzer et al. 1973).  Since the dense, massive clouds which are the source of molecular radio emission lines have well over five magnitudes of visual extinction, they are totally opaque in the ultraviolet; our knowledge of these regions must be derived from radio frequency transitions of rarer molecules.

Clearly favored for a galactic survey of molecular regions is the $J = 1 \rightarrow 0$ emission of CO at 2.6-mm.  In every extended source of radio frequency molecular transitions the CO line is most intense; often it is found to maintain a high intensity even to the edge of molecular clouds where the density falls below $10^3$ cm$^{-3}$ and other transitions have become unobservable.  The emission is optically thick in all dark clouds and often thermalized in the line center.  Since emission in the rotational line ultimately results from collisions of $H_2$ and CO, CO observations yield information on the $H_2$ density $n_{H_2}$ and temperature $T_k$. We have made and report here a preliminary survey of CO emission from the galactic plane.  Combining the results of this survey with the known properties of well mapped clouds (in CO and $^{13}$CO Penzias et al.

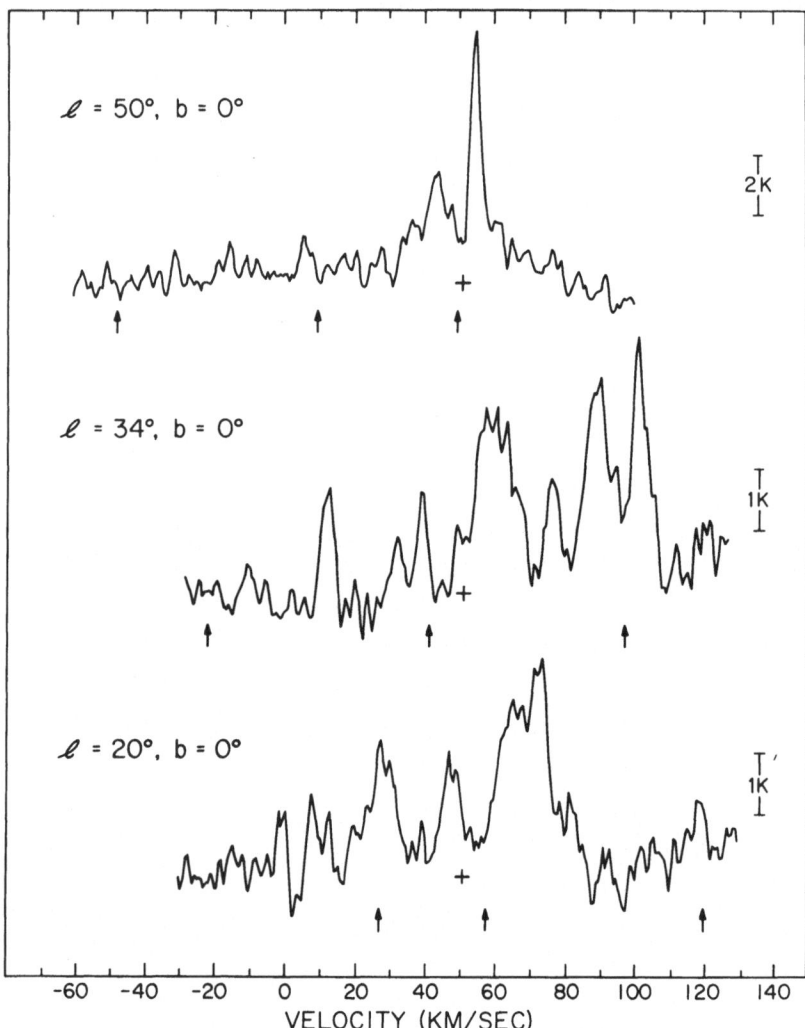

Fig. 1   CO spectra are shown for three typical positions crossing the
inner region of the galaxy.  At each position many distinct
features are seen.  The velocities of local maxima in 21-cm
spectra at the same position are indicated by arrows (Kerr 1969).
The widths of the 21-cm are typically at least 30 km/sec.  The
intensity units are °K, $T_A$ corrected for atmospheric attenuation
but not for telescope beam efficiency ($\eta_B$ = 0.6).

1971 and Lizst 1973), we are able to obtain an idea of the distribution
and temperature of $H_2$ throughout the galaxy.

CO spectra were taken along the galactic equator over the longitude range $348° \leq \ell \leq 90°$. In this initial effort, we randomly sampled
the gas clouds throughout the galaxy, avoiding the selection bias of
most previous molecular observations towards compact HII regions or
nearby dark clouds which appear on photographs. The observations were
separated by 1° in longitude; selected longitudes ($\ell = 20°$, $30°$, $41°$,
and 44°) were also mapped perpendicular to the galactic plane from
$b = -1°$ to $+1°$ with a spacing $\Delta b$ of 0.25°. The velocity coverage was
varied with longitude so as to contain most of the gas indicated by HI
surveys (see Figure 5) and expected from galactic rotation.

The CO data for $|\ell| \lesssim 3°$ were taken from our more detailed study
of the galactic nucleus (Scoville et al. 1974). The new observations
for the remainder of the galactic plane were obtained in October 1973
with the 36-foot antenna (HPBW = 1.2) and front-end receiver of the
NRAO* Kitt Peak, Arizona. Spectral resolution was provided by a 256-
channel filter bank covering 166 km/sec with a resolution of 0.65 km/sec
at the CO frequency. A chopper wheel was employed for noise calibration
and all antenna temperatures were corrected for atmospheric attenuation
(Davis and Vanden Bout 1973). The instrumental baseline was partially
removed by switching to reference positions 3° above the galactic plane
for alternate 30 sec periods during a 10 minute observation. The telescope control system did not permit switching a greater distance. We
expect that the observations of gas near the sun, at zero velocity, are
unreliable since in many instances the reference position at $b = 3°$
will also contain this gas (e.g., Fig. 1 $\ell = 20°$, V = 0 km/sec). Some
spectra also contained a residual curvature of amplitude 1°K most often
near the edge of the filter bank. Such curvature was judged to be instrumental and has been eliminated from Figure 2 when it did not recur
in other spectra at the same position.

The galactic plane CO emission is displayed in Figure 2 with shad-
ing representative of its intensity. Figure 3 depicts the longitude
variation of the integrated intensity $\int T_A dv$.

---

*The NRAO is operated by Associated Universities, Inc., under contract
with the National Science Foundation.

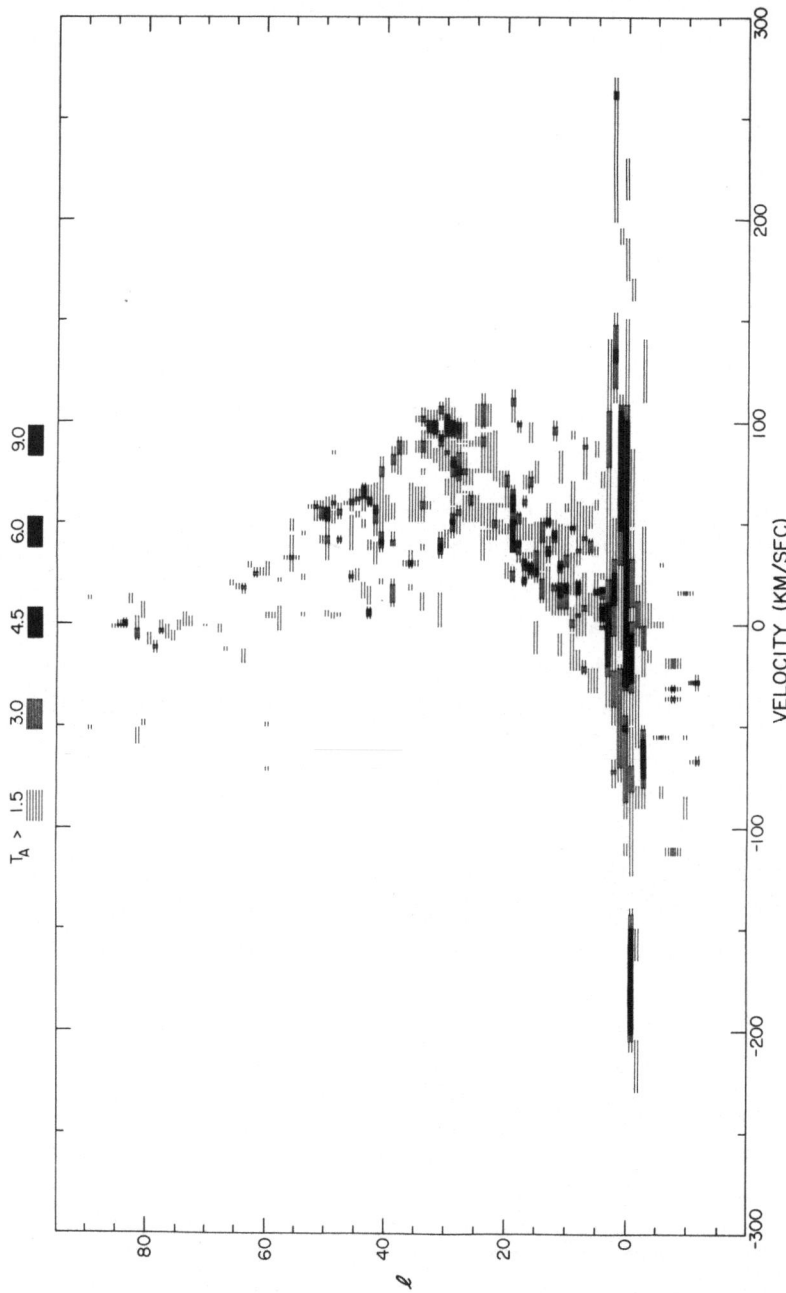

Fig. 2 (a)

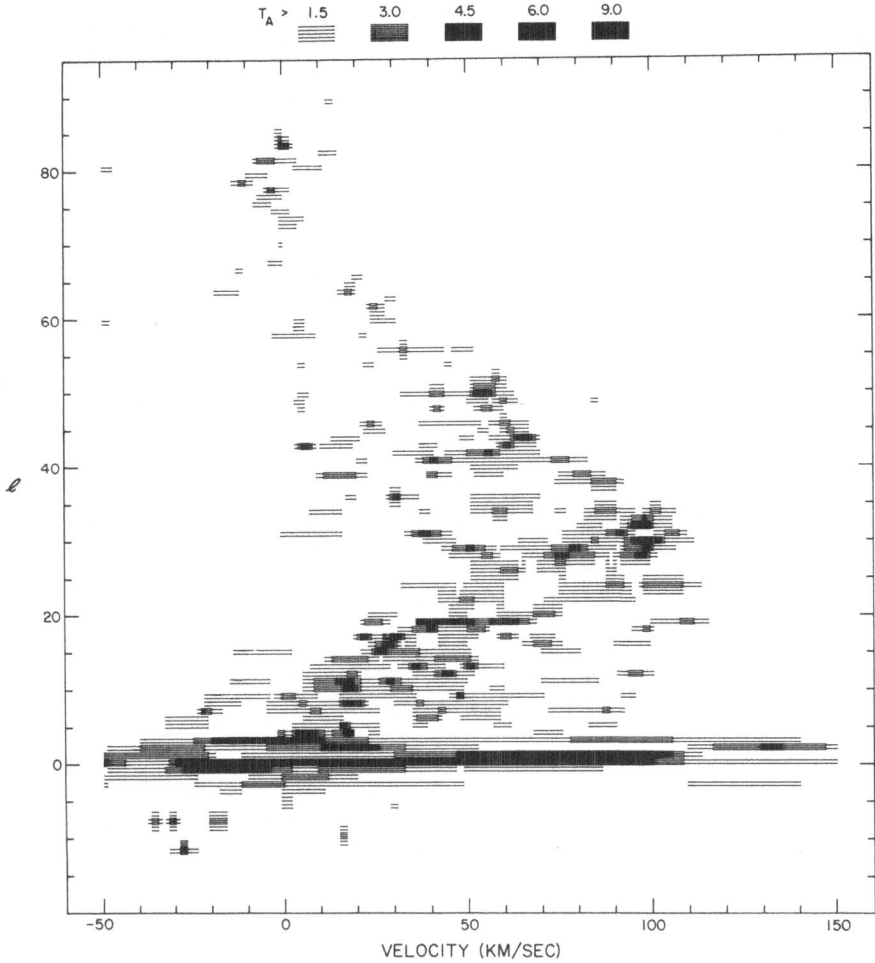

Fig. 2   The intensity of CO emission along the galactic equator is
         shown as a function of longitude and velocity.  Molecular
         emission tends towards lower longitudes and more positive
         radial velocities as compared with 21-cm (see Kerr 1969)
         indicating that the molecules are concentrated towards the
         center of the galaxy.  Figure 2a includes more velocities
         (±300 km/sec) in order to show the full range found in the
         galactic center.

## II.  DISTRIBUTION OF CO EMISSION

A brief glance at the data reveals that emission lines are distributed non-uniformly in the spiral arms and galactic disk.  The molecules in the galactic nucleus, between the nucleus and the solar circle, and those exterior to the sun are easily distinguished in Figure 2.  Essentially all the emission at $|\ell| \lesssim 4°$ is from the nucleus; both the intensities and radial velocities are high in this region.  The remainder of the CO at $\ell < 4°$ may be shown to be inside (or outside) the solar circle where the radial velocities are positive (or negative) assuming only that angular velocity increases toward the center.*

Major characteristics in the distribution of molecular clouds apparent from the data in Figures 2 and 3 are:  1) the extremely strong emission originating from the galactic nucleus at $|\ell| < 4°$; 2) a maximum at $\ell = 30°$; and 3) a sharp drop off in integrated intensity beyond $\ell = 45°$.  The fall off of integrated intensity with increasing longitude points to an inverse relationship between molecular emission and distance from the galactic center.

## A) Galactic Nucleus

The most spectacular emission is found at longitudes less than 4° corresponding to galactic radii $\bar{\omega} < 0.8$ kpc.  The very high velocities (Fig. 2a) were emphasized in the first HI observations of the galactic nucleus (Rougoor and Oort 1960); the kinematics of molecules in the nucleus are a complex mixture of both radial and orbital motions with the radial motions occurring at smaller galactic radii ($\bar{\omega} \approx 300$ pc) than was evidenced by the 21-cm observations (see Scoville et al. 1974 for more discussion).  Special also is the extremely high abundance of molecular gas - many times the atomic hydrogen there.  We have estimated in the previous CO survey that the mass of $H_2$ is $5 \cdot 10^7$ $M_\odot$ as compared with $4 \cdot 10^6$ $M_\odot$ in HI (Rougoor 1964).

---

*Excepting the 3 kpc expanding arm.  This feature crosses $\ell = 0°$ at -53 km/sec and presumably continues to a tangency point at $\ell = 20°$, $v \approx 100$ km/sec.

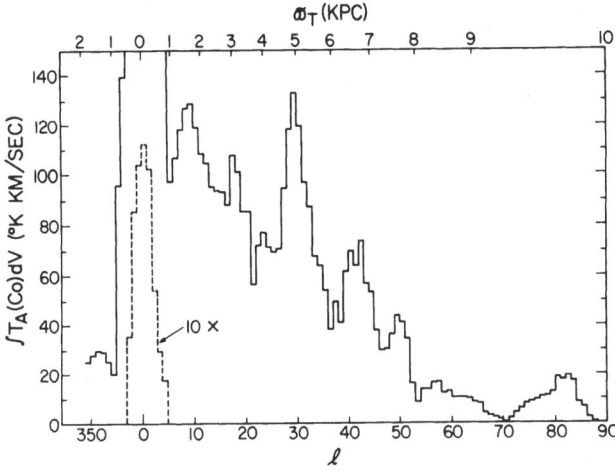

Fig. 3   The distribution of the integrated CO intensity $\int T_A dv$ in
$^\circ K \cdot km \cdot sec^{-1}$ is shown as a function of longitude. The
tangential radius on each line-of-sight is also given on
the top.  The extraordinarily strong emission from the galactic center may be seen.


## B) Galactic Disk

For the gas outside the galactic center where circular motion is
a good approximation we use a galactic rotation curve (Schmidt 1965)
to derive unambiguously the galactic radii which correspond to each
observed $\ell$, V point in Figure 2.  For the data at $\ell \geq 10°$ the mean
CO intensity as a function of $\varpi$ is shown in Figure 4.  Moving out in
the galactic plane we discover that <u>the emission rises sharply to a</u>
<u>maximum at 5.5 kpc; beyond this peak a dramatic fall of is seen; and</u>
<u>exterior to the sun there are few features.</u>

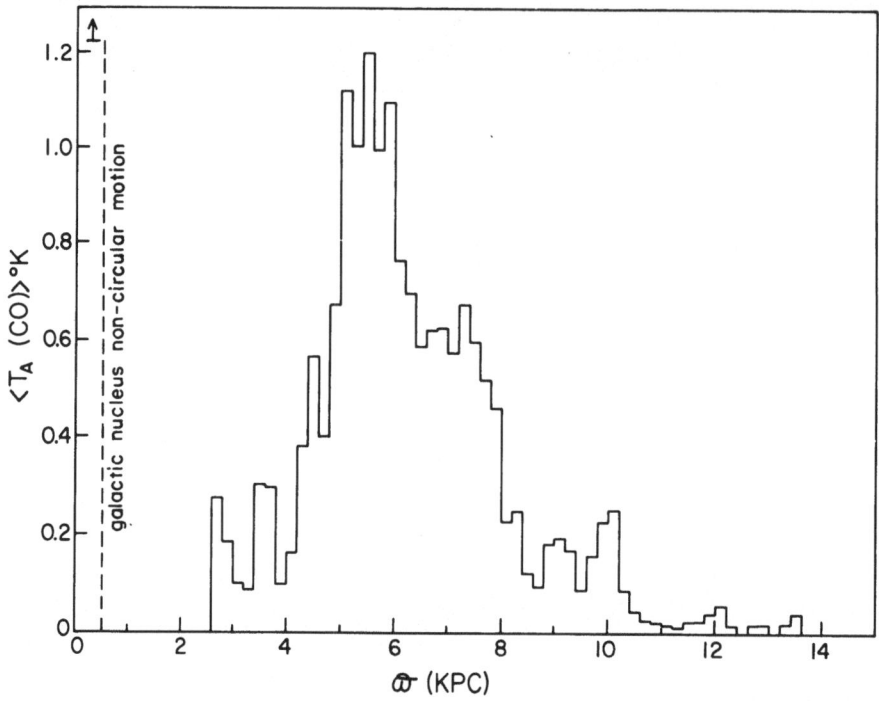

Fig. 4 The mean CO antenna temperature as a function of radius in the galaxy ϖ was calculated using the Schmidt (1965) rotation law to transform ℓ, v in Figure 2 to ϖ. We use only data at ℓ ≥ 10° in order to exclude the galactic center where much of the gas clearly is not in pure rotation. Note the sharp peak in CO at radius of 5.5 kpc and dramatic fall off towards the sun and beyond.

We judge that this distribution is real and cannot be explained as observational bias given by the threshold for detection in the present survey ($3\sigma = 1.5°K$). The major limitation to our results is the possibility of distributed low intensity emission which is uniform in some region of the galaxy but always less than our detection limit. This possibility could not be easily discounted using the present data alone because the $<T_A>$ in Figure 4 are less than $1.5°K$. However, more recent observations with five times the present sensitivity indicate that this problem is not significant (Scoville and Solomon 1975).

To some extent CO emission exterior to the sun could be underrepresented since at low longitudes this gas must be observed on the far side of galaxy, 20 kpc distant. But since the 1.2' beam still subtends only 6 pc as compared with typical 10 pc sizes of clouds the resulting beam dilution is probably not severe. 21-cm observations suggest that more serious errors occur because the galactic plane is bent and its scale height increases at large distances. However, the little mapping we have done in galactic latitude (Scoville and Solomon 1975) shows no strong evidence of systematic increase in the negative velocity emission as one moves above or below the galactic plane. Moreover, where the exterior gas would have been much closer, in the longitudes greater than $50°$, no strong emission was seen.

A scarcity of molecules between 1 and 4 kpc is implied by both Figure 4 and Figure 3 where the total integrated emission is shown as a function of longitude. In circular orbit this gas would appear at high radial velocities ($80 < V < 250$ km/sec) in the longitude range $5°$ to $25°$. Thus one might be inclined to argue that its absence in Figure 4 is caused by non-circular motions which make the transformation em-' ployed there, based on the Schmidt rotation law, inapplicable. Nevertheless provided the emission is not fortuitously blocked by a foreground cloud at the same radial velocity, any significant features would show up in Figure 3 as an increase of $\int T_A dv$ regardless of the gas kinematics. Since the peak at $\ell = 8°$ is no higher than that at $\ell = 30°$ (corresponding to $\bar{\omega}_T = 5$ kpc), we conclude there is no evidence of additional emission when the line-of-sight passes closer to the center than the tangential point at $\bar{\omega} = 5$ kpc (excepting $\ell < 5°$).

## III. COMPARISON WITH HI AND HII

The galactic plane 21-cm emission shown in Figure 5 (Kerr 1960 and Mezger et al. 1970) is more pervasive in both space and velocity than the CO: there is little tendency to concentrate at $\ell < 40°$ and fairly strong emission is found at negative velocities ($\bar{\omega} > R_o$). The 21-cm lines are much broader $\Delta V \approx 30$ km s$^{-1}$ and most local maxima show little correspondence with those of CO (the arrows in Figure 1 indicate the

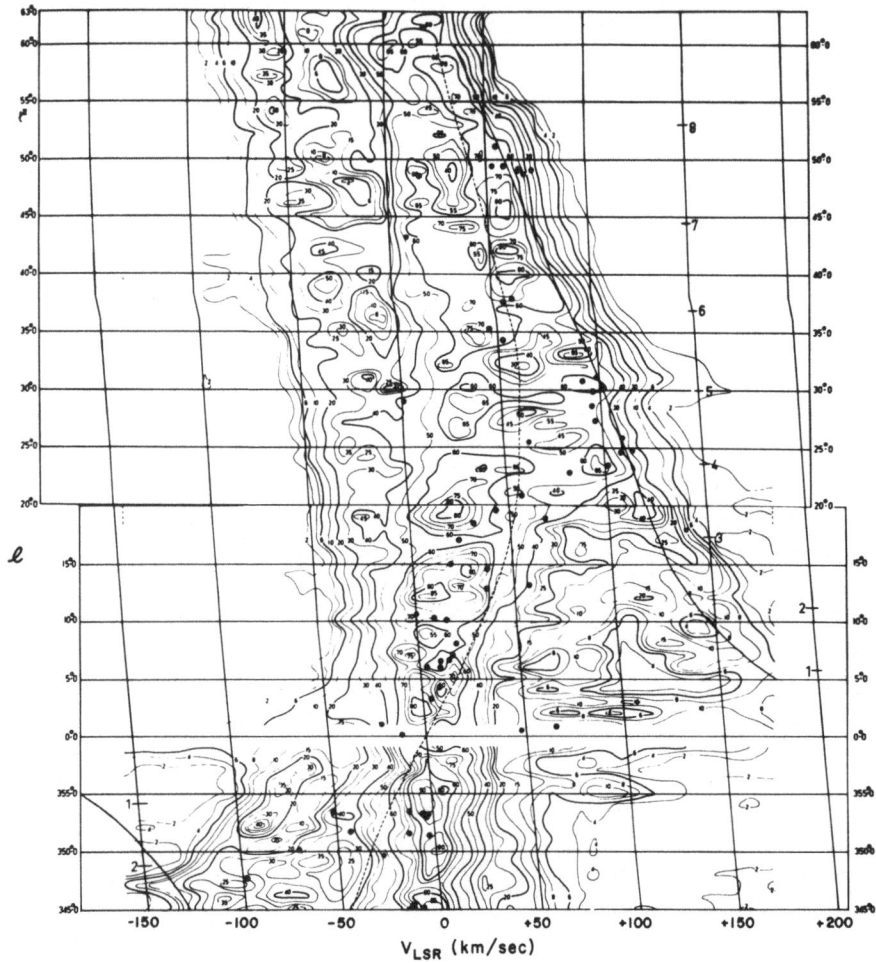

Fig. 5   Radial velocities of compact HII regions are shown as dots
         superimposed upon the 21-cm emission of HI (Mezger et al. 1970
         and Kerr 1969).   The solid and dotted lines are the maximum
         and half-maximum radial velocities in the Schmidt model.   HI
         contour unit is $1.75^{\circ}$K, $T_B$.

21-cm maxima). The spectra at $\ell = 20°$ and $34°$ thus show four and seven separate CO peaks in a velocity range showing only one HI maxima. With this many discrete features in each spectrum it would be foolhardy to attempt to trace spiral features from one spectra to the next. If spiral features can be traced perhaps in HI, the CO peaks might be interpreted as small condensations within the arms. That the peak CO emission occurs primarily on the high velocity wings of the 21-cm emission may be due to molecular clouds occurring preferentially on the inner parts of the spiral arms.

Outside the galactic nucleus there is close correspondence between the surface density of HII regions (Westerhout 1958; Mezger 1970) and the CO emission intensity (Figure 6). Both show a peak at 5.5 kpc and decrease a factor of 6 going to 10 kpc, with striking similarity in the detailed dependence. This correlation is precisely what is expected since dense HII regions result from formation of massive stars inside molecular clouds. We guess that the lack of correlation between CO emission and HII regions in the galactic nucleus (Figure 6) is explained either by the relatively low radio luminosity of HII regions formed in dense molecular clouds* or more likely by the youth and chaotic state of the molecular clouds in the galactic center where the clouds may be only $2 \cdot 10^6$ years old.

The HII-CO correlation does not prove that all molecular clouds have HII regions but does demonstrate that randomly sampled molecular clouds and dense HII regions have the same large scale distribution. The spatial extent of the individual molecular clouds is far greater than that of HII regions. Virtually every dense HII region is embedded in or on the edge of a molecular cloud, but there are many molecular clouds without dense HII regions.

## IV. $H_2$ DENSITY IN THE GALAXY

Given the limited amount of observational data we estimate the mass of molecular hydrogen contained in dense clouds in two ways: 1) simply

---

*At densities greater than $10^5$ cm$^{-3}$, dust grains may absorb most ionizing photons, and if an HII region does form it will initially be optically thick at radio wavelengths.

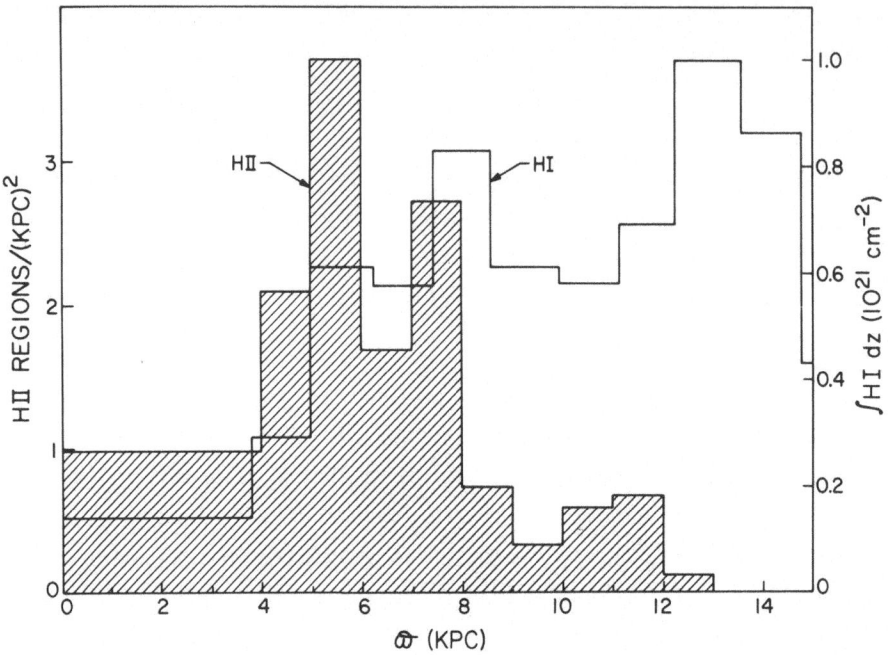

Fig. 6  The surface density giant HII regions (shaded area Mezger 1970)
and free-free continuum radiation (see Figure 16 in Westerhout
1958) show a remarkable similarity to the radial distribution
of CO (Figure 4). In contrast the HI surface density varies
little with galactic radius (Van Woerden 1965). The giant HII
regions are defined to be intrinsically as luminous as Orion A
= M42.

by assuming that the clouds observed here are akin to dark nebulae in
the solar neighborhood, or more accurately 2) by application of CO excitation and line formation theory.

We note that a typical line-of-sight crossing the inner part of
the galaxy ($10° < \ell < 50°$) shows 5 well-defined features with typical
width 8 km/sec and $\overline{T}_A = 2°K$ (i.e., $\int T_A dv = 80°K \cdot$ km/sec). The antenna
temperature corrected for antenna efficiency ($\eta_B = 0.6$) and atmospheric
attenuation is therefore approximately 3.3°K which is equivalent to a
true Planck brightness temperature of 6.6°K. This is strikingly similar
to CO intensities in nearby dark clouds; it is a factor of ten weaker

than the emission from the vicinity of strong infrared sources and compact HII regions such as W3, W51, NGC6334, and the Kleinman-Low nebula.
If the CO levels are nearly thermalized (a hypothesis which is reasonable but certainly not proven), then the average gas kinetic temperatures are only 7°K.

The relevant path length along the line-of-sight where these features are found is typically 10 kpc mostly in the vicinity of the 5.5 kpc
peak. The mean line-of-sight distance $\ell$ between clouds is therefore
2 kpc at $4 < \bar{\omega} < 7$ kpc in contrast to $\ell = 10$ kpc in the solar neighborhood. Thus there are approximately five times as many dense clouds at
5 kpc as at 10 kpc. Whether each feature is a single cloud or a complex
is impossible to determine at present.

The "smoothed out" hydrogen density $\bar{n}_{H_2}$ may now be determined from
the density within clouds $n_{H_2}$ and the cloud size r. The fraction of
space filled by clouds is $f = 4r/3$ and

$$\bar{n}_{H_2} = \frac{4r}{3\ell} n_{H_2} \qquad (1)$$

From studies of nearby dark nebulae (for example the cloud complex near
Orion B and A - Tucker et al. 1973 and Wannier 1975) we take $5 \leq r \leq 15$ pc
and $n_{H_2} = 300$ (a minimum density estimated to exist in molecular clouds)
to 1000 $cm^{-3}$. Combining these numbers gives a smoothed out molecular
hydrogen density of $\bar{n}_{H_2} = 1$ to 10 $cm^{-3}$ at $\bar{\omega} = 4$ to 7 kpc.

Alternately we may attempt to treat the line formation in greater
detail (see Goldreich and Kwan 1974 and Scoville and Solomon 1974). In
the notation of Scoville and Solomon we find that

$$n_{H_2} = \beta \left( \frac{E}{2\epsilon} \frac{\Delta v}{r} \right)^{\frac{1}{2}} \qquad (2)$$

where

$$\beta = \left( \frac{8\pi}{3\lambda^3} \frac{1}{<\sigma v>} \right)^{\frac{1}{2}} , \qquad (3)$$

and $\epsilon$ is the $CO/H_2$ abundance ratio. For the CO transition $\beta = 4 \cdot 10^6$
and the degree of excitation parametrized by $E = \alpha n_{H_2} <\sigma v> n_{CO}$ is equal
to 100 for thermalized CO emission and $^{13}CO$ at 1/3 the intensity of CO.

This $^{13}CO/CO$ intensity ratio of ~1/3 is found for the lines in Figure 1
(Scoville and Solomon 1975). Setting $\Delta v/r = 1$ km/sec/pc and $\varepsilon = 5 \cdot 10^{-5}$
(approximately 10% of all C in CO) we obtain $n_{H_2} = 730$ cm$^{-3}$ and $\bar{n}_{H_2} = 2.4$
to 7.3 cm$^{-3}$ in the 4 to 7 kpc region.

## V. COMMENTS

The major result of this preliminary survey is that in contrast to
atomic hydrogen the molecular hydrogen shows a strong density gradient
with galactic radius. We find a maximum $H_2$ density at 5.5 kpc and estimate that most of the interstellar medium there is molecular. The region between 0.5 and 4 kpc is a density minimum in both molecular and
atomic gas. The analysis presented here, taken along with our previous
work on the galactic center, indicates that much of the interstellar
medium in the interior of the galaxy $\bar{\omega} \lesssim 7$ kpc is molecular: a maximum
at $\bar{\omega} \lesssim 400$ pc and a secondary peak at $\bar{\omega} = 5.5$ kpc. Whether the outer
maximum is ring shaped (i.e., fully axially symmetric) we cannot deter-
mine from the present data. It does seem clear, however, that the maximum is not merely the result of a single spiral arm being coincidentally
tangential to our line of sight at $\bar{\omega} = 5.5$ kpc. Figures 3 and 4 indicate that the width of the peak is at least 2 kpc at half-intensity
points; this of course is much greater than the thickness of spiral arms.

At $4 < \bar{\omega} < 7$ kpc, a typical molecular cloud is found to have a
radius of 10 pc, density ($H_2$) of 500, and temperature 7°k. If its mass
of $9 \cdot 10^4$ M$_\odot$ were smoothed out over its neighboring volume, the resulting hydrogen atom density would be about 5 cm$^{-3}$.

We think it reasonable that the observed clouds should end up forming primarily into OB associations rather than bound clusters or T Tauri
associations. Their internal velocity dispersion ( 8 km $\cdot$ s$^{-1}$) is similar to that observed for stars in OB associations but much greater than
the dispersion within T Tauri associations (only 1-2 km s$^{-1}$, Herbig 1962).
Using the above densities and distribution we find the total mass in
molecular clouds is about $1 \rightarrow 3 \; 10^9$ M$_\odot$, mostly concentrated inside
$\bar{\omega} = 8$ kpc. If the lifetime of a cloud is $3 \cdot 10^6$ yr (= $1/\sqrt{G\rho}$ with

$n_{H_2} = 500$ cm$^3$), then star formation with an efficiency of 1% will give a current rate of star formation in the galaxy of 3 M$_\odot$ yr$^{-1}$.

# REFERENCES

Davis, J.H. and Vanden Bout, P. 1973, Ap. Lett., 15, 42.

Goldreich, P. and Kwan, J. 1974, Ap.J., 189, 441.

Herbig, G.H. 1962, Adv. Astrm. Astrophys., 1, 47.

Kerr, F.J. 1969, Aust. J. Phys. (Astrophysical Supplement), 9, 1.

Lizst, H. 1973, unpublished Ph.D. thesis Princeton University.

Mezger, P.G. 1970, I.A.U. Symposium, 38, 107.

Mezger, P.G., Wilson, T.L., Gardner, F.F., and Milne, D.K. 1970,
  Astrm. and Astrophys., 4, 96.

Penzias, A.A., Jefferts, K.B., and Wilson, R.W., 1971, Ap.J., 165, 229.

Rougoor, G.W. 1964, B.A.N., 17, 381.

Rougoor, G.W. and Oort, J.H. 1960, Proc. Natl. Acad. Sci. U.S.A., 46, 1.

Schmidt, M. 1965, Stars and Stellar Systems (University of Chicago Press),
  5, 513.

Scoville, N.Z. and Solomon, P.M. 1974, Ap.J. (Letters), 187, L67.

Scoville, N.Z., Solomon, P.M., and Jefferts, K.B. 1974, Ap.J. (Letters),
  187, L63.

Scoville, N.Z. and Solomon, P.M. 1975 (in preparation).

Spitzer, L., Drake, J.F., Jenkins, E.B., Morton, D.C., Rogerson, J.B.,
  and York, D.G. 1973, Ap.J., 181, L116.

Tucker, K.D., Kutner, M.L., and Thaddeus, P. 1973, Ap.J. (Letters), 186,
  L13.

Van Woerden, H. 1965, Trans. I.A.U., 12A, 789.

Wannier, P.G. 1975 unpublished Ph.D. thesis Princton University.

Westerhout, G. 1958, B.A.N., 14, 215.

RADIO OBSERVATIONS OF HII REGIONS IN EXTERNAL GALAXIES


F.P. ISRAEL

Sterrewacht, Leiden




1. INTRODUCTION


       Because of their large distances even very bright extragalactic
HII regions appear to terrestrial observers as objects much smaller
and weaker than galactic HII regions. Why should one then want to
observe these distant regions rather than the close ones in the home
Galaxy? One can think of a number of reasons other than the lure of
exotic lands.
(i) By observing extragalactic HII regions, one can study their
       distribution over a whole galaxy. The HII regions in almost
       all galaxies are effectively at the same distance. In our own
       Galaxy, HII regions have vastly different distances which are
       often hard to determine. Accordingly, it is extremely difficult
       to obtain a complete and homogeneous sample of galactic HII regions
       over the whole Galaxy.
(ii)By comparing the HII regions content from galaxy to galaxy, one
       can hope to determine whether the HII region population varies,

and whether population differences are correlated with other parameters of the galaxies in question. One also might hope to learn something about the frequency of star formation in galaxies, and about the birth rate of early type stars.

(iii) Extragalactic observations enable one to observed the properties of very large and very bright HII regions which appear to be absent in our Galaxy. The core region of NGC 5461 in the galaxy M101 for instance has a $\lambda$ 6 cm flux-density of 13 m.f.u. at a distance of 7 Mpc, and has a linear size of 205 pc (Israel, Goss and Allen, 1975). Placed inside the Galaxy at a distance of 20 kpc, such an object would show up as an HII region or group of HII regions emitting 1600 f.u. in an area only 35 arcmin across. Only if placed very close to the Sun it might have been missed by galactic surveys. In that case, however, its massive cluster of exciting stars would have been observed long ago. Clearly, no object like the core of NGC 5461 is present in the Galaxy.

Granted the importance of observing extragalactic HII regions, it would seem easier to observe them only optically than to try and detect them also at radio wavelengths where angular resolution and sensitivity is poorer. However, there are several advantages in making radio observations.

(iv) It is easier to obtain absolute total fluxes in the radio region than in the optical region.

(v) A comparison of radio observations with available optical measurements yields information on the extinction of the HII regions, and hence their dust content. As will be shown in section 5, extragalactic HII regions seem to contain appreciable amounts of dust, so that optical fluxes cannot be used to determine reliable physical parameters.

2. PRESENT STATUS OF OBSERVATIONS

In order to observe extragalactic HII regions in the radio regime one requires aperture synthesis telescopes for the following reasons

(i)  One needs a sensitivity of the order of milliflux–units[1].

For example, Orion A has a λ 21 cm flux-density of about 500 f.u.
at a distance of 0.5 kpc. Seen at the distance of M33 (D = 720 kpc)
this source would have a flux-density of only 0.25 m.f.u. The
giant HII region 30 Dor in the Large Magellanic Cloud (D = 55 kpc)
would only show up as a source of about 4 m.f.u. at the distance
of M101 (D = 7 Mpc).

(ii) One needs an angular resolution in the order of arcseconds.

For example, one arcsec corresponds to a linear size of 5 pc
at a distance of 1 Mpc (which is the limiting distance of Local
Group galaxies) and 50 pc at a distance of 10 Mpc (which is the
limiting distance of observable HII regions on the basis of the
sensitivity of present-day telescopes).

Only HII regions in the Magellanic Clouds, and a few of the brightest
regions in some galaxies like M33 and M101 can be observed with large
single dish radiotelescopes, preferably at high frequencies.
With respect to angular resolution and sensitivity, the best
observations of extragalactic HII regions at the present time are
comparable to the best observations of galactic HII regions with
radiotelescopes in the fifties. Tables 1 and 2 summarize the available
data.

TABLE 1A – PUBLISHED RADIO DATA ON EXTRAGALACTIC HII REGIONS

| Name | Number of Observed Radio sources identified with HII regions | Weakest Observed HII region (m.f.u.) | Wavelength (cm) | Authors |
|------|------|------|------|------|
| LMC  | 24 | 180 | 6, 11, 75 | McGee and Newton |
|      | 8  | ——  | 6 (line) | McGee et al (1974) |
| SMC  | 6  | 170 | 75 | Mills and Aller (1971) |
| M33  | 67 | 1.5 | 21 | Israel and Van der Kruit |
| M101 | 8  | 2.7 | 6, 21, 49 | Israel, Goss and Allen (1975) |

1) 1 m.f.u. = $10^{-3}$ f.u. = $10^{-29}$ Wm$^{-2}$ Hz$^{-1}$

TABLE 1B - UNPUBLISHED RADIO DATA ON EXTRAGALACTIC HII REGIONS

| Name | Number of Observed Radio Sources identified with HII regions | Weakest Observed HII region (m.f.u.) | Wavelength (cm) | Authors |
|------|------|------|------|------|
| NGC 2403 | 9 | 1.7 | 6, 21 | Israel |
| NGC 6946 | — | —— | 6, 21, 49 | Allen |
| NGC 4631 | 11: | 1.0: | 6, 21 | Israel |
| M51 | 5: | 2: | 21 | Israel |

: uncertain


TABLE 2 - STUDIES OF INDIVIDUAL EXTRAGALACTIC HII REGIONS

| Name | Wavelength (cm) | Resolution (arcmin) | Authors |
|------|------|------|------|
| 30 Dor (LMC) | 75 | 2.8 x 3.4 | Le Marne (1968) |
| | 75, 21, 11 | 48, 14, 7.5 | Mathewson and Healey (1964) |
| | 6 | 4 | Mezger et al (1970) |
| | 6 | 4 | Huchtmeier and Churchwell (1974) |
| NGC 604 (M33) | 21 | 1.5 x 3.0 | Wright (1971) |
| | 21,6 | 0.4 x 0.75 0.1 x 0.2 | Israel and Van der Kruit (1974; unpublished) |
| | 11,3 | 0.15, 0.05 | Spencer(1973) |
| NGC 595 (M33) | 11,3 | 0.15, 0.05 | Spencer (1973) |
| IIZw40 | 21 | 0.4 x 6.0 | Jaffe (1972) |
| M51 VS53/55 | 11,3 | 0.15, 0.05 | Spencer (1973) |
| NGC 5447 (M101) NGC 5455 NGC 5461 NGC 5462 NGC 5471 | 49, 21, 6 | 1.0 x 1.2 0.4 x 0.5 0.1 x 0.1 | Israel, Goss and Allen (1975) |
| NGC 2403 VS44 | 21,6 | 0.4 x 0.45 0.1 x 0.11 | Israel (unpublished) |

## 3. DISTRIBUTION OF HII REGIONS

Only in the case of the LMC (McGee and Newton, 1972) and M33
(Israel and Van der Kruit, 1974) is the number of detected HII regions
large enough to study their distribution both in space and in intensity.
The HII region content of the LMC and M33 are remarkably similar
in several respects, in spite of the fact that the two galaxies differ
in mass by a factor of 10, and are of different type (Ir and ScIII
respectively).

(i)     In both galaxies the most massive HII regions are found at about
        one third of the optical radius (i.e. at about 1.5 kpc in the
        LMC, taking the HI rotation center as the center of that
        galaxy, and at about 2 kpc in M33).

(ii)    Both galaxies have two large and bright HII regions (30 Dor and
        N11 in the LMC, NGC 595 and NGC 604 in M33) that are much
        brighter than the other regions.

(iii)   The integrated number of HII regions, brighter than a certain
        limiting flux-density $S_o$ plotted logarithmically as a function
        of $S_o$ defines a straight line in both galaxies with a slope of
        about $-2.2$ (see Figure 1).

(iv)    In a $n_e - d$ diagram, the distribution of HII regions in both
        galaxies shows an upper envelope to the excitation
        parameter $u = 210$ pc $cm^{-2}$, excluding the regions mentioned
        under (ii). This limiting value to the excitation parameter
        corresponds to excitation by $3 - 5$ O5 stars or their equivalent.

(v)     Extrapolating the flux-density distribution mentioned in (iii)
        down to HII regions comparable to Orion A, one predicts for
        both galaxies a thermal flux-density which is consistent with
        the observed spectrum and total flux-densities of M33 and the
        LMC as a whole.

In contrast, the few observed HII regions in M101 (Israel, Goss and
Allen, 1975) do not define a similar flux-density distribution. They
by themselves contain an appreciable amount of the total thermal flux-
density of the galaxy, perhaps as much as 50 percent. They are very
much brighter than the HII regions seen in M33 and the LMC, and also
very much brighter than most other HII regions in M101 itself.

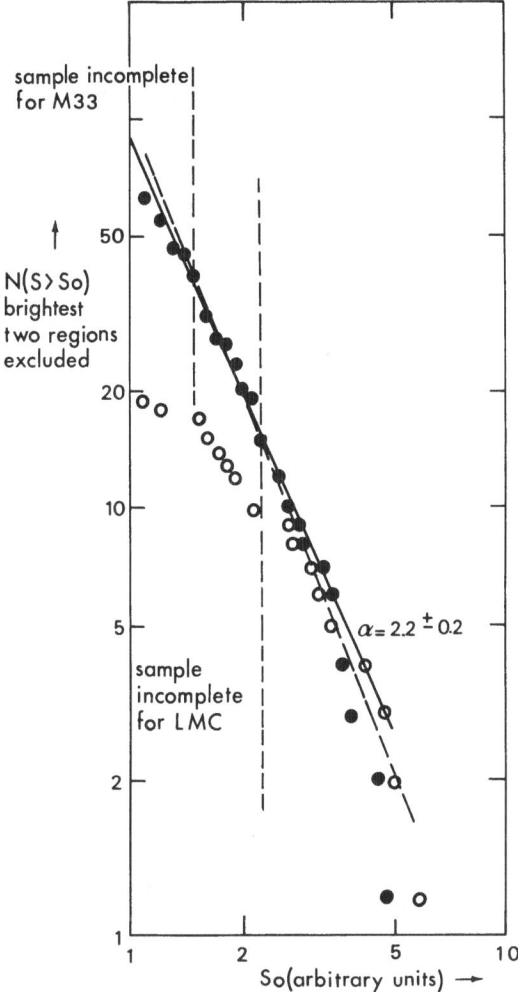

Figure 1  Integrated number N(S) of HII regions brighter than
a limiting flux-density $S_0$ plotted as a function of
$S_0$. Open circles denote regions in the LMC, filled
circles denote regions in M33. The HII regions in
the LMC are plotted as if their distance is 720 kpc
instead of 55 kpc in order to make a meaningful
comparison with HII regions in M33 possible.
The horizontal scale thus indicates flux-densities
in m.f.u. at a distance of 720 kpc.

Two suggestions can now be made that need to be confirmed by more observations.

The first suggestion is that Sc and Ir galaxies have something like a standard distribution of HII regions irrespective of other factors. Observationally, this would translate into an equal radio surface brightness of the thermal emission from galaxies. This in turn would mean that the thermal contribution to the total flux-density is in principle determined only by the size of the galaxy. One may predict the thermal contribution for a large sample of Sc and Ir galaxies (about 40) using the inferred thermal flux-density of M33 and for instance the maximum Hohmberg size of the galaxies. The predicted thermal flux-density at $\lambda$ 21 cm is then indeed lower than the observed total flux-density for all these galaxies. Thus the first suggestion is not contradicted by the available information. As expected the relation does not hold for Sb galaxies, where in most cases a thermal contribution would be predicted much larger than the observed flux-density. The second suggestion is, that some galaxies also contain an additional population of very large and very bright HII regions superimposed on the standard population, (cf. M101; Israel, Goss and Allen, 1975). Observationally, this has a number of consequences. In the first place an excess of large and bright HII regions will appear in a number of galaxies. In the second place, in these galaxies the few brightest HII region will account for an appreciable percentage of the thermal part of the flux (cf. M101). Although the statistics for individual galaxies are poor, the effect will show up clearly in a large sample of galaxies.

## 4. SPECTRUM AND STRUCTURE OF EXTRAGALACTIC HII REGIONS

In the LMC, McGee and Newton (1972) find six cases where a thermal source, identified with an HII region is found close to a nonthermal source, presumably a supernova remnant. A good example is 30 Dor, whose $\lambda$ 21 cm flux-density is 30 percent nonthermal.

In contrast to this, the M33 regions NGC 604 and NGC 595; the M101
HII regions NGC 5447, NGC 5455, NGC 5461, NGC 5462 and NGC 5471, and
the NGC 2403 region VS44 all seem to be optically thin and thermal
to at least 90 percent of their flux-density at λ 21 cm.
Therefore, giant starclusters containing up to a thousand 05 stars
or their equivalent are needed to explain the excitation for these
regions. As a consequence, these HII regions may approach irregular
dwarf galaxies in size and mass. The object  IIZw40  (Jaffe, 1972)
looks like something between a super HII region without galaxy and a
peculiar dwarf galaxy.
Most of the resolved extragalactic HII regions are best described in
terms of a core-envelope model, sometimes with multiple cores
(Mathewson and Healey, 1964; Israel and Van der Kruit, 1974; Israel,
Goss and Allen, 1975). This is especially clear for 30 Dor (Figure 2,
Table 3), but also for the brightest regions in M33 and M101
(Table 4, Figures 3 and 4).

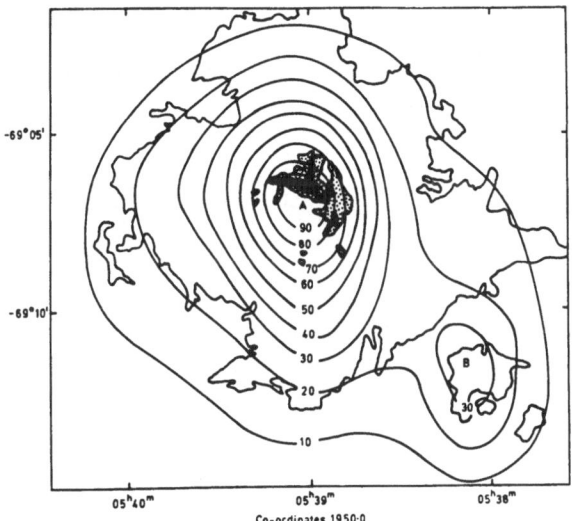

Figure 2 Contourmap of 30 Dor in the LMC(taken from Le Marne,
1968).Contours at λ 75 cm are superimposed on a sketch
of the optical nebulosity.

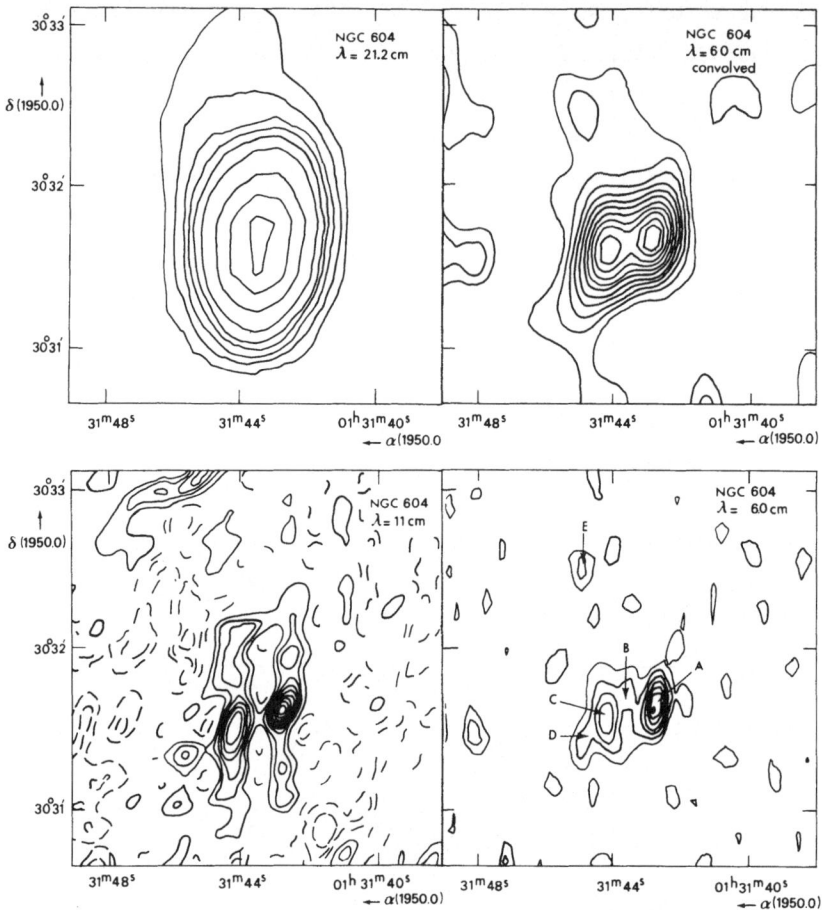

Fig. 3 Contour maps of NGC 604 in M33 at λ 21.2 cm (taken from Israel
and Van der Kruit, 1974), at λ 11 cm (taken from Spencer, 1973)
and at λ 6 cm (Israel, unpublished).

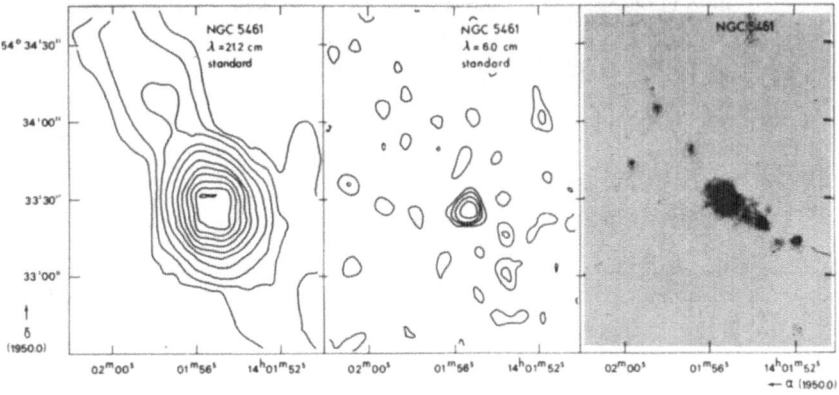

Fig. 4 Contourmaps of NGC 5461 in M101 at λ 4.2 cm and at λ 6.0 cm
compared with an Hα photograph of the region (taken from Israel
Goss and Allen, 1975).

The cores have r.m.s. electron-densities 10 - 30 times higher than the
envelopes, which have typical densities of 1 - 3 cm$^{-3}$; the core
sizes are in the order of 10 - 25 percent of the envelope sizes and core
masses are similar by two orders of magnitude. Since optical observations (Searle, 1971; Smith, 1975) show that regions denser than
500 cm$^{-3}$ do not significantly contribute to the fluxes, a useful
upper limit of about 100 is found for the clumping factor
$(n_e/<n_e^2>^{\frac{1}{2}})^2$ in the cores listed in Table 4.
Because the envelope-densities are comparable to the density of neutral
hydrogen at least in the case of M101 (Israel, Goss and Allen, 1975)
it is tempting to speculate that the envelopes consist of spiral arm
interstellar medium ionized by photons escaping from the density-bounded
cores. Finally it should be noted that in the LMC, M101 and M33 the
detected HII regions have a strong tendency to exist at places of high
HI surface density on a typical scale of 1 kpc. The same conclusion
was drawn for M31 by Emerson (1974). Emerson also remarks that HII
region cluster around peaks of HI emission rather than coincide with
those. The same is found for M33 by Boulesteix et al (1975). Only in

the case of the brightest HII regions in M101 do the HI and the HII peaks coincide (see also Allen, 1975).

TABLE 3 - OBSERVATIONS OF 30 DOR

Model Mathewson and Healey (1964) - 1410 MHz

| Size of Component (arcmin) | S(f.u.) | Nature | R.m.s. Electron-density |
|---|---|---|---|
| 4 | 40 | Thermal | 60 |
| 24 | 25 | Nonthermal | — |
| 45 | 20 | Thermal | 1 |

Model Le Marne (1968) - 408 MHz

| | | | | |
|---|---|---|---|---|
| A | 4x5 | 30 | Thermal | 40 |
| B | 1 | 4 | Nonthermal | — |
| C | 26x13 | 44 | Nonthermal | — |

Nonthermal emission not centered on 30 Dor; $A_v$ $2^m.0$.

Recombination Lines H109α, He109α

| $T_e$ (K) | He/H | $V_{turb}$ (km s$^{-1}$) | $V_{rad}$ (km s$^{-1}$) | Authors |
|---|---|---|---|---|
| 9400±1500 | 17% | 40 | 267 | Mezger et al (1970) |
| 11900±1800 | 5-8% | 40-60 | 253 | Huchtmeier and Churchwell(1974) |
| ——— | 8% | 40-60 | 265 | McGee et al (1974). |

TABLE 4 - CORE ENVELOPE STRUCTURE

(Taken from Israel, Goss and Allen, 1975; Israel, unpublished)

| Name | $S_{4995}$ (m.f.u.) | $\theta$ (arcmin) | d (pc) | $n_e$ $(cm^{-3})$ | M $(10^5 M_\theta)$ |
|---|---|---|---|---|---|
| **Core Parameters** | | | | | |
| NGC 604 A | 19.5 | 0.12 | 30 | 75 | 0.4 |
| C | 19.0 | 0.21 | 53 | 30 | 0.9 |
| NGC5455 | 5.5 | 0.10 | 245 | 15 | 47 |
| NGC5461 | 13 | 0.09 | 205 | 35 | 55 |
| NGC5471 | 3x2.7 | 3x0.05 | 3x120 | 3x35 | 3x12 |
| **Envelope Parameters** | | | | | |
| NGC 604 | 8 | 0.75 | 190 | 3 | 4 |
| NGC5455 | 1 | 0.45 | 1095 | 0.7 | 190 |
| NGC5461 | 9 | 0.67 | 1665 | 1.2 | 1060 |
| NGX5471 | 4 | 0.33 | 810 | 2.3 | 240 |

## 5. PRESENCE OF DUST IN EXTRAGALACTIC HII REGIONS

Mills and Aller (1971) observed that a comparison of radio and optical measurements in the LMC and the SMC indicate a higher extinction of the observed HII regions than one would expect from optical observations alone. The same effect is found when the radio results on M33 and M101 are combined with Searle's (1971) optical data (Table 5).

TABLE 5 - COMPARISON OPTICAL AND RADIO DATA

(Adapted from Israel and Van
der Kruit, 1974; Israel,
Goss and Allen, 1975)

| Name | Galaxy | A(Hβ) optical | A(Hβ) radio |
|------|--------|---------------|-------------|
| IC 131 | M33 | negative | $1\overset{m}{.}1$ |
| IC 132 | | $0\overset{m}{.}26$ | 2.0 |
| IC 142 | | 0.52 | 2.2 |
| NGC 604 | | 0.14 | 1.9 |
| NGC 5455 | M101 | 0.14 | 1.5 |
| NGC 5471 | | 0.14 | 1.4 |
| VS 53 | M51 | 0.20 | 3.0 |
| VS 71 | | 0.48 | 3.5 |

In all galaxies on which data are available, the reddening is
relatively small, and the reddening curve seems to be normal
(Smith, 1975); in the same galaxies however the observed HII regions
suffer extinction up to 2 - 3 magnitudes. The most plausible
explanation is the presence of large amount of dust within the
boundaries of the ionized region. This dust must be present in both
cores and envelopes in order to explain the observed optical fluxes and
radio flux-densities.

6. BRIGHTEST HII REGIONS IN GALAXIES

Recently, Sandage and Tammann (1974a, 1974b) have found a relation
between the luminosity class of Sc and Ir galaxies and the mean linear

core/envelope size of their largest HII regions. Does such a relation
also exists between the type of galaxy and the monochromatic power
of the HII regions? In Table 6 the monochromatic power of the largest
and brightest HII regions in a number of galaxies is given, together
with the luminosity class of the galaxies. Galactic HII regions are
included only for reference purposes.

Looking at Table 6, the power emitted by the brightest HII regions
in M33 and M101 suggests a strong correlation with the luminosity
class of those galaxies, in the sense that there is a difference
of a factor 35 between the intrinsic brightness of the brightest

TABLE 6 - MONOCHROMATIC POWER OF BRIGHTEST HII REGIONS

| Name | HII region | $P_{1415}$ $(10^{18}W\ Hz^{-1})$ | D (kpc) | Type of Galaxy |
|---|---|---|---|---|
| Galaxy | W51 | 3.6 | 6.5 | — |
| LMC | 30 Dor(nucl.) | 9.1 | 55 | Ir |
|  | N11 | 1.2 |  |  |
| SMC | N66 | 2.2 | 80 | Ir |
|  | N19 | 1.7 |  |  |
| M33 | NGC 604 | 3.6 | 720 | ScII-III |
|  | NGC 595 | 1.2 |  |  |
| M31 | AB74 | 0.6 | 720 | Sb |
| M101 | NGC 5461 | 111.7 | 7000 | ScI |
|  | NGC 5462 | 70.6 |  |  |
| NGC 2403 | VS44 | 31.3 | 3300 | ScIII |
|  | VS42 | 10.1 |  |  |
| NGC 6946 | —— | ∿40 | 9000 | ScI-II |
| NGC 4631 | —— | ∿30 | 6500 | ScIII |
| M51 | VS55 | ∿25 | 9000 | ScI |

HII region in M33 and that in M101. A further look however shows that
the brightest region in M51 (another ScI galaxy), is intrinsically
weaker by a factor of 4 than the one in M101, and that the brightest
regions in NGC 2403 and NGC 4631, both classified as ScIII galaxies
are about 8 times stronger than NGC 604 in M33. This suggests that
the brightest HII regions in Sc galaxies have about equal
monochromatic power at radio wavelengths, and that both M101 and M33
are notable exceptions.

Unfortunately, the data available at present do not allow one to go
further than noting tendencies. A larger sample is needed before one
can draw more definite conclusions.

Table 6 also shows clearly that in Local Group galaxies no HII regions
are found comparable in power to giant complexes like NGC 5461 in M101
or VS 44 in NGC 2403. The lack of Sc galaxies with brightest HII regions
of the same intrinsic power as Local Group HII regions at larger
distances is due to observational selection at these distances. It is
difficult to detect the intrinsically much weaker HII regions like the
ones found in the Local Group.

CONCLUSIONS

Only in recent years has it become possible to observe
extragalactic HII regions at radio wavelengths. Radio observations
yield physical parameters that make a comparison of HII regions in
different galaxies possible.

The first results indicate that the Large Magellanic Cloud and M33 have
similar HII region populations, leading to the suggestion that perhaps
Sc and Ir galaxies possess standard HII region populations. Superimposed
on such a standard population one may find a few very large and very
bright HII regions, like the brightest HII regions found in M101.
These very large HII regions and HII region complexes must be excited
by large numbers of early 0 stars. They even may show some of the

characteristics of irregular dwarf galaxies.

Extragalactic HII regions are generally found near positions of maximum density in the neutral hydrogen distribution; they appear to contain large amounts of internal dust.

So far, no clear correlation is found between the monochromatic power of the brightest HII regions and the luminosity classification of their parent galaxy.

Most of these conclusions are preliminary, since they are based only on a small sample of HII regions in a few galaxies. More observations are necessary, and will be done in the near future. Only when these results are available will it be possible to study the nature of the brightest HII regions in galaxies in detail.

REFERENCES

Allen, R.J. 1975, to be published in: CNRS Intern. Coll. Dynam. Spiral
        Gal., Inst. Hautes Et. Scient. Bures-sur-Yvette.
Boulesteix, J., Courtès, G., Lavel, A., Monnet, G., Petit, H., 1975
        Astron. Astroph. submitted.
Emerson, D.T., 1974, Mon. Not. R.A.S. 169, 607.
Huchtmeier, W.K., Churchwell, E., 1974 Astron. Astroph. 35, 417.
Israel, F.P., Goss, W.M., Allen, R.J., 1975 Astron. Astroph. submitted.
Israel, F.P., Van der Kruit, P.C., 1974 Astron. Astroph. 32, 363.
Jaffe, W.J. 1972 Astron. Astroph. 20, 461.
Le Marne, A.E. 1968 Mon. Not. R.A.S. 139, 461.
Mathewon, D.S., Healey, J.R., 1964 I.A.U. - U.R.S.I. Symposium 20
        "The Galaxy and the Magellanic Clouds", Canberra, p.283.
McGee, R.X., Newton, L.M., 1972 Austr. phys. 25, 619.
McGee, R.X., Newton, L.M., Brooks, J.W., 1974 Austr. J. Phys. 27, 729.
Mezger, P.G., Wilson, T.L., Gardner, F.F., Milne, D.K., 1970, Astroph.
        Lett. 5, 117.
Mills, B.Y., Aller, C.H., 1971 Austr. Phys. 24, 609.
Sandage, A., Tammann, G.A. 1974a Astrophys. J. 190, 525.
                          1974b Astrophys. J. 194, 223.
Searle, L., 1971 Astrophys. J. 168, 327.
Smith, H.E., 1974 thesis Univ. Cal. - Berkeley.
Spencer, J.H., 1973 thesis Mass. Inst. Techn., - Cambridge, Mass.
Wright, M.C.H., 1971 Astrophys. L. 7, 209.

OPTICAL OBSERVATIONS OF H II REGIONS IN EXTERNAL GALAXIES

E. M. BURBIDGE and G. R. BURBIDGE

Department of Physics
University of California, San Diego
La Jolla, California 92037 U.S.A.

I.    INTRODUCTION

H II regions in the spiral arms of external galaxies and in irregular galaxies are often complexes of objects that cannot be studied individually, as can many H II regions in our Galaxy, because of resolution problems. The giant H II regions like 30 Doradus in the Large Magellanic Cloud and NGC 5471 in M101 are clearly more complex than nearby compact H II regions like Orion in our Galaxy, and must have a complicated temperature, density, excitation, and ionization structure throughout which averaging procedures must be carried out. The kinematic properties are also more complicated (cf. Smith and Weedman 1971, 1972).

There do exist, however, general properties that are correlated with morphological type of galaxies in which they are found. Sersic (1960, 1964) showed that the diameters of H II regions are well correlated with galaxy type and hence with that proportion of the galaxy mass which consists of uncondensed gas, so that diameters of H II regions could be used as distance indicators for galaxies. That the diameters depend not only on morphological type (evolutionary stage, uncondensed gas mass/total mass) but also on luminosity of the galaxy (and thus on its total mass) complicates the problem of using H II regions as distance indicators, but Sandage and his co-workers have set up a calibrated relation which provides a basic step in the determination of the distance scale of the Universe (Sandage and Tammann 1974, and references given therein).

H II regions can be used also as a basic means of investigating the evolution of galaxies, since a careful study of their physical conditions enables abundances of certain key elements to be determined, and thus information is provided on the chemical composition and chemical evolution of the uncondensed matter in galaxies. It is with this latter use of H II regions that we shall be concerned in this paper.

## II.     ABUNDANCES IN EXTRAGALACTIC H II REGIONS - GENERAL DISCUSSION

There have been several reviews of abundances from optical observations in extragalactic H II regions, of which the most recent is by Peimbert (1975). The observations involve the same problems as encountered in studies of galactic H II regions: only certain elements can be observed, the effective temperature of the ionizing radiation, the degree of ionization of the gas, and its electron temperature and density must be determined, and the dust content and reddening must be evaluated. In addition, observations necessarily integrate over large volumes of gas within which these parameters may vary considerably.

However, results have become available for a number of galaxies of a range of types, and we shall present and discuss these, drawing on Peimbert's and earlier reviews, and on recent results by Smith (1975).

The elements that can be studied are H, He, N, O, Ne, S. Since the key element O must be studied in more than one stage of ionization, observations must cover at least the region 3727-6731 Å ([O II] to S II). The Balmer line intensities, combined with the usual theoretical Balmer decrement (e.g. Brocklehurst 1971) and Whitford's (1958) reddening law, can be used to correct for extinction if it is assumed that the reddening properties of dust in external galaxies are similar to those in our Galaxy.

The determination of $T_e$ is best done by measuring the [O III] line intensity ratios, $\lambda 4363/\lambda\lambda 4959, 5007$; fluctuations in $T_e$ within the region observed can be taken into account following Peimbert (1967), although the uncertainties involved in the estimation of these fluctuations can be a source of error. Further, it is rarely possible to resolve the [O II] $\lambda 3727$ doublet, so that upper limits to electron densities have to depend on the more widely separated [S II] $\lambda\lambda 6715, 6731$ lines, but it appears that collisional depopulation is not important. Since H and He, the most abundant elements, control the Strömgren spheres, the H II regions can be divided into ionized and neutral helium zones. However, the fact that in extragalactic work we are necessarily dealing with complex H II regions ionized by aggregates of O, B stars must be an additional source of error, though Shields (1974) has shown that this effect may not introduce as large uncertainties as it was at one time feared.

## III.    THE HELIUM ABUNDANCE

Helium lines are produced by recombination, and most abundance determinations depend on measures of He I $\lambda 5876$, although $\lambda\lambda 4472, 6678$ may also be measurable. Until recently, measures in different galaxies have not been sufficiently accurate to look for variations in helium abundance from galaxy to galaxy, or in different regions within a galaxy. Although the most recent determinations are still subject to errors, e.g. in estimating the mean square temperature variation (Peimbert and Torres-Peimbert 1974), and in making allowance for the amount of neutral and doubly-ionized helium, they

are a considerable improvement on the earlier values and now show
that there are variations from galaxy to galaxy (cf. Section IV). Table
1 of Peimbert (1975) shows a range from one of the two peculiar
Zwicky compact galaxies studied by Searle and Sargent (1972a) with
He/H = 0.056 (II Zw 40), through values around 0.08 (0.077 for mean
SMC results (Dufour 1974), 0.08 for I Zw 18 (Searle and Sargent 1972a)
and 0.081 for mean LMC results (Dufour 1974; Peimbert and Torres-
Peimbert 1974) to values of about 0.1 or greater for the remaining
galaxies in the table and Orion Nebula. Peimbert (1975) and Smith
(1975) both give formulae for deriving He/H:

$$He/H = Y = Y^o + Y^+ + Y^{++}$$

where

$$Y^i = \frac{\int N_e \, N(He^i) \, dV}{\int N_e \, N(H^+) \, dV}$$

and

$$Y^+ = 0.089 \, T_o^{0.23} \frac{I(5876)}{I(H\beta)} = 0.74 \, T_o^{0.11} \frac{I(4472)}{I(H\beta)} = 0.26 \, T_o^{0.25} \frac{I(6678)}{I(H\beta)}$$

and

$$Y^{++} = 2.08 \times 10^{-2} \, T_o^{0.15} \frac{I(4686)}{I(H\beta)} \quad .$$

A combination of Peimbert's Table 1 and Table 4 from Smith (1975)
gives the results shown in Figure 1 and Table 1.

As will be shown in the next section, these results indicate that
there is a slight tendency for low He/H to be correlated with low Z/H,
although the range in the He/H ratio is very much smaller than that for
the heavier elements. As we shall see, lower O/H abundances result
in less cooling of H II regions, and the resulting higher temperatures
lead to more accurate He/H determinations.

## TABLE 1

### He/H Abundances By Number

| OBJECT | TYPE | FRACTIONAL RADIAL DISTANCE | PEIMBERT | SMITH |
|--------|------|----------------------------|----------|-------|
| Galaxy: Orion Nebula | | | 0.101 | |
| M82 | Irr-Sc | | 0.130 | |
| NGC 7679 | SBO | | 0.090 | |
| M101: NGC 5461 | Scd I | 0.4 | 0.090 | 0.080 |
| NGC 5455 | | 0.5 | 0.073[a] | 0.083 |
| NGC 5447 | | 0.6 | | 0.074 |
| Searle 12 | | 0.8 | | 0.085 |
| NGC 5471 | | 1.0 | 0.099 | 0.074 |
| M33: NGC 604 | Scd II-III | 0.5 | 0.112 | 0.13 |
| NGC 588 | | 0.8 | | 0.10 |
| IC 132 | | 1.0 | | 0.10 |
| NGC 6822 | Irr IV-V | | 0.090 | 0.074 |
| NGC 4449 | Irr III | | 0.088 | |
| ⟨LMC⟩ | Irr III-IV | | 0.084 | |
| ⟨SMC⟩ | Irr IV-V | | 0.081 | |
| I Zw 18 | Irr pec | | 0.085[a] | |
| II Zw 40 | Irr pec | | 0.056[a] | |

[a] Searle and Sargent, adjusted by Peimbert.

Figure 1 shows several interesting results:

1.     The average He/H ratio in the galaxies for which
       results are available is below the previously accepted
       value of ~ 0.1 by number.

2.     There is a slight correlation between He/H and mor-
       phological type.  The lowest He/H ratios occur in the
       irregular galaxies I Zw 18, II Zw 40, and the
       Magellanic Clouds.  It should be noted that these H II
       regions are metal-poor, therefore hotter, and the
       amount of neutral H and the consequent error in
       He/H ratio is considered by Peimbert to be negligible.
       M 33 appears to have a slightly higher average He/H
       ratio than M 101 (Smith 1975).

3.     There is no detectable gradient in the He/H ratio
       with position in M 33 and M 101, the only galaxies for
       which determinations in several regions have been
       made so far.

## IV.    HEAVY ELEMENT ABUNDANCES; OVERALL RESULTS WITHIN GALAXIES AND GRADIENTS IN COMPOSITION

It is easier to determine abundances of N, O, Ne, and S relative
to each other than to determine their absolute abundances relative to
H, since the latter depend on the accuracy with which the temperatures
can be determined.  Despite the difficulty of the observations, there
are a number of results available now.  More than thirty years ago
Aller (1942) noted that there was a gradient in the [O III]/H$\beta$ line
intensity ratios with distance from the center in M 33.  Many accurate
determinations of such line intensities are now available, by Searle
(1971), Rubin et al. (1972), Benvenuto et al. (1973), Warner (1973),
Comte and Monnet (1974), Smith (1975), and from these detailed
studies of this effect have been made.

A gradient in [N II]/Hα with distance from the center in the inner regions of many galaxies, first noted by Burbidge and Burbidge (1962), was shown by Peimbert (1968) to be due to a gradient in the N/H abundance ratio. Searle (1971) subsequently showed that N/H ratio decreases outward in M33 and M101, the numerical value of the ratio decreases by a factor ~ 10 between inner and outer parts. Aller's effect in the [O III] lines was shown by Searle and confirmed by others to be due to a gradient in the O/H abundance, in a sense opposite to what one might at first suppose. Since the main cooling mechanism for H II regions is infrared fine-structure transitions in oxygen, or in regions with low oxygen abundance, the [O III] nebular lines themselves, a decreased O abundance gives rise to a hotter H II region and a larger [O III]/Hβ intensity ratio. Smith (1975) finds a steeper O/H abundance gradient than that obtained by Searle: for M101 he found a radial decrease by a factor ~ 7 between H II regions 10 kpc and 24 kpc from the center. Because temperature and dust corrections become more severe for regions nearer the center, where H II temperatures may be as low as ~ 5000 °K, Smith gives only a lower limit for the innermost regions in M33 and M101, namely $O/H \geq 1.5 \times 10^{-3}$, while Shields (1974) using more elaborate models found a gradient from $O/H = 2 \times 10^{-4}$ in the outermost H II regions to $O/H = 3 \times 10^{-3}$ for the innermost ones, i.e. a factor of 15.

Smith found a decrease in the N/O ratio by a factor of about 4 on going from the innermost regions to those out at about 60% of the Holmberg radius. Beyond this, this ratio stayed roughly constant, with a value of about half that in the Orion Nebula. His resultant N/H variation from central to outermost regions was thus a factor ~ 40. He found no detectable radial variations in the ratios S/O or Ne/O, so that the ratios S/H and Ne/H presumably reproduce the O/H variation.

These results are collected together in Table 2 where average results are also shown for the SMC and LMC and for II Zw 40 and I Zw 18. It can be seen that there is a general correlation in heavy

## TABLE 2

### Heavy Element Abundances by Number

| OBJECT | TYPE | (O/H) (x $10^4$) | (N/O) (x $10^2$) | (S/O) (x $10^2$) | (Ne/O) |
|--------|------|------|------|------|--------|
| Galaxy: | | | | | |
| Orion Nebula[a] | | 6.7 | 12 | 0.5 | 0.19 |
| M101:[b] | Scd I | | | | |
| Inner | | 6.9 | 8.2 | 1.4 | 0.15 |
| Outer | | 1.4 | 5.1 | 2.8 | 0.21 |
| All | | 4.4 | 5.0 | 1.6 | 0.22 |
| M33:[b] | Scd II-III | | | | |
| Inner | | 5.3 | 5.2 | 1.1 | 0.21 |
| Outer | | 1.2 | 5.5 | 2.2 | 0.24 |
| All | | 3.7 | 4.6 | 1.4 | 0.25 |
| NGC 6822[b] | Irr IV-V | 2.8 | 3.5 | 1.5 | 0.21 |
| ⟨LMC⟩ | Irr III-IV | | | | |
| a) | | 3.8 | 3.3 | -- | 0.23 |
| c) | | 3.3 | 2.1 | 0.5 | 0.17 |
| ⟨SMC⟩[c] | Irr IV-V | 1.2 | 2.6 | 1.2 | 0.14 |
| I Zw 18[d] | Irr pec | 0.2 | -- | -- | 0.2 |
| I Zw 40[d] | Irr pec | 1.3 | -- | -- | 0.3 |

[a] Peimbert et al.

[b] Smith

[c] Dufour

[d] Searle and Sargent

element/hydrogen abundance ratio with type of galaxy, in that the compact peculiar irregulars I Zw 18 and II Zw 40, and also the SMC, show extreme metal-poorness, which, as described in the preceding section, tends to go with a small He deficiency. The LMC and NGC 6822 are intermediate in composition between these objects and the Orion Nebula. Meanwhile, the existence of the very pronounced abundance gradients found in M 101 and M 33 means that their outermost H II regions are as metal-poor as the intermediate galaxies, and they approach the values found in the extreme metal-poor objects.

## V. IMPLICATIONS FOR NUCLEOSYNTHESIS AND GALACTIC EVOLUTION

### 1. Heavy Elements

The heavy-element abundance ratios provide a very beautiful verification for the general view that these elements are synthesized in stars. Even with the small body of evidence now available, we see that there is a correlation between:

a)    Metal content and M(gas)/M(total) [but not necessarily between metal content and M(total)].

b)    Metal content and presumed evolutionary stage.

c)    Metal content and the location in a galaxy.

These can all be related to the fact that matter in certain irregular galaxies and the outer parts of normal spirals has undergone a lesser amount of nuclear processing in stars than the matter in early-type galaxies and the inner parts of galaxies. Mixing in M 33 and M 101 and indeed in most spirals must be far from complete. It is worth noting that Smith found that M 51, although it shows a similar heavy-element gradient to M 33, has all of its abundances higher by a factor ~ 5. However, according to Searle, Sbc galaxies do in general show a less steep heavy-element gradient than Scd galaxies.

The fact that the gradient of the N/H ratio is much larger than the O/H, Ne/H, and S/H ratios is probably a direct consequence of the fact that O, Ne, S are all produced in explosive stellar nucleosynthesis, while N is produced in CNO reactions in normal H-burning. Thus, N is produced in "secondary" nucleosynthesis while O, Ne, S can be produced in explosive events in earlier generations of stars. Smith (1975) believes that the relative abundance gradients suggest that some nitrogen may have been produced along with O, Ne, S, but this proposal must await more detailed abundance determinations in H II regions throughout many more galaxies, and more detailed models of the late evolutionary stages of stars in which C, O, Ne, S generated in the cores pass through hydrogen shells.

2.      Helium

It is well known that, while helium is synthesized in stars, the observed value of the He/H ratio in different places in our own Galaxy, and the values given for different galaxies here, are too high to be accounted for by the normal processes of stellar evolution (Burbidge 1958). There are several possible explanations. They are:

1)      that the overall abundance ratio for the Galaxy is low because the bulk of the matter is tied up in low-mass stars which have a very small He/H ratio;

2)      that the helium is primordial and was made in a big bang;

3)      that the helium was made in the early phase in the life of a galaxy. In such a phase the galaxy must be much more luminous than it is at present or in some way the energy which is released, $\sim 10^{61}$ erg for a galaxy with a mass of $10^{43}$ gm and a He/H ratio of 0.1 by number, must be radiated by thermal or nonthermal processes or expelled in ejection of gas, etc.

In principle it should be possible to decide between these alternatives as follows:

If (1) were correct, we would expect to find many old stars or nebulae in the Galaxy where the He/H ratio is very small, perhaps pure hydrogen stars, and we would also expect to find very considerable variations between He/H ratios in different parts of a galaxy.

If (2) were true, we would expect to find a constant and uniform value for the He/H ratio with only very small but variable increases above the primordial value.

If (3) were correct, we would expect to find that the He/H ratio varies between different galaxies but is approximately the same throughout a given galaxy.

What are the observed facts? Different versions of them have been discussed in recent years by Burbidge (1969) and by Searle and Sargent (1972b), while a critically important point relating to (2) has been made by Fowler (1970).

The evidence from the old stars in our own Galaxy is conflicting. Arguments based on the interior structures and evolution suggest that the oldest stars have He/H ratios close to the value obtained for H II regions and young stars in our Galaxy. However, spectroscopic analyses of some hot old stars show very small abundances of helium. Since many of these stars show anomalies in other elements, it has been argued quite strongly (e. g., Searle and Sargent 1972b) that these stars are showing the results of internal nucleosynthesis so that the helium abundances in them are not representative of the abundance in the Galaxy when these stars formed. Thus, such results have been considered irrelevant to the problem and it has generally been argued that the true He/H ratio is the higher value obtained from internal structure arguments. Thus, explanation (1) has been excluded, and we are left with either (2) or (3).

Proposal (2) is the one which is currently very popular. Its observational basis is firstly that there is direct evidence for a big bang -- the form of the microwave background radiation spectrum -- and secondly that there is an approximately constant value of the He/H ratio in all galaxies observed.

The first of these arguments is reasonably strong, as far as simplistic cosmological arguments go, but the second is on more shaky ground. We see from Table 1 and Figure 1 that there are apparently variations in this ratio among different galaxies. It has been generally argued that the differences are due to uncertainties in the determinations, and lately Peimbert and Torres-Peimbert (1974) have argued that stellar nucleosynthesis at different rates in different galaxies will give rise to variations in the He/H ratio above a base value of He/H generated in a big bang. However, it is not obvious that this is numerically plausible.

Given the range of values of the He/H ratio so far discovered, it appears to be equally plausible to argue that (3) is correct, i. e., that the helium is largely synthesized in the early lives of galaxies through "little big bangs" (Wagoner, Fowler, and Hoyle 1967) with significantly different values in different galaxies. Many have rejected this possibility, either because the bandwagon effect is at work, i. e., many observers have been led to believe that a constant value of this ratio is expected, so that given the uncertainties and errors involved, they tend to settle upon values close to the "canonical" one, or because they believe that there was a big bang and conclude that therefore helium must have been made in it.

However, this latter argument is not correct. We briefly summarize the argument of Fowler (1970). Wagoner et al. (1967) showed that little or no helium results from big-bang nucleosynthesis if either neutrinos or antineutrinos are sufficiently degenerate. Since, for this to be the case, a very large ratio of the lepton to baryon ratio is required so that the helium (and deuterium) production is suppressed, this argument has not been considered to be very plausible. However, Fowler then shows that given the very reasonable possibility that, following meson and hadron decay in the big bang, the number of photons was approximately equal to the number of neutrinos, there will be effectively no big bang nucleosynthesis and hence no "cosmological helium." Of course, the ratio $n_\gamma/n_b$ will still be quite high, $\sim 10^7 - 10^9$, but this may be what is to be expected from energy considerations since baryons have large rest masses.

We conclude that the observed values of the He/H ratio in H II regions in external galaxies are compatible with the idea that the helium was made in little big bangs, giving rise to different helium abundances in different galaxies. Clearly, more determinations of the He/H ratio which extend the range of galaxies observed are required.

We are very grateful to Dr. H. E. Smith for providing data in advance of publication, and for many helpful discussions. Extragalactic research at University of California, San Diego, is supported in part by grants from the National Science Foundation and the National Aeronautics and Space Administration.

## REFERENCES

Aller, L. 1942, Ap. J., 95, 52.
Benvenuto, P., D'Odorico, S., and Peimbert, M. 1973, Astron. & Astrophys., 28, 447.
Brocklehurst, M. 1971, Mon. Not. Roy. Astron. Soc., 153, 471.
Burbidge, E. M., and Burbidge, G. R. 1962, Ap. J., 135, 694.
Burbidge, G. R. 1958, Publ. Astron. Soc. Pacific, 70, 83.
Burbidge, G. 1969, Comments Astrophys. & Space Phys., 1, 101.
Comte, E., and Monnet, G. 1974, Astron. & Astrophys., 33, 161.
Dufour, R. J. 1974, unpublished Ph. D. Thesis, University of Wisconsin.
Fowler, W. A. 1970, Comments Astrophys. & Space Phys., 2, 134.
Peimbert, M. 1967, Ap. J., 150, 825.
Peimbert, M. 1968, Ap. J., 154, 33.
Peimbert, M. 1975, Ann. Rev. of Astron. & Astrophys., Vol. 13, in press.
Peimbert, M., and Torres-Peimbert, S. 1974, Ap. J., 193, 327.
Rubin, V. C., Kumar, C. K., and Ford, W. K. 1972, Ap. J., 177, 31.

Sandage, A., and Tammann, G. A. 1974, Ap. J., 190, 525.

Searle, L. 1971, Ap. J., 168, 327.

Searle, L., and Sargent, W. L. W. 1972a, Ap. J., 173, 25.

Searle, L., and Sargent, W. L. W. 1972b, Comments Astrophys. & Space Phys., 4, 59.

Sersic, J. L. 1960, Zs. f. Ap., 50, 168.

Sersic, J. L. 1964, Zs. f. Ap., 58, 259.

Shields, G. A. 1974, Ap. J., 193, 335.

Smith, H. E. 1975, Ap. J., in press.

Smith, M. G., and Weedman, D. W. 1971, Ap. J., 169, 271.

Smith, M. G., and Weedman, D. W. 1972, Ap. J., 172, 307.

Wagoner, R. V., Fowler, W. A., and Hoyle, F. 1967, Ap. J., 148, 3.

Warner, J. W. 1973, Ap. J., 186, 21.

Whitford, A. E. 1958, Astron. J., 63, 201.

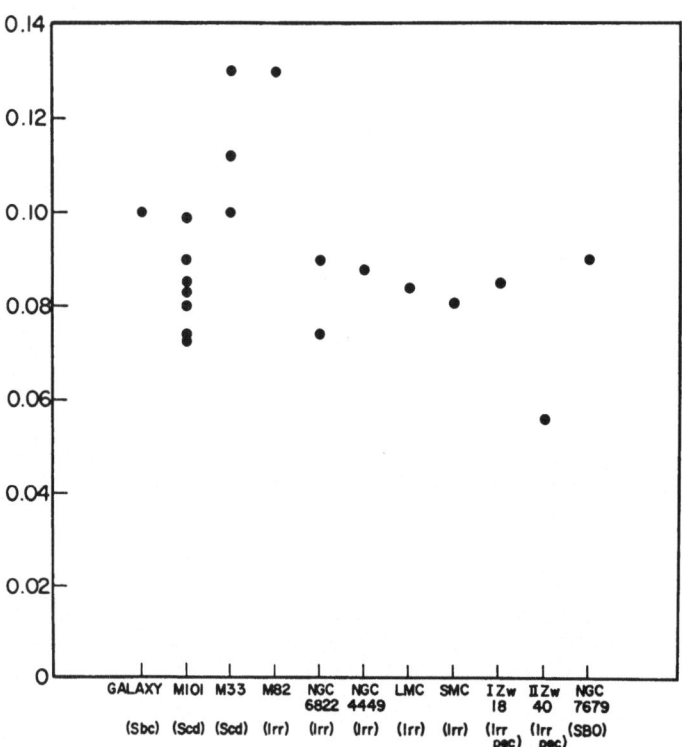

Figure 1. A plot of values of the He/H ratio taken from Table 1.

# IONIZED GAS IN NGC 4151

K. J. FRICKE

Astronomische Institute der Universität Bonn,
Bonn, FRG

## ABSTRACT

The spatial and radial velocity distributions of the ionized gas in the inner region ($<4$ kpc) of NGC 4151 are determined from the $[O\ III]\lambda 5007$ emission. The data imply a projected rotation axis in P.A. $120^o$ and radial motions predominantly in the NE and SW directions.

21 cm measurements of the velocity field in NGC 4151 obtained by
Davies(1973) suggest a projected rotation axis of the system in P.A.
120°. This is nearly perpendicular to the minor axis of the dominant
optical structure which had been identified earlier with the direction
of the rotation axis. Davies' results have recently been confirmed by
the Nancay group (Lequeux 1975, private communication). From the optical and the radio data for the outer region of this galaxy an inclination angle i < 34° has been inferred. Anderson(1974) has pointed
out that the velocity field within 400 pc of the nucleus as found by
Ulrich(1973) from the $[O\ III]\lambda\lambda$ 4959,5007 emission may be due to pure
rotation about the revised rotation axis rather than to radially outward motions in the NE - SW direction.

Fricke and Reinhardt(1975) obtained 11 untrailed spectra of NGC
4151 in 8 different position angles with the image tube spectrograph
at the Palomar 60-inch telescope. The dispersion was 144 A/mm and the
slit length corresponded to 124" or ≈ 12 kpc at the distance of the
of the galaxy. The extent of the $[O\ III]\lambda 5007$ line perpendicular to
the dispersion has been measured in all slit positions. In the NE - SW
directions the emission can be traced out to 4 kpc while in the SE - NW
directions only out to 1 kpc. The radial velocities are positive in
the NE sector and negative in the SW sector and of order 200 km/s. The
distribution of the ionized matter is much more elongated along the
NE - SW axis than could be explained by a rotating disk having inclination i < 34° and its rotation axis in P.A. 120°. Basically the same
result follows from Ulrich's spectra of the innermost region having
radius ≈ 400 pc. The radial velocity field can only be explained by
assuming a superposition of rotational and radial motions. The rotational contribution within 400 pc is ≈ 0.2 km/s/pc (corresponding to
a maximum central mass ≈ $2\ 10^9\ M_\bullet$) with rotation axis in P.A. 120°.
Radial outflow takes place in the NE - SW and E - W directions out to
at least 4 kpc distance from the center. These motions do not increase
linearly with distance from the center but are decelerated. The braking is either due to the gravitational field of the disk material or
due to ram pressure if the matter is partly injected into the plane of
the galaxy. The latter picture is also suggested by the irregular
structure of the spiral arms in the directions of the ejection.

The results indicate that NGC 4151 exhibits mass motions of a similar kind as observed in the inner regions of NGC 1068 and NGC 4258.

## REFERENCES

Anderson, K. S. 1974, Ap. J., 187, 445.

Davies, R. D. 1973, M.N.R.A.S. , 161, 25P.

Fricke, K. J. , and Reinhardt, M. 1975, Astr.Ap.(in press).

Ulrich, M.-H. 1973, Ap.J. 181, 51.

# ON THE IONIZATION CONDITIONS IN THE INTERARM GAS OF M 33

P. Benvenuti and S. D'Odorico

Osservatorio Astrofisico di Asiago
Università di Padova

## ABSTRACT

Deep spectroscopic observations of the galaxy Messier 33 in the spectral region $\lambda\lambda 4300$-$6700$ Å show the presence of the $\lambda 5007$ Å emission line of [O III] in the interarm region. We derive for the log $I(5007)/I(H_\beta)$ a preliminary value of .12, which indicates moderate ionization conditions in the interarm gas.

## INTRODUCTION

The spiral galaxy M 33 is the best case where it is possible, beside bright H II regions, to study the emission from the area between the arms, the so called interarm or disc emission. This is due to the large angular size, to the small inclination with respect to the line of sight, to the high relative surface brightness and to the morphological type of the galaxy. It is important to measure emission line strengths in the interarm region because they can give a clue to the ionization mechanism thus providing a better understanding of the phenomena in the interstellar medium. The interarm emission in M 33 was discovered, studied and discussed in several papers by the Marseille group (Carranza et al., 1968; Deharveng and Pellet, 1970; Monnet, 1971). More recently Benvenuti et al. (1973) measured the $H_\alpha/[N\ II]$ and $H_\alpha/[S\ II]$ intensity ratios at several interarm positions in the galaxy. Comte and Monnet (1974) presented an estimate of the $H_\beta/[O\ III]$ ratio based on Fabry-Perot observations.They concluded that the interarm gas is likely to be ionized by Lyman continuum photons escaping from the spiral arms, the other possibilities being ruled out mainly by the high value of the $H_\beta/[O\ III]$ ratio.

Within the program of extragalactic observations of H II regions carried out at Asiago in collaboration with M. Peimbert of the University of Mexico, we decided in October 1974 to extend our spectroscopic observations of the disc of M 33 to the blue spectral region. This had became possible after the mirrors of the 122 cm telescope were re-aluminized, with an increase by a factor of two of the speed of the system.

## OBSERVATIONS

The grating image-tube spectrograph of the Asiago Observatory, attached to the newtonian focus of the 122 cm reflector, was used for the observations. The dispersion is 125 Å mm$^{-1}$; the resolution 10 Å and the scale normal to the dispersion 127 arcsec mm$^{-1}$ on the plate. The spectral response of the image-tube photocathode peaks at $\lambda$4500 Å, the grating is blazed at $\lambda$ 7000 Å with a resultant, maximum sensitivity at about $\lambda$5800 Å .

We obtained nine spectra of M 33 in the spectral region $\lambda\lambda$4300-6700 Å. Exposure times up to 250 minutes were needed to record the fainter H II regions and the disc emission. The spectra show the following emission lines : H$_\alpha$, $\lambda$6584 Å of [N II] , $\lambda$4959 and 5007 Å of [O III] , H$_\beta$ . Fig.1 is an H$_\alpha$ photograph of the central region of M 33 kindly made available by G. Courtès: the observed H II regions and interarm zones are indicated. From our spectra it will be possible to estimate emission line relative intensities for 15 H II regions and 3 disc positions. Details of the method to derive line intensity ratios will be given in a forthcoming paper. Spectra and calibration plates are traced with an automatic microphotometer; the data come out in the form of punched paper tape and are reduced with an HP 2100 computer. Eventually, plots and intensity tables of the spectra are produced.

## PRELIMINARY RESULTS FOR THE DISC EMISSION

All spectra were obtained in November and December, 1974 and the reductions are still in progress. We report here a first, indicative measurement of the disc emission. The [O III] $\lambda$5007 line is visible in the interarm in all the deep

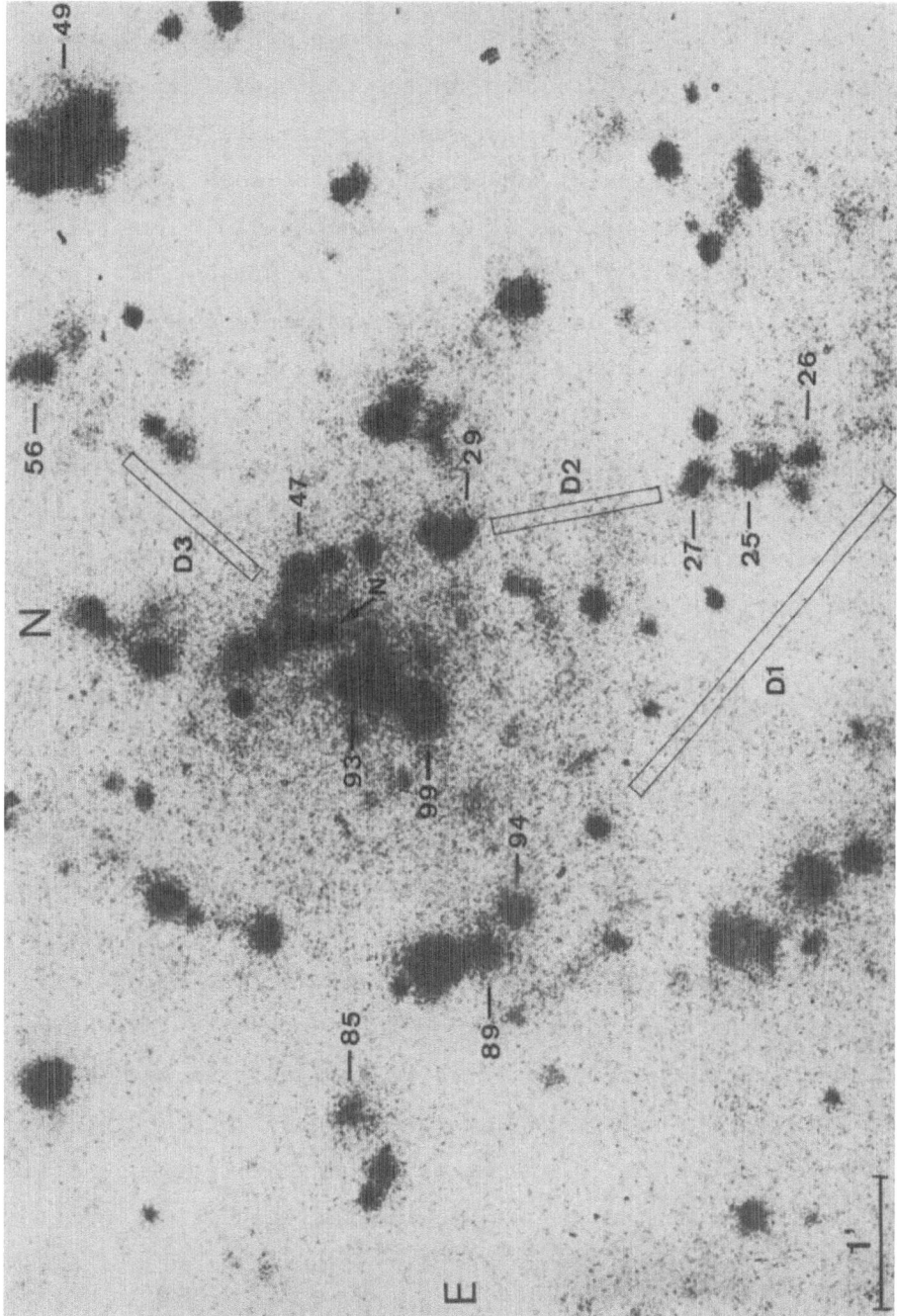

Fig. 1 : An H$_\alpha$ photograph of the central region of M 33
(courtesy of G. Courtès).Positions where interarm emission
at $\lambda$ 5007 Å has been detected are indicated by a D letter.
Numbers of H II regions after Courtès and Cruvellier (1965).

exposed interarm spectra. In some cases also $\lambda4959$ Å barely
appears at a visual inspection of the plates. The lines occur at the same wavelength than the corresponding emissions
from the H II regions on the same spectra. In spectra S2441
and S2450 ( position D1 and D3) $H_\beta$ is not seen in a visual
inspection of the plates,while in spectra S2466 and S2488
(position D1 and D2) a faint emission is present. So far
only spectrum S2441 has been fully reduced and an intensity
tracing is shown in Fig. 2. A feature is present within a
broad absorption at the $H_\beta$ ·wavelength. It is not clear whether it results from a structure within the absorption line
or from the presence of a superposed emission. In this latter hypothesis we can derive a value of log I $(5007)/I(H_\beta)$
=.12 ,once the correction for the spectral response is
applied. The uncertainty is ±.15 in the logarithm and it
will be improved when the other spectrum taken at the same
position will be reduced. A visual inspection of the other
plates shows that the above ratio could be lower in the
positions D2 and D3,but it is unlikely that it is negative.
In conclusion,the observed regions in the disc show a moderate excitation emission spectrum.
Comte and Monnet (1974) reported I( 5007)/I(H$_\beta$)< .25 in a
region of the disc north-east of the center. At this stage
of the reduction of the data it is difficult to say whether
this difference indicates an inconsistency between the two
observations or the existence of a gradient in the condition of the gas in the disc. Some indications will be provided
from the comparison of the results on the disc with those
of H II regions of different surface brightness. It is clear
however that the ionization mechanisms in the interarm will
have to be re-discussed taking into account these new data.

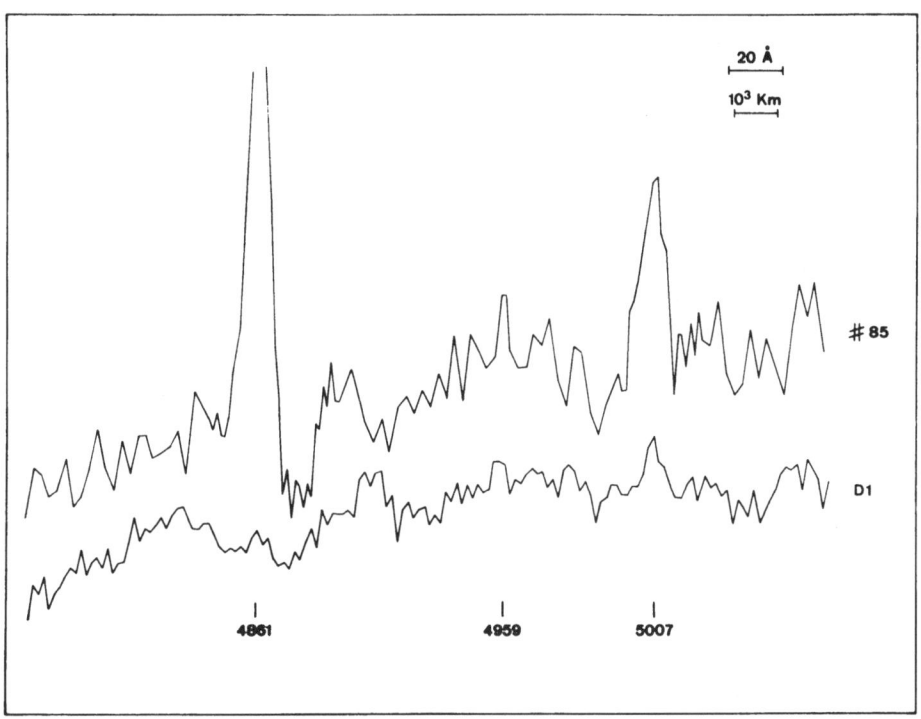

Fig. 2 An intensity tracing of the observed spectrum of the
interarm region D1 and of the low excitation H II
region #85. Emission at $H_\beta$, $\lambda 4959$ and $\lambda 5007$ Å of
[O III] are indicated.

From our observations we will also compute the $H_\alpha/H_\beta$,
$H_\alpha/$[N II] intensity ratios for the positions in the disc
and the H II regions. The $H_\alpha/H_\beta$ intensity ratios, which up
to now were known only for 8 H II regions studied by Searle
(1971) in the outer part of the galaxy, will be useful to
estimate the reddening in the innermost regions. Furthermore, with Searle's absolute measurements of emission line
fluxes for bright H II regions as a calibration, we will
estimate the absolute flux at $\lambda 5007$ Å from the disc areas.

## REFERENCES

Benvenuti, P., D'Odorico, S., and Peimbert, M. 1973, Astr.
     and Ap., 28, 447.
Carranza, G., Courtès, G., Georgelin, Y., Monnet, G., and
     Pourcelot, A. 1968, Ann. d'Astrophys., 31, 68.
Comte, G., Monnet, G. 1974, Astr. and Ap., 33, 161.
Courtès, G., Cruvellier, P. 1965, Ann. d'Astrophys., 28, 683.
Deharveng, J.M., Pellet, A. 1970, Astr. and Ap., 9, 181.
Monnet, G. 1971, Astr. and Ap., 12, 379.
Searle, L. 1971, Ap. J., 168, 327.

# OBSERVATIONS OF THE EXCITED HYDROGEN RADIO LINE H56α IN SEVERAL H II REGIONS

J.J. BERULIS, G.T. SMIRNOV, R.L. SOROCHENKO

Lebedev Physical Institute,
Moscow, USSR

## ABSTRACT

Measurements of the excited hydrogen radio line H56α have been carried out in Orion A, W3, W12, W51 and DR 21. The electron temperatures obtained are lower than those deduced from the non-LTE analysis of observations at longer wavelengths. For the Orion nebula, detailed consideration gives $T_e$ = 7000 - 7300 K.

Observations of the excited hydrogen radio lines give important information about physical conditions in H II regions. The line measurements give the possibility of obtaining information about electron temperature, density and velocity of the internal macroturbulent motions.
From a wide range of wavelengths where observations of the radio line
are carried out now, the high frequency measurements at $\nu > 15 - 20$ GHz
are of special importance.

Observations in this range have following advantages: i) the optical
depth even of the brightest H II regions is small enough that the increase of line intensity due to the maser effect is negligible; ii)
Stark broadening is absent; iii) observations at high frequency have
better angular resolution.

However, there are very limited hydrogen radio line measurements at
high frequencies. The radio line H56$\alpha$ (36.46 GHz) has been observed in
the Omega (Sorochenko et al. 1969) and Orion nebulae (Sorochenko and
Berulis 1969), the radio line H65$\alpha$ (23.40 GHz) in Orion (Churchwell et
al. 1969), the radio line H66$\alpha$ (22.36 GHz) in five H II regions including Omega and Orion (Waltman and Johnson 1973) and the H73$\alpha$ line
(16.56 GHz) in four sources (Papadopoulos et al. 1972). Recently, in
the Orion nebula, the presently shortest wavelength radio line, H42$\alpha$,
(85.69 GHz) was observed (Waltman et al. 1973).

In this paper the results of the new observations of the radio line
H56$\alpha$ in the Orion nebula and also of observations in the sources W3,
W12, W51 and DR 21 are presented. The program of observations was based
on the 8.2 mm survey of thermal radio sources (Berulis and Sorochenko
1973).

## EQUIPMENT AND METHOD OF OBSERVATIONS

The radiometer with an 8.2 mm maser and symmetrical beam switching
was used (Sorochenko et al. 1969). The separation between beams was 23'
and beam width 1.9. System noise temperature at the time of observations
was 250 - 300 $^{\circ}$K. The spectrometer had a 32-channel filter bank with a

filter bandwidth of 500 kHz The information was recorded and analyzed
on line by a TPA-I computer (Kuzenko et al. 1974). The observations were
carried out in 15 min cycles. The source has been tracked by one and
then the other beam for 5 and 10 min, respectively. In the second part
of the cycle, 4 min were spent on calibration by the noise signal of
5.5 $^{\circ}$K. One minute transition time was allowed. The intensity of each
channel in units of the antenna temperature for each cycle was calculated by the computer.

## REDUCTION OF OBSERVATIONS AND RESULTS

The continuum $T_c$ and line $T_L$ intensities were separated in a measured
spectrum. The continuum level was defined by 16 - 18 channels displaced
from the line. Use of the symmetrical method of observations (Gudnov
and Sorochenko 1967) and the low continuum intensity of the sources
under investigation have caused effective absence of the baseline effect:
the continuum was separated by a straight line which had no slope
relative to the frequency axis within the limits of rms error. After the
continuum level had been defined, the ratio $\frac{T_L}{T_c}$ for each channel was determined, and all spectrograms for each source were averaged. Two blended spectral lines: H56$\alpha$ (36466.32 GHz) and H80$\gamma$ (36467.97 GHz) were
within the frequency range of the spectrograms. The separation of these
lines and determination of their parameters was made by fitting of two
Gaussians of equal dispersion and with amplitude ratio 0.12. The profiles of the observed lines with the fit to H56$\alpha$ and H80$\gamma$ are shown in
Figure 1.

The results of the measurements are presented in Table 1. Table 1
lists the source name, galactic coordinates, integration time and continuum antenna temparature, corrected for atmospheric extinction from the
8.2 mm survey.

In last three columns are given the ratio of the antenna temperature
at maximum line to the continuum temperature $\frac{T_L}{T_c}$ , half power line width
$\Delta\nu$ and the radial velocity with respect to the L.S.R. $V_{LSR}$.

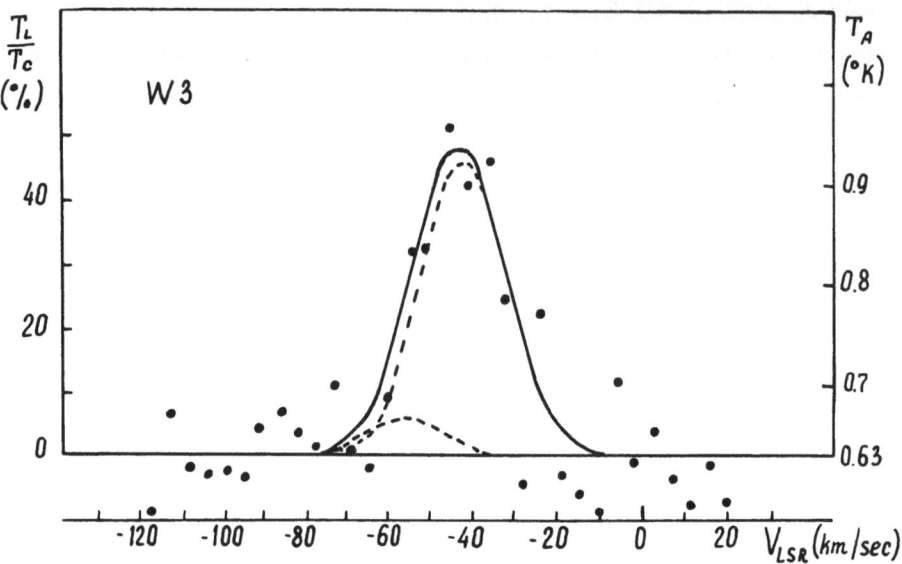

Fig. 1 Profiles of the radio line H56α. Right hand scale – antenna temperature corrected for atmospheric extinction.

a) W3

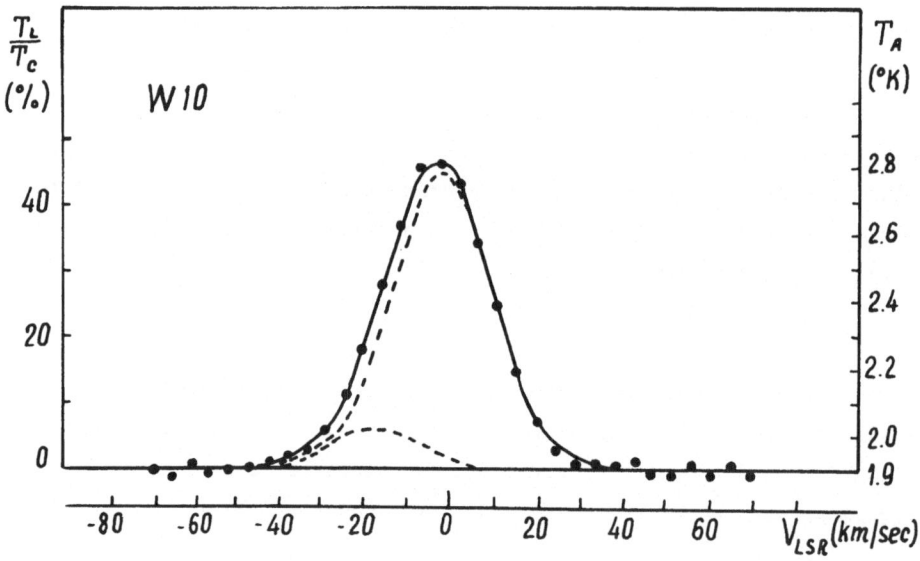

b) W10 (Orion A)

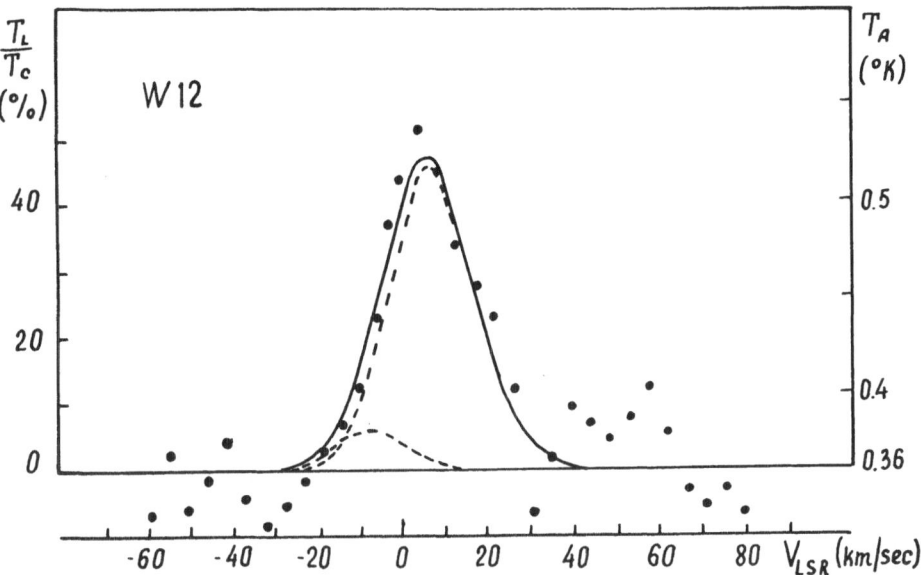

Fig. 1  c) W12 (NGC 2024)

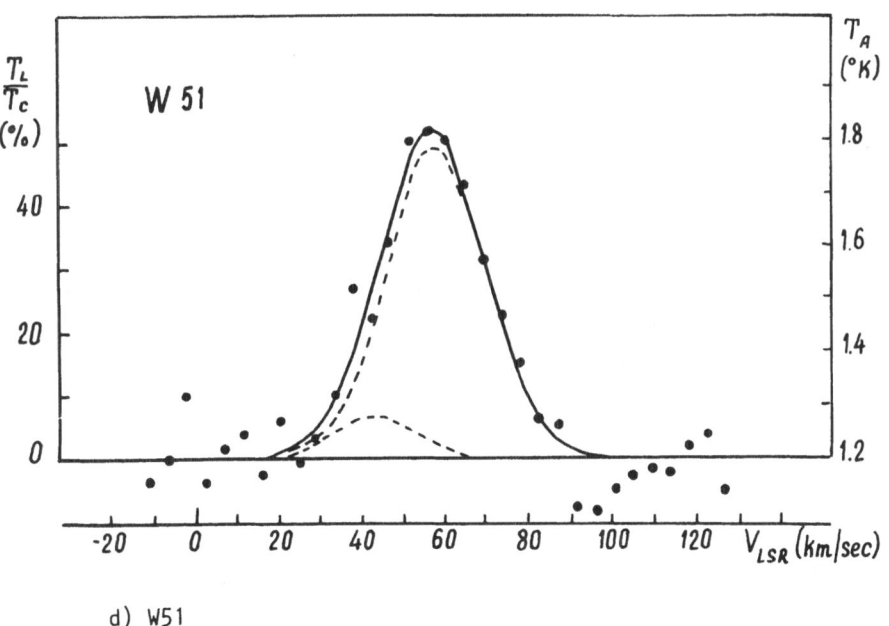

d) W51

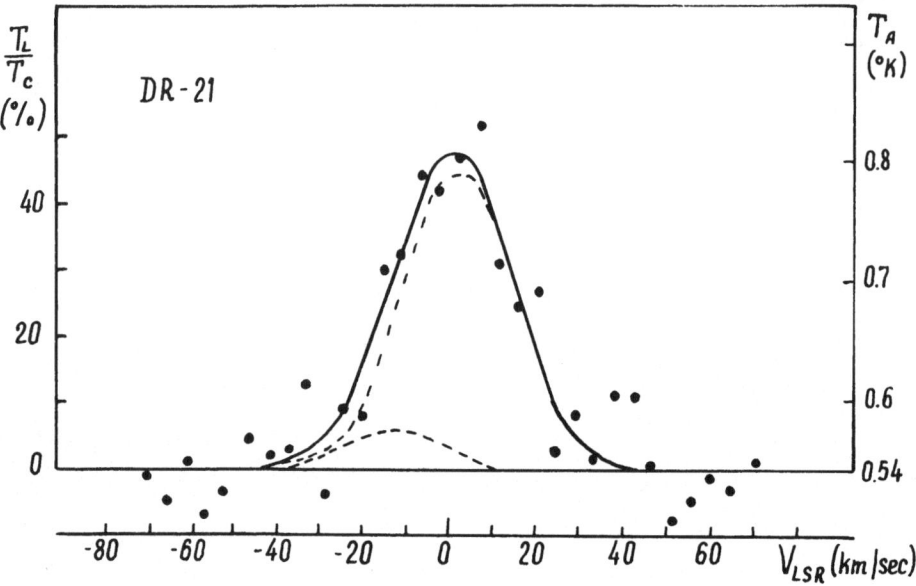

Fig. 1  e) DR 21

## Table 1

| Source | G | Time integration (minutes) | $T_c$ (°K) | $\frac{T_L}{T_c}$ (%) | $\Delta\nu$ (kHz) | $V_{LSR}$ (km/sec) |
|---|---|---|---|---|---|---|
| W 3 | 133.71 + 1.22 | 20 | 0.63 | 46±3 | $(2.98\pm0.2)\cdot10^3$ | -42.2±1.6 |
| W 10 | 209.01 - 19.38 | 40 | 1.9 | 45±2 | $(3.34\pm0.1)\cdot10^3$ | -2.6±0.8 |
| W 12 | 206.54 - 16.35 | 25 | 0.36 | 46±3 | $(2.93\pm0.25)\cdot10^3$ | 5.9±2 |
| W 51 | 49.49 - 0.39 | 30 | 1.2 | 49±3 | $(3.45\pm0.2)\cdot10^3$ | 57.2±1.6 |
| DR-21 | 81.67 + 0.53 | 25 | 0.54 | 45±5 | $(3.73\pm0.22)\cdot10^3$ | 1.4±1.8 |

Measured parameters of the radio lines give the possibility to derive the electron temperature in investigated H II regions. Since at the wavelength 8.2 mm, the optical depth $\tau \ll 1$, and the increase in line intensity due to the maser effect is negligible for the calculation of the electron temperature, one can use the expression:

$$T_e = \frac{b_n \cdot 1.27 \cdot 10^{-10} \nu^2}{\frac{T_L}{T_c} \cdot \Delta\nu \cdot \ln A \left(1 + \frac{N_{He}}{N_H}\right)} \tag{1}$$

This expression differs from the analogous one (Sorochenko and Berulis 1969) by including the relative abundance of helium and hydrogen $\frac{N_{He}}{N_H}$. In the expression (1), $\ln A = \ln (24.3\cdot10^{14}\ T_e^{3/2}\ \nu^{-2})$, the $b_n$ coefficient (less than 1) shows the deviation of the population of the n-th level from local thermodynamic equilibrium (LTE) and $\nu$ is the frequency of the line.

The results of the calculation of $T_e$ for LTE conditions and an assumed value $\frac{N_{He}}{N_H} = 0.08$ are presented in Table 2, column 3. In the second column of the Table are measured values of $\Delta\nu \frac{T_L}{T_c}$ with rms errors of measurement.

*Table 2*

| Source | $\frac{T_L}{T_c} \cdot \Delta\nu$ ($\kappa Hz$) | $T_e$ (°K) | | |
|--------|--------|--------|--------|--------|
| | | $H56\alpha_{LTE}$ | $H66\alpha_{LTE}$ | $non - LTE$ |
| W 3 | $(1.37 \pm 0.13) \cdot 10^3$ | $8260 \pm 600$ | $8100 \pm 1700$ | $11000 \pm 1100$ |
| W 10 | $(1.50 \pm 0.1) \cdot 10^3$ | $7670 \pm 500$ | $7800 \pm 800$ | $10000 \pm 1000$ |
| W 12 | $(1.34 \pm 0.14) \cdot 10^3$ | $8400 \pm 650$ | $7000 \pm 1300$ | |
| W 51 | $(1.69 \pm 0.14) \cdot 10^3$ | $6960 \pm 500$ | $7450 \pm 1000$ | $9000 \pm 900$ |
| DR-21 | $(1.68 \pm 0.21) \cdot 10^3$ | $7020 \pm 700$ | | |

In the last two columns of Table 2 are values of $T_e$ obtained from the H66$\alpha$ radio line measurements with the assumption of LTE (Waltman and Johnson 1973) and values of $T_e$ obtained on the basis of non-LTE analysis of the data of radio line measurements at longer wavelengths with use of higher order lines (Hjellming and Davies 1970). One can see from the Table that the values of $T_e$ obtained from the lines H56$\alpha$ and H66$\alpha$ agree well, but are lower than values obtained by the non LTE method.

## DISCUSSION

In order to find out the reason for the difference in the values of $T_e$, we have analyzed the results for the Orion nebula. For this nebula, there are available the largest number of measurements of the excited hydrogen radio lines, and several interpretations were proposed

(Hjellming and Davies 1970; Sorochenko and Berulis 1970; Pjatunina 1973; Brocklehurst and Seaton 1972; Guljaev and Sorochenko 1974).

Fig. 2 shows the values $\Delta\nu \dfrac{T_L}{T_c}$ obtained in the Orion nebula from different radio lines.

The data on $\alpha$-lines, collected by Hjellming and Gordon (1971) were used with the addition of recent and more exact measurements of H42$\alpha$ (Waltman et al. 1973), H66$\alpha$ (Waltman and Johnston 1973), H85$\alpha$ (Doherty et al. 1972), H104$\alpha$ (Gudnov et al. 1968), H166$\alpha$ (Pedlar and Davies 1972), H183$\alpha$ (Cato 1973), H192$\alpha$ and H220$\alpha$ (Pedlar and Davies 1972). To exclude the strong frequency dependence, the ordinate is $\dfrac{\Delta\nu T_L}{\nu^2 T_c}$.

The dashed line is a calculated curve obtained with the non-LTE method for the model of nebula with $T_e = 10^4 \, ^\circ K$, $\langle N_e \rangle = 1.7.10^4 \, cm^{-3}$ and emission measure $\langle E \rangle = 1.9.10^7 \, cm^{-6} \, pc$ (Hjellming and Davies 1970). As seen from the figure, this curve well approximates the results of measurements for $n \geq 73$, but differs from them for $n < 73$.

This difference, which was already noticed in the first observations of the radio line H56$\alpha$ in Orion (Sorochenko and Berulis 1969) became even larger with the present observations. The results for the line H56$\alpha$, which well correlate with observations of the lines H66$\alpha$, H65$\alpha$ and H42$\alpha$, made difficult the interpretation of the results with electron temperature $T_e = 10^4 \, ^\circ K$.

As follows from expression (1), the measurements of the high frequency radio lines allow determination of a direct value of $T_e$ with errors (besides errors of measurements $\dfrac{\Delta\nu T_L}{T_c}$ which depend only on the accuracy of $b_n$. From the $b_n$ calculation (Brocklehurst 1970) and possible limits of $N_e$ (upon which $b_n$ depends), the error of $T_e$ due to the $b_n$ factor is not more than 6 - 8 %.

At the same time, the interpretation of the low frequency radio lines, which necessarily takes into account the maser effect, allows different models depending on $N_e$ and E. With some assumptions about a model of the nebula, the data for $n > 85$ may be reconciled with $T_e =$

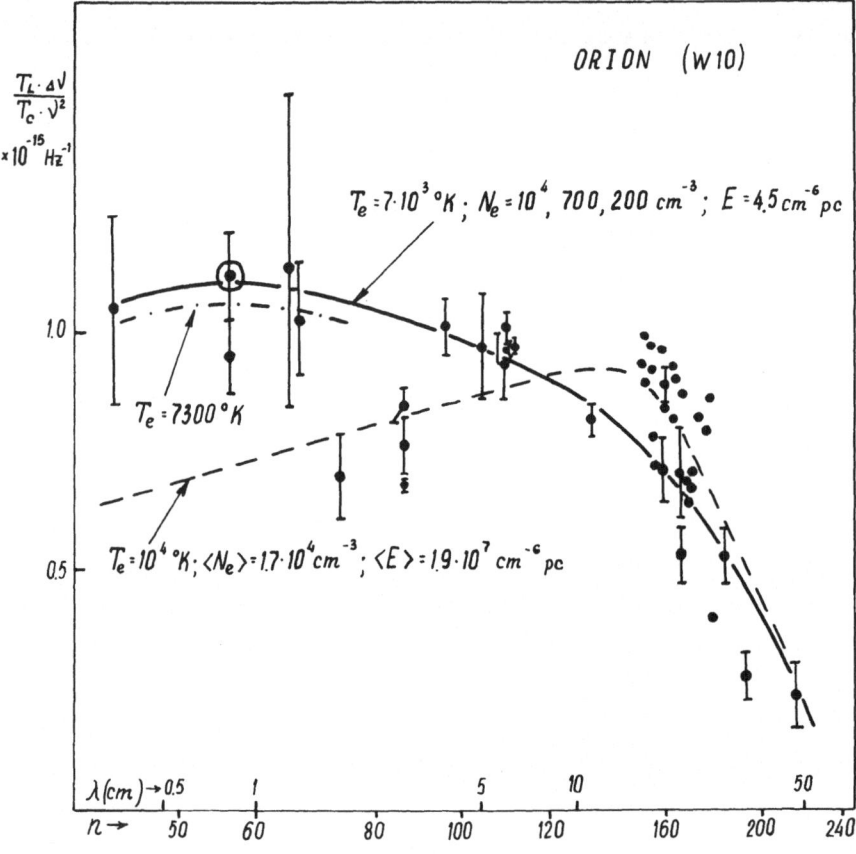

Fig. 2 Comparison for the Orion nebula of measured values

$$\frac{\Delta v \cdot T_L}{v^2 \cdot T_c}$$

with those predicted by the different theoretical models of the nebula. Double circle - present observations.

7000 °K. This is shown by the solid line on Fig. 2 which corresponds to $T_e$ = 7.10$^3$ °K and a model of the Orion nebula consisting of clouds with $N_e$ = 10$^4$ cm$^{-3}$ in the nucleus of nebula and a rarefied envelope (Pikel'ner and Sorochenko 1973).

In the low frequency part of the spectrum for $n \geq 85$, calculations made with a computer have taken account of the maser effect and also the difference of the beam averaging of the nebula corresponding to the difference of beamwidths of radio telescopes used at different wavelengths (Guljaev and Sorochenko 1974). It was assumed that the envelope has two layers with $N_e$ = 700 cm$^{-3}$ and $N_e$ = 200 cm$^{-3}$ and emission measure in the direction to the center of nebula E = 4.5.10$^6$ cm$^{-6}$ pc, very near to the measured value.

At $n < 85$, the calculated values of $\frac{\Delta\nu}{\nu^2} \cdot \frac{T_L}{T_c}$ were obtained directly from the expression 1. The value of $T_e$ = 7300 °K is in better agreement for these lines (broken line), but consideration of models with such $T_e$ has not been carried out. The calculations of $b_n$ by Brocklehurst (1970) have been used everywhere.

If one gives to the results on the lines H149$\alpha$ -H178$\alpha$ (solid circles on Fig. 2), (obtained in a single work) the statistical weight of one measurement, one can say that the solid curve agrees well with experimental results except with the data for H73$\alpha$ and H85$\alpha$.

Fig. 3 shows the comparison of values $\frac{\Delta\nu}{\nu}$ measured in the different lines. In the range from H42$\alpha$ to H85$\alpha$, the agreement is very good, indicating high accuracy of measurements of line width at these wavelengths. The increase of $\frac{\Delta\nu}{\nu}$ for $n > 85$ can be explained both by increase of beam width and the Stark broadening (Cato 1973). Therefore the difference of the results for the lines H73$\alpha$ and H85$\alpha$ on the one side and higher frequency lines on the other side is connected only with the difference in measurements of the ratio $\frac{T_L}{T_c}$. The reason for such a difference is not clear, but one can note that, at higher frequencies, the accuracy of measurements of the ratio must be better. It is connected both with the higher value of the ratio (45 % for H56$\alpha$ and 9 % for H85$\alpha$) and lower baseline distortion caused by interference of the signal which

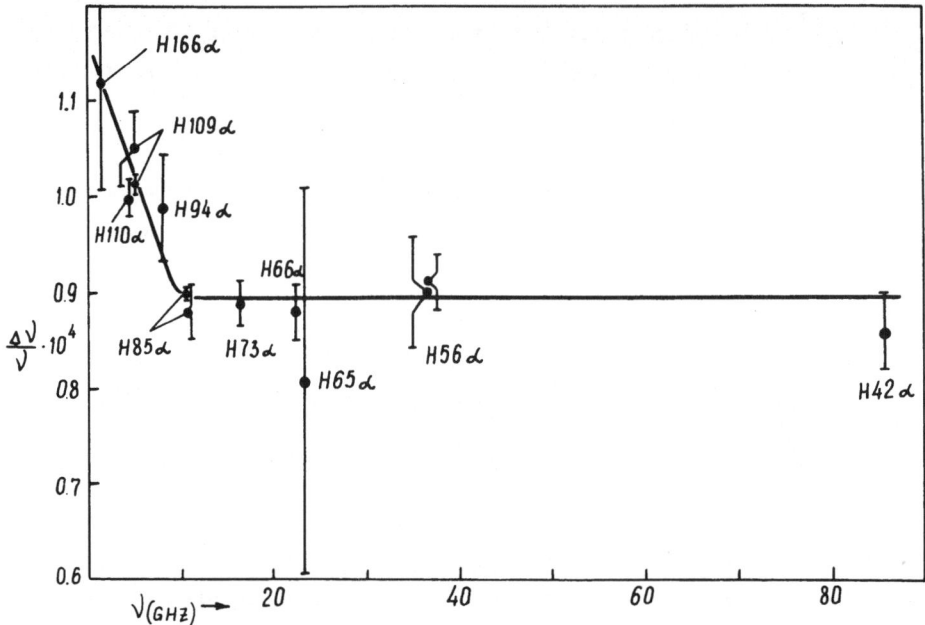

Fig. 3  Comparison of values $\frac{\Delta\nu}{\nu}$ measured in the different radio
lines.

is    proportional to $T_c$.

In the present interpretation, the results from higher order lines, as used in the non-LTE analysis which gives for Orion $T_e = 10^4$ °K (Hjellming and Davies 1970), have not been taken into account. However, the experimental results of decrease of energy of higher order lines may have another interpretation.

In the model of the Orion nebula with an inhomogeneous density distribution, the lines with n > 140 - 160 must have very extended wings (Brocklehurst and Seaton 1972). The increase of the energy in the wings is evident also from the measurements made by Cato (1973). One can assume that Stark broadening must be stronger for higher order lines than for $\alpha$-lines. When separating the line from the base level, es-pecially if subtracting a fourth-order polynomial curve, wings may not enter the profile providing the effect of a decrease of line energy which enhances as the order of the line increases.

Therefore the value of electron temperature in the Orion nebula 7000 - 7300 °K, obtained from the measurements of the H56$\alpha$ radio line and other high frequency lines seems to be highly plausible and in agreement with most measurements of hydrogen radio lines.

The same analysis for other H II regions where the radio line H56$\alpha$ was measured is now in process.

REFERENCES

Berulis, J.J. and Sorochenko, R.L. 1973, Astron. Zh. 50, 270.
Brocklehurst, M. 1970, M.N.R.A.S. 148, 417.
Brocklehurst, M. and Seaton, M.J. 1972, M.N.R.A.S. 157, 179.
Cato, B.T. 1973, Research Report 114, Onsala Space Observatory,
    Chalmers University of Technology.
Churchwell, E., Mezger, P.G., Reifenstein, E.C., Rubin, R.H., and
    Turner, B. 1969, Astrophys. Letters 10, 89.
Doherty, L.H., Higgs, L.A., and MacLeod, J.M. 1972, Astrophys. Letters
    12, 91.
Gudnov, V.M. and Sorochenko, R.L. 1967, Astron. Zh. 47, 1001.
Gudnov, V.M., Zotov, V.V., Nagorniih, L.M. and Sorochenko, R.L. 1968,
    Astron. Zh. 45, 942.

Guljaev, S.A. and Sorochenko, R.L. 1974, Astronom. Zh. 51.

Hjellming, R.M. and Gordon, M.A. 1971, Astrophys. J. 164, 47.

Hjellming, R.M. and Davies, R.D., Astron. and Astrophys. 5, 53.

Kutzenko, A.B., Polociantz, B.A., Smirnov, G.T., Sorochenko, R.L. and
     Teriokhin, S.A. 1974, preprint FIAN, 84.

Menon, T.K. and Payne, J., 1969, Astrophys. Letters 3, 25.

Papadopoulos, G.D., K.Y.Lo., Rosenkrantz, P. and Chaïsson, E.J. 1972,
     Astrophys. Letters 10, 89.

Pedlar, A. and Davies, R.D., 1972, M.N.R.A.S. 159, 129.

Pikel'ner, S.B. and Sorochenko, R.L. 1973, Astron. Zh. 50, 693.

Pjatunina, G.B. 1973, Astron. Zh. 50, 955.

Sorochenko. R.L., Puzanov, V.A., Solomonovich, A.E. and Shteinshleger,
     V.B. 1969, Astrophys. Letters 3, 7.

Sorochenko, R.L. and Berulis, J.J. 1969, Astrophys. Letters 4, 173.

Sorochenko, R.L. and Berulis, J.J. 1970, Astron. Zh. 47, 850.

Waltman, E.B. and Johnston, J.J. 1973, Astrophys. J. 182, 489.

Waltman, E.B., Schwartz, P.R., Johnston, K.J., and Wilson, W.J.
1973, Astrophys. J. 185, L135

# FAR INFRARED OBSERVATIONS OF H II REGIONS

D. A. HARPER

University of Chicago

## ABSTRACT

Far infrared observations of H II regions are reviewed. Special emphasis is placed on recent observations having high angular resolution.

In 1970, Low and Aumann (1970) reported the detection
of large far infrared fluxes from M42 and M17. A short
time later, Harper and Low (1971) showed that intense far
infrared emission was a common, perhaps universal, characteristic of "compact H II regions". In fact, most of the
power radiated from the sources they observed was emitted
at wavelengths between 40 and 750μ. Subsequent observations
can generally be categorized either as surveys (e.g. Hoffmann et al. 1971, Emerson et al. 1973, Furniss et al. 1974,
Olthoff 1974) or detailed studies of individual sources
(e.g. Harper 1974, Fazio et al. 1974). Since Richard
Jennings has discussed the former in some detail in a preceding talk, I will concern myself primarily with the
latter.

It seems likely that most, if not all, galactic far
infrared sources can be explained in terms of dust heated
by stars. In current far infrared data, the limited sensitivity of far infrared telescopes has introduced a strong
bias toward objects powered by one or more O-stars. In
fact, one of the primary uses of the available observations
is to give lower limits to the luminosities of OB stars or
clusters which are heavily obscured by local dust clouds.
In the near future, however, it should be possible to obtain data on many lower luminosity sources which may be
associated with recently formed stars of lower mass. There
is some evidence that the very luminous far infrared sources
in the nuclei of some galaxies (e.g. M82, NGC 253, NGC 1068)
may be related to star formation similar to, but occurring

on a much larger scale than, the phenomena associated with galactic H II regions. However, there are other indications — principally the presence of strong, non-thermal, radio sources — which suggest that these sources may differ qualitatively from a simple assembly of normal H II regions (Harper and Low 1973).

If we are, indeed, seeing thermal re-radiation from heated dust, there are a number of questions in addition to those related purely to energetics to which far infrared data may be relevant. We would like to know the composition of the dust, where it is located, and how the composition and grain size distribution vary in space and time. We would also like to understand how the presence of dust alters the transfer of radiant energy in and around H II and how it may affect such things as abundance determinations (e.g. the observed He/H ratio) and measurements of the properties of the existing stars (e.g. total luminosities and Zanstra temperatures). Some information may be derived from "total flux" measurements with large beams (e.g. Mezger et al. 1974), but it is becoming increasingly evident that a rather broad range of observations, at many wavelengths and sensitive to a wide range of angular scale sizes, are required for a complete understanding of what can be very complicated sources.

Radio-frequency observations have shown that compact H II regions are often associated with complex systems containing large amounts of mass in neutral clouds with molecular hydrogen densities of $10^3$-$10^8$ cm$^{-3}$. It is important to realize that the physical conditions in such systems may change appreciably over distances of $\sim$0.1 pc, and that observations of sources at different distances may represent different sorts of averages over a variety of physically distinct phenomena. Until the structure of at least a few far infrared sources can be observed with a spatial resolution better than $\sim$0.1 pc, the results of lower resolution observations of more distant objects must be treated with a great deal of caution. At the same time, it must be remembered that current far infrared observations are almost always performed by beam-switching to a reference position only a few beam diameters from the observed position. Therefore, valid comparisons between nearby and distant sources will often be impossible without good, low angular resolution observations of the closer objects.

This point is illustrated by Figure 1, where I have plotted the range of angular frequencies which have been sampled in far infrared measurements of several of the better-studied sources against their distances. Several points should be noted. First, even the most thoroughly observed objects have not yet been measured with angular resolution equal to the best which is currently available. Second, present measurements of nearby objects are not sensitive to large, low surface brightness source components which could make a significant contribution to the total fluxes from more distant objects. In particular, it would seem desirable to make a systematic search for galactic objects with scale sizes of the order of 100 pc, since this

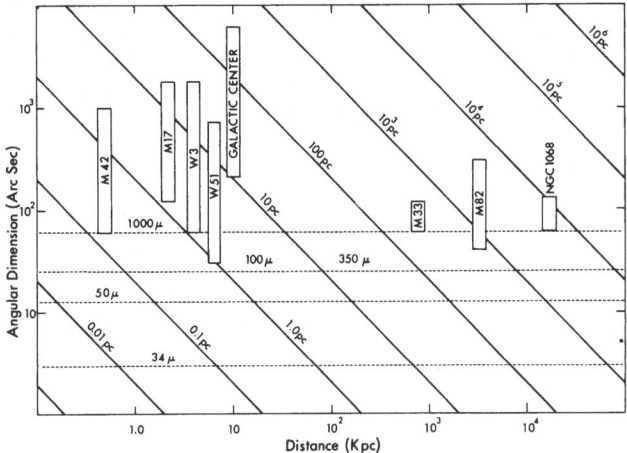

Figure 1:   Ranges of angular scale sizes sampled by current
            far infrared observations of H II regions and re-
            lated objects.

may represent a "natural" upper limit for the size of in-

dividual H II regions (this corresponds roughly to the

thicknesses of spiral arms or the galactic disk).   Such

measurements would be especially useful for comparison with

observations of giant H II regions in other galaxies.   For

example, they might help to explain the inability of Strom

et al. (1974) to detect extragalactic H II regions at 10μ.

Finally, the maximum angular resolution at which measure-

ments can be obtained over the full range from visible and

near infrared wavelengths through 1000μ is presently limited

by diffraction to ∿1'.   Coincidentally, the maximum resolu-

tion for present millimeter-wavelength molecular line

measurements is also ∿1'.   Thus it should at least be possi-

ble to make consistent comparisons between far infrared and

molecular observations.   At 34μ, measurements can be made

through a marginally transparent atmospheric window using

large, ground-based telescopes with beam sizes as small as

∿3" (Low et al. 1973). However, it has been shown (Harper et al. 1975, Werner et al. 1975) that the spatial distribution of radiation at wavelengths ≳100μ may differ significantly from that in the interval 20-50μ. Therefore care must be exercised when extrapolating the results of 34-μ observations to longer wavelengths.

Perhaps the most significant recent instrumental development in far infrared astronomy is the development of ∿1-m aperture telescopes for use in the 30-300-μ range where the far infrared sources commonly emit most of their energy. The two such instruments now in operation are the 91-cm NASA Airborne Infrared Observatory Telescope (Cameron et al. 1971) and the Harvard-Smithsonian Astrophysical Observatory 40-inch balloon-borne telescope (Fazio et al. 1974). For the first time, these telescopes permit 30-300μ observations with beam sizes similar to those used for ground based observations at shorter and longer wavelengths.

Figure 2 shows a set of preliminary multiband photometry which my associates and I have obtained with the NASA 91-cm telescope. Airborne 20-300-μ data have been combined with ground-based data at 10 and 20-μ to determine the characteristic spectral distribution of the sources. (Data on M8, M17, and M42 were taken with 2'-4' beam sizes using the NASA 30-cm telescope). One of the most striking features of the spectra is that a large percentage have very similar spectra, peaking sharply at a wavelength of 50-70μ. Those which do not fit this "standard" pattern generally fall into two categories — those with maxima at shorter or longer

wavelengths (DR 21 OH, OMC-2, and NGC 7027) and those with
extremely broad spectra.

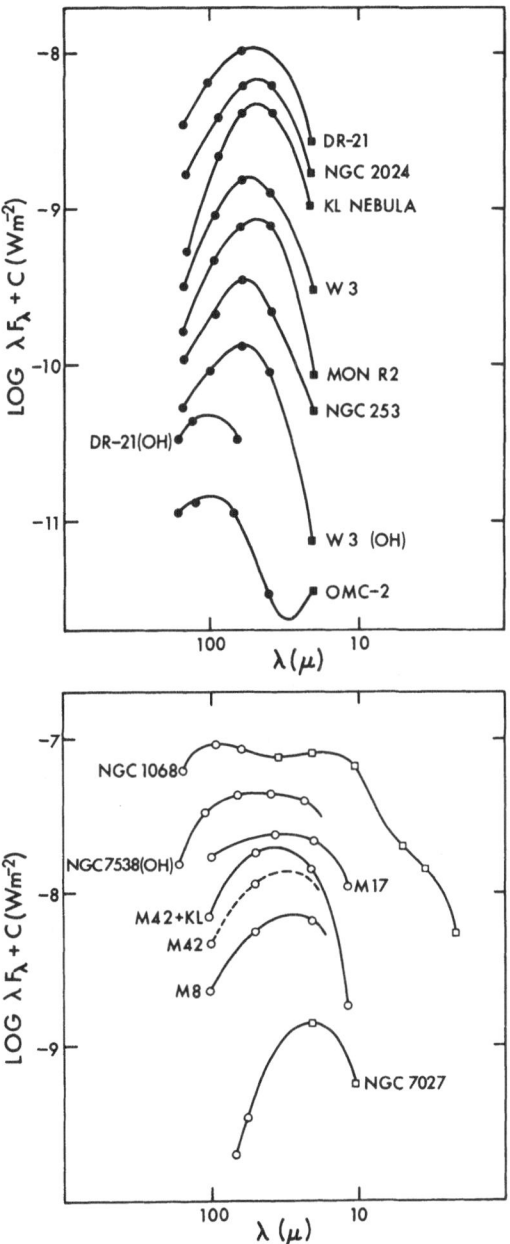

Figure 2: Far infrared spectral distributions for selected
sources.

The physical conditions in the sources represented in Figure 2 cover rather wide ranges.  For example, the gas densities vary from $10^3$-$10^8$ cm$^{-3}$, the sizes from $\sim$0.1-200 pc, and the observed ratio of infrared luminosity to Lyman $\alpha$ luminosity from $\sim$2.5 to >$10^2$.  The luminosity range for various types of objects are shown schematically in Figure 3. Of course it is possible that some of the larger and more powerful sources may be composites formed from an assembly of more or less independent objects of lower luminosity. However, even subclasses such at the "compact" ($\gtrsim$0.5 pc diameter) sources range in luminosity over at least three orders of magnitude.

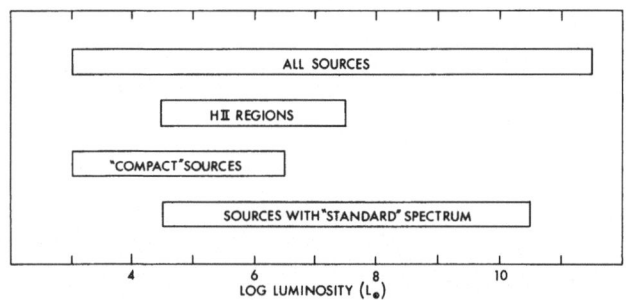

Figure 3:  Ranges of luminosities observed for different classes of far infrared objects.

Sources with "standard" spectra, as defined above, have their peak flux densities in roughly the same wavelength interval as 50-80°K black bodies.  However, the spectra are typically steeper toward long wavelengths and less steep at short wavelengths than Planck curves.  The shapes of the observed spectra depend on such factors as the characteristics of the dust, the geometry of the nebula, and the spectrum and intensity of the exciting radiation. In general, we would expect the average grain temperature in a source to vary roughly as

$$T^4 \propto \rho^2 L \; Qa/Qe$$

where $\rho$ is the dust density, L is the luminosity of the exciting star (or stars) and Qa and Qe are Planck-averaged grain emissivities appropriate to the spectrum of the absorbed and emitted radiation, respectively. Since both $\rho$ and L vary over many orders of magnitude among the various sources, the similarity of many of the spectra seems to require an explanation. It is possible, of course, that the effect is fortuitous. However, it is also possible that one or more types of thermostating mechanisms may be important. For example, most realistic grain models have rapidly falling emissivities at long wavelengths, which would keep their temperatures from dropping too low. Alternately, the grains could evaporate rapidly if heated above a given temperature, preventing them from becoming too hot. Local and "interstellar" extinction may also play a role in determining the spectrum of distant sources or those associated with extensive molecular cloud complexes. This effect could help to explain, for example, the very low color temperature (Harper 1974) of Sgr B2. It is important to remember, however, that any explanations of the spectra of far infrared objects must contain either explicitly or implicitly, assumptions about the spatial structure of the emitting region. This is especially true in light of recent evidence that many of the sources may be highly asymmetric (e.g. the model for the

Orion Nebula proposed by Zuckerman 1973).

Several explanations have been suggested for the very broad spectra of the type observed in M17. Lemke and Low (1972) attributed the effect to a mixture of "hot" and "cold" grains mixed with the ionized gas. The temperature difference between the two types of grains in their model results from differences in their emissivities at infrared and/or ultraviolet wavelengths. Wright (1973), however, has suggested that the hot and cold grains are separated in space with the hot dust inside and the cold dust outside of the ionized zone, respectively. His model requires a high degree of dust depletion within the H II region in order to produce high intensities for trapped Lyman $\alpha$ radiation and thus a sharp change in equilibrium grain temperature across the boundary of the H II region. New observational data (Harper et al. 1975) do not support the detailed predictions of either of the simple models outlined above. Instead, they suggest that the proper interpretation of the spectrum of M17 may involve the details of the geometry of the nebula as well as grain parameters which may vary in a complex manner with time and position. In particular, the structure of the ionization fronts separating the H II region from dense, adjacent molecular clouds may play a crucial role in determining the observed properties of the infrared source.

High angular resolution far infrared ($\sim$40-300$\mu$) data now exist for a number of sources. These include:

1) Maps of M42, NGC 2024, and M17 at a resolution of $\sim$2$\overset{.}{!}$2 (Harper 1974, Harper et al. 1975)

2) Maps of M42 at ∿1' resolution (Fazio et al. 1974, Werner et al. 1975)

3) A map of the W3 area at ∿1' resolution (Fazio et al. 1975)

4) A 1' resolution map of W58 (Harper 1975)

5) A 30" resolution map of W51 (Thronson et al. 1974)

6) Photometry of areas of W51 at ∿15" – 2' resolution (Harvey et al. 1975)

7) The 1' resolution photometry shown in Figure 2 (Harper 1975).

Much of this material is either incomplete or in a very preliminary form. However, several general comments may be useful at the present time. First, there appear to be significant differences between the spatial distributions of the radiation at 50, 100, and 350μ. For example, although the 45-150-μ brightness distribution of W5J (Thronson et al. 1974) is similar to the distribution of radiofrequency free-free emission, the 350μ emission (Rieke et al. 1973) may be in better overall agreement with the distribution of molecular emission. Maps of the core of the Orion Nebula at 50μ and 100μ having ∿1' angular resolution (Werner et al. 1975) display a similar tendency toward a better correlation between longer wavelength infrared emission and molecular line intensity. Second, in the 50 and 100-μ maps of Orion mentioned above, there are a number of features which can be associated with ionization fronts. Further studies of such regions can be expected to yield important information on the interactions between young main-sequence stars and dense molecular clouds. Finally, there are a number of objects (e.g. the Kleinmann-Low Nebula, DR21(OH), W3(OH)) for which there is evidence for large optical depths at far infrared wavelengths. Such objects are of great interest since they may be examples of very early stages of star formation. However, additional observations at the highest possible spectral and angular resolution are essential for further progress in understanding the nature of the exciting stars and the details of the radiation transfer process in very thick circumstellar clouds.

## REFERENCES

Cameron, R. M., Bader, M., and Mobley, R. E. 1971, Appl. Opt., 10, 2001.

Emerson, J. P., Jennings, R. E., and Moorwood, A. F. M. 1973, Ap. J., 184, 401.

Fazio, G. G., Kleinmann, D. E., Noyes, R. W., Wright, E. L., Zeilik, M. II, and Low, F. J. 1975, preprint.

Fazio, G. G., Kleinmann, D. E., Noyes, R. W., Wright, E. L., Zeilik, M. II, and Low, F. J. 1974, Ap. J., 192, L23.

Furniss, I., Jennings, R. E., and Moorwood, A. F. M. 1974, Proc. 8th ESLAB Symposium "H II Regions and the Galactic Center"

Harper, D. A. 1974, Ap. J., 192, 557.

Harper, D. A., and Low, F. J. 1973, Ap. J. 182, L89.

Harper, D. A., and Low, F. J. 1971, Ap. J., 165, L9.

Harper, D. A., Rieke, G. H., Thronson, H. A., and Low, F. J. 1975, Yerkes Observatory Preprint #151.

Harvey, P. M., Hoffmann, W. F., and Campbell, M. F. 1975, preprint.

Hoffmann, W. F., Frederick, C. L., and Emery, R. J. 1971, Ap. J., 170, L89.

Lemke, D. and Low, F. J. 1972, Ap. J., 177, L53.

Low, F. J., and Aumann, H. H. 1970, Ap. J., 162, L79.

Low, F. J., Rieke, G. H., and Armstrong, K. R. 1973, Ap. J., 183, L105.

Mezger, P. G., Smith, L. F., and Churchwell, E. 1974, Astr. and Ap., 32, 269.

Olthof, H. 1974, Astr. and Ap., 33, 471.

Rieke, G. H., Harper, D. A., Low, F. J., and Armstrong, K. R. 1973, Ap. J., 183, L67.

Strom, S. E., Strom, K. M., Grasdalen, G. L., and Capps, R. W. 1974, Ap. J., 193, L7.

Thronson, H. A., Harper, D. A., Loewenstein, R. F., and
    Telesco, C. M. 1974, B. A. A. S., 6, 443.
Werner, M., Gatley, I., Becklin, E., Cheung, L., Harper, A.,
    Loewenstein, R., Telesco, C., and Thronson, H. 1975,
    presented at the 145th meeting of the American Astro-
    nomical Society, Bloomington, Indiana, 23-26 March
    1975.
Wright, E. L. 1973, Ap. J., 185, 569.
Zuckerman, B. 1973, Ap. J., 183, 863.

# INTERSTELLAR ISOTOPIC RATIOS, STELLAR MASS LOSSES AND GALACTIC EVOLUTION

J. AUDOUZE[+][*] and J. LEQUEUX[+]

[+] Département de Radioastronomie, Observatoire de Meudon
Meudon, FRANCE

[*] also Laboratoire René Bernas
Orsay-Campus, FRANCE

## ABSTRACT

A study of presently available data shows that the $^{12}C/^{13}C$ ratio in the interstellar matter is about 40, rather than 89 as in the solar system, which has the isotopic composition of the interstellar matter $4.6 \times 10^9$ years ago. Mass loss by evolved stars and possibly novae is responsible for relative enrichment in $^{13}C$. Isotopes of CNO seem to be very promising tools for the study of galactic evolution

Recently, a number of CNO isotopic ratios in the interstellar gas have been determined mostly through observations of interstellar molecules (for references previous to 1974, see Bertojo et al, 1974). In particular, it has been suspected for some time that the interstellar $^{12}C/^{13}C$ may be smaller than the corresponding terrestrial ratio of 89. However optical depth effects have made difficult the interpretation of most observational results. An extensive study of the CO isotopic species has just been performed by Wannier (1974), in which optical depth effects have been carefully discussed : all the CO observations concerning 14 molecular clouds distributed all throughout the galactic disk are consistent with a $^{12}C$ $^{18}O/^{13}C$ $^{16}O$ ratio of 0.080 instead of the terrestrial 0.178. Comparison with observations of other molecules shows that this is due to a smaller $^{12}C/^{13}C$ ratio in the interstellar medium, rather than to a larger $^{16}O/^{18}O$ ratio. This is confirmed by a recent measurement of the $^{12}C/^{13}C$ ratio in $HC_3N$ by Gardner and Winnewisser (1975) which gives $^{12}C/^{13}C = 36 \pm 5$ in Sgr B2, and by a very recent determination of the ratio of unsaturated $^{12}C^{18}O, ^{13}C^{18}O$ lines in NGC 2024 giving $^{12}C/^{13}C \simeq 50$ (preliminary value ; Lucas, private communication). Surprisingly enough, there is no evidence for a variation in the interstellar $^{12}C/^{13}C$, ratio with distance of the cloud to the galactic center.

Data concerning other interstellar isotopes are not so extensive. The $^{16}O/^{18}O$ interstellar ratio seems to be roughly equal to that in the solar system, but $^{16}O/^{17}O$ may be smaller by a factor 2, and $^{14}N/^{15}N$ also smaller by a factor 2.

Isotopic separation in the interstellar matter seems to be unimportant (except for deuterium/hydrogen) ; the best evidence is that the same ratios are obtained from different molecules .

A simple interpretation for the differences between
the interstellar matter and the solar system isotopic ratios
is that the interstellar medium is continuously enriched in
rare isotopes with respect to the main ones : the isotopic
composition of the solar system reflects that of the interstellar matter at the time of its formation, $4.6 \times 10^9$ years
ago. $^{13}C$, $^{17}O$, $^{18}O$ and $^{15}N$ as well as $^{14}N$ are products of
secondary nucleosynthesis, being manufactured from $^{12}C$ and
$^{16}O$, themselves produced in an earlier-generation large mass
stars. Partial mixing between the hydrogen and the helium
zones inside red giant stars can produce $^{13}C$, $^{17}O$ and also
$^{14}N$, while novae can also generate these isotopes as well
as $^{15}N$. The corresponding enrichments are directly observed
at the surfaces of red giants (see e.g. Lambert and Tomkin,
1974) where $^{12}C/^{13}C \simeq 6$, and also observed to some extent
in novae (for which most information is rather of theoretical
nature : Starrfied et al, 1974) By using an estimate of the
present rate of mass ejection in novae explosions, we suggest
that they can produce all the $^{15}N$ and a part of the $^{17}O$.
We have also estimated that the present rate of mass loss
by evolved stars must be about 0.4 to 0.8 M$\odot$/year in the
Galaxy to account for the change of the $^{12}C/^{13}C$ ratio. This
mass loss may take place during the steady red giant phase,
the Mira phase or the planetary nebula phase ; it should be
stressed that the rate of mass loss by Miras as estimated
by Gehrz and Woolf (1971) is sufficient to account for the
increase in $^{13}C$. The change in the $^{16}O/^{17}O$ ratio can also be
accounted for by these mass losses ; the situation is not
settled for $^{18}O$.

It is obvious that the study of the interstellar
isotopic ratios offers a new powerful method for studying
mass loss by different types of stars and galactic evolution
in general. We are presently investigating these problems,

and show in particular that the uniformity of the $^{12}C/^{13}C$ ratio throughout the galaxy is not inconsistent with the $^{14}N/^{16}O$ gradient with galactocentric distance observed in spiral galaxies (Audouze et al, 1975).

Finally, we remark that since the comets seem to have a terrestrial $^{12}C/^{13}C$ ratio similar to that of the solar system (Danks et al, 1974), they probably do not originate from outside the solar system.

## REFERENCES

Audouze, J., Lequeux, J., Vigroux, L., 1975, submitted to Astron. Astrophys.

Bertojo, M., Chui, M.F., Townes, C.H., 1974, Science, 184, 619.

Danks, A.C., Lambert, D.L., Arpigny, C., 1974, Astrophys.J, 194, 745

Gardner, F.F., Winnewisser, G. 1975, this Symposium and Astrophys. J. Letters, in press.

Gehrz, R.D., Woolf, N.J., 1971, Astrophys. J., 165, 285

Starrfield, S., Sparks, W.M., Truran, J.W., 1974, Astrophys. J. Suppl., 28, 247

Wannier, P.G., 1974. Ph D thesis, Princeton University.

# "REVIEW" OF ORION A AND ORION B

B. ZUCKERMAN

University of Maryland
College Park, Maryland, U.S.A.

## ABSTRACT

Some current problems in our understanding of the Orion A and Orion B regions are discussed. About 3 times more time is devoted to Orion A than to Orion B.

## ORION A

The Orion A region is remarkable because it contains the highest surface brightness HII region visible from the Earth and, only 1' away, a cluster of extremely bright infrared sources. Ever since the discovery of the IR sources, astronomers have wondered if this near coincidence of position is simply a coincidence or if there is a physical relationship between the two regions. Some of the contributed papers given in this session suggest that we may be able, at last, to make a tentative choice between these two possibilities. I return to this problem at the end of my talk. This talk will be structured according to 11 questions that we may hope to answer that are related to the Orion Nebula (Orion A) and nearby regions. These questions, listed in Table 1, are mainly concerned with the ages (evolutionary state) of and distances to astronomical objects. This, I think, is a reasonable way to proceed since the purpose of much of the time and effort expended by astronomers since the development of the telescope has been to answer the twin questions: how old and how far away?

---

### TABLE 1
Some Questions Concerning Orion A (The Orion Nebula)

1) What is the role of dust in determining the optical appearance of the Orion Nebula?

2) What is the meaning of the optical morphology?

3) What is the age of the Orion Nebula?

4) Where is the Trapezium relative to the molecular cloud?

5) What is the origin of the infrared radiation?

6) What is the evolutionary state of the BN-KL infrared complex?

7) Where is BN relative to the Orion Nebula?

8) Where is KL relative to the Orion Nebula?

9) Where is the region of origin of the molecular ridge ("spike") emission relative to the Orion Nebula?

10) Where is the region of origin of the molecular "plateau" emission relative to the Orion Nebula?

11) What about magnetic fields?

---

## Question 1

We first consider the optical HII region, the Orion Nebula. I think

that God must have wanted us to understand HII regions or He wouldn't
have allowed us to view the exciting stars of a young, high density,high
emission measure, dusty HII region.  Thus we must energetically strive
to understand this object.  A problem that has received much attention
is: what is the role of dust in determining the optical appearance of the
Orion Nebula?  Although, as we mention below, it now seems generally accepted that the Nebula is embedded in a dusty neutral region the role
played by dust embedded within the Nebula itself is still not clear.
Schiffer and Mathis (1974) have made an ambitious attempt to determine
the distribution of (scattering) dust in the ionized and nearby neutral
gas.  They attempt to fit the distribution of scattered light to a two
component dust model.  One component is a shell of dust and ionized gas
that surrounds the Trapezium and has an optical depth at $H_\beta$ of 1.5 - 2.0.
The inner radius of this shell is perhaps 0.2 - 0.3 pc and the ionized
region inside of this radius contains little dust.  Thus the Trapezium
is surrounded by a "hole" that contains only a little dust that scatters
visible light effectively.  That is, if neutral condensations exist in
the ionized gas near the Trapezium they do not scatter light well.  Observations of [SII] emission by Dr. T. Gull at KPNO also imply a limited
number of neutral condensations distributed throughout the core of the
Nebula.  Besides the ionized shell, the calculations of Schiffer and
Mathis suggest that a substantial portion of the scattered light may
originate in the dense neutral cloud that lies behind the HII region. Dr.
Gull's interpretation of certain polarization data suggests that the neutral slab may be responsible for some of the scattered radiation observed
in the outer ($\gtrsim 10'$ from the Trapezium) regions of the Nebula.  Further
polarization studies may help to determine better the relative fractions
of the light scattered in the neutral and in the ionized gas.

Another interesting problem related to the dust is the location,rel-
ative to the ionized gas, of the dark bays to the East and North-East
of the Trapezium.  Some workers place the bulk of the dust in front of
the emitting layers whereas others believe that the dust and gas are well
mixed but with a higher dust/gas ratio than in the brighter regions of
the nebula (see, e.g., Perinotto and Patriarchi 1974).

Question 2
We know that a large molecular cloud exists behind the Orion Nebula.
Photographs of the Nebula suggest that it is embedded in the molecular

cloud which seems to envelop the ionized gas on all sides. But does it? Arguments based on kinematics and carbon recombination line maps strongly imply that the answer is "yes", the Nebula is ionization bounded on its rear side, and perhaps some other sides as well, by the molecular cloud (see Zuckerman and Palmer (1974) for a summary of the arguments). Therefore, pending further observational evidence, we accept this model. Then flows off of the dark cloud can replenish the ionized gas in times less than the dynamical expansion time ($\sim 10^{4.5}$ years) of the Nebula.

Important flows suggest prominent ionization fronts. The most obvious front is a "bar" that runs between $\theta^1$ and $\theta^2$ from the northeast to the southwest. Dr. T. Gull has noted that this bar is nearly as bright in lines of [OI] as the brightest part of the Nebula. Such ionization fronts are most prominent in relatively low excitation species such as [SII], [OII], [NII], [OI] and OI studied by Gull, Meaburn, and other observers. Gull finds that [SII] lines display seeing limited (1" - 2") structure. Thus, detailed measurements of the spatial and velocity structure near ionization fronts in low ([SII]) and high ([OIII]) excitation lines should lead to a clearer observational understanding of the interaction between the neutral and ionized gas. Information on the velocity structure near the fronts might also be obtainable via observations of high frequency radio recombination lines with high angular resolution ($\lesssim 30"$). Deeper theoretical studies of these fronts than have hitherto been carried out are also desirable. The bar has recently been observed in the far IR by a joint Univ. of Chicago - Cal. Tech. team, as will be reported in a contributed paper. Ionization fronts in Orion are also indicated in a new aperture synthesis map made at Cambridge and Owens Valley and these important results, which are discussed more fully in another contributed paper, probably enable a definitive placement of the Kleinmann-Low infrared cluster along the line of sight (discussed below - Question 8).

Another interesting problem worthy of further optical study is the high velocity features long reported, in a rather hazy and contradictory fashion, to exist in Orion and other HII regions. A recent paper by Traub et al. (1974) presents convincing evidence for highly blue shifted ($v = -60$ to $-100$ Km/s) H$\alpha$ emission between the Trapezium and the bar. The origin of this high velocity gas, which comprises $\lesssim 0.5\%$ of

the mass of the nebula, is still uncertain. Two conceivable explanations might be (1) free expansion from the dense molecular cloud into a
"vacuum" although it might be difficult to accelerate so much mass to
such high velocities, or (2) gas and dust accelerated by radiation pressure or stellar winds from the Trapezium (suggested to me by Dr.
Appenzeller). Gas velocities of 50 - 100 Km/s have been observed near
young stars (Strom et al. 1974) and the Kleinmann-Low nebula (Kuiper et
al. 1975). Complete maps in Hα should be sufficient to decide between
these two possibilities (both of which might be incorrect). Mapping
the high velocity gas by means of radio recombination lines is probably
not possible (owing to the weakness of the features).

## Question 3

What is the age of the Orion Nebula? I think that no one knows
($10^5$ years is a canonical guess). Conti and Leep (1974) have noted that
of a list of 130 O stars, θ'C is the only one that displays an inverse
P Cygni profile that might be interpreted as due to infalling material
suggesting extreme youth. However, such profiles might have nothing to
do with infalling material (Strom et al. 1975).

## Question 4

The 4th question concerns the distance "d" between the Trapezium
and the front edge of the background molecular cloud. There are at
least four methods for estimating d. The first uses an excitation parameter (100 $cm^{-2}$ pc) for the Trapezium and an assumed density (3 x $10^4$
$cm^{-3}$) for the Nebula and one obtains d ∿ 0.1 pc. (The assumed density is
a rough geometric mean between the rms density in the central portions
of the Nebula and the density of the background molecular cloud.). From
the scattering of light by the dust near the Trapezium one derives d ≲
0.26 pc (Schiffer and Mathis 1974). From the carbon line map of Balick
et al. (1974) one estimates d ∿ 0.4 pc and from the extent of IRe 3
(Harper 1974), d ∿ 0.3 pc. Thus d = 0.3 pc is probably not too far off.
If so, the average density between the Trapezium and the background
neutral cloud is ∿ $10^4$ $cm^{-3}$. Since this is probably a full factor of
10 or more below the density of the neutral cloud a model is suggested
whereby the dense emitting layers of the Nebula (of thickness ≲ 0.1 pc)
lie behind the Trapezium and near the neutral cloud. Such a model is
consistent with the picture advocated by Wurm and Perinotto who, based

on measurements of HeI 3889 Å, place the stars closer to the observer than the main emitting masses of the HII region.

## Question 5

So far we have been discussing the ionized gas in Orion. What else is there? We see prominent infrared and radio molecular sources. The strongest infrared source is the Kleinmann–Low infrared cluster (KL) which contains the Becklin – Neugebauer "point" source (BN). There is also an extended far IR source that appears to be due in part to radiation from the molecular cloud and in part to radiation associated with the Trapezium stars (e.g., Harvey et al. (1974); Harper 1974). Associated with these infrared sources are a variety of molecular structures alluded to in questions 9 and 10 that I will not have time to discuss (see Zuckerman and Palmer 1974 and 1975 for further details).

## Questions 6, 7 and 8

What is the evolutionary state of the BN–KL complex? Most astronomers believe that it is probably a cluster of "protostars" although a few prefer a model where we are seeing a cluster of highly reddened evolved stars (BN is envisaged as an F supergiant with a 20μ excess in this model). Although there are already problems with the latter model that make it doubtful, a definitive choice between the two could be made if we could place KL and BN along the line of sight. In the protostar model the IR cluster is embedded within the molecular cloud probably somewhere near the front edge, whereas in the most recent version of the evolved supergiant model (Allen and Penston 1974) the cluster is located behind the molecular cloud (whose size is ∿ 10 pc). Placement of the cluster along the line of sight is also important in choosing between models of the massive molecular cloud behind the Orion Nebula. If a systematic radial motions (e.g. collapse) model (Liszt et al. 1974) is appropriate then KL is probably near the center of the extended molecular cloud (i.e. many pc behind the HII region). On the other hand, if KL is near the HII region (Zuckerman 1973) then "turbulent" models (e.g. Zuckerman and Evans 1974) for the molecular cloud appear tenable. Thus, for a number of reasons, it is important to know the location of KL relative to the Orion Nebula along the line of sight.

The first good evidence for a real physical connection between the HII region and the infrared cluster is suggested by the contributed papers in this session. The Cambridge-Cal. Tech. aperture synthesis map of S. Gull and A. Martin shows a probable ionization front cutting through KL from the northeast to the southwest. Since this synthesis map (and optical maps) show many ionization fronts in the Nebula this positional coincidence is not, by itself, especially significant. However, a map made with the 100-meter Bonn antenna by R. Hills, and discussed in a contributed paper, shows that the $H_2O$ maser sources near KL lie in a line that is parallel to the ionization front cutting through KL. Observations of other compact HII regions reported at this meeting suggest that $H_2O$ maser sources are often (perhaps almost always?) located in neutral gas just outside of an ionization front. Thus, it is not unreasonable to suggest that the $H_2O$ maser sources in Orion lie just outside of the ionization front in the Orion Nebula that cuts through the position of KL. In the general vicinity of Orion A both KL and the $H_2O$ masers studied by Hills are unique sources. Because of their positional coincidence there can be little doubt that they are physically associated. Thus we conclude that the ionization front that cuts through KL is also located near KL along the line of sight.

If we accept that KL and the Trapezium are located within a pc of each other and remember that OMC-2 (Gatley et al. 1974) is also within a pc or so of the Trapezium, a scenario for fragmentation of massive molecular clouds is suggested. A large molecular cloud (containing $\sim 10^5 \, M_\odot$) fragments into clusters of $\sim 10^3 \, M_\odot$ blobs of gas and dust with 1 pc as a typical separation between the individual blobs (this separation seems to hold for other regions, e.g. M 17 and NGC 6334, besides Orion). Each blob is envisaged as an incipient (KL) or fully formed (Trapezium) star cluster. This clustering of star clusters might be related to the Jeans length in the cloud (e.g. Phillips et al. 1974) or to pressure from the HII region which helps to induce star formation in neighboring regions. Thus, at the present time, active star formation seems to be confined to a relatively small fraction ($\sim 1 \, pc^3$) of the total volume ($\sim 1000 \, pc^3$) of the Orion molecular cloud. By the time these clusters of star clusters have evolved sufficiently to be called a classical open cluster it is likely that their internal velocity dispersions will have smeared them out into a single entity.

## ORION B

The structure of Orion B (NGC 2024) is not as well understood as
that of Orion A probably because the central parts of Orion B are totally obscured by a "bar" of dark material that is elongated in the north-
south direction (e.g. Figure 2 in Grasdalen 1974). Thus the exciting
star(s) of Orion B is (are) still not identified with certainty. Because there is a substantial amount of neutral gas both in the bar and
in a background molecular cloud, the location of the gas responsible for
mm- wavelength molecular emission and narrow cm-wavelength recombination
lines from carbon and hydrogen is still uncertain. Furthermore, there
is still some observational uncertainty regarding the location of the
peak of radio continuum emission as a function of wavelength. The most
ambitious attempt to model Orion B is due to Grasdalen (1974).

Grasdalen discovered, behind the bar, a $2\mu$ point source that he
called NGC 2024 #2. Because a visible star, NGC 2024 #1, located near
the edge of the bar is believed to be able to account for neither the
radio nor the infrared fluxes, Grasdalen postulated that 2024 #2 is the
exciting star for the nebula. By measuring the relative colors of 2024
#1 and #2 he deduced that #2 suffers 32 magnitudes of visual extinction.
Consideration of the ultraviolet flux necessary to ionize the nebula
then implies that 2024 #2 is a star of luminosity $1.6 \times 10^6$ $L_\odot$ with a
surface temperature of 23,000 K (the surface temperature must be fairly
low since there is very little ionized helium in Orion B). A star of
this luminosity and surface temperature must have already left the main
sequence and its age would be $\sim 5 \times 10^5$ years. From the optical size
($\sim 30'$) and an expansion velocity of 10 Km/s one finds an age for the
nebula of $4 \times 10^5$ years in agreement with the deduced age of the star.
The optical nebula is the result of gas flows escaping to the north of
2024 #2. To the south, where there is an extended $8\mu$ continuum source
and a compact HII region, the nebula is ionization bounded.

Just as Napoleon met Wellington at Waterloo so Grasdalen's model
meets Harper (1974) in the far infrared. That is, a problem arises in
that the luminosity of 2024 #2 given above is $\sim 40$ times the total infrared luminosity of the NGC 2024 region. Where is all the ultraviolet
energy disappearing to? One possibility that might narrow the discrepancy is to accept, as suggested by Grasdalen, that NGC 2024 is pri-

marily density bounded. (The low value of the ratio $L_{IR}/L_\alpha$ for NGC 2024 independently suggests a density bounded nebula [Harper 1974].) Still it seems unlikely that the entire discrepancy can be due to this effect. Perhaps the model can be patched up by some minor adjustment such as a lower extinction for 2024 #2. On the other hand, a more substantial modification may be required, possibly of the sort suggested in the model outlined by Dr. Higgs in his contributed paper. Another slightly different problem arises in the relative placement of the compact HII region south of 2024 #2 and the extended $8\mu$ source. The degree to which these two sources do or do not coincide is dependent on the still somewhat uncertain position of the radio peak.

## REFERENCES

Allen, D. A. and Penston, M. V. 1974, Nature, 251, 110.
Balick, B., Gammon, R. H. and Doherty, L. H. 1974, Ap. J., 188, 45.
Conti, P. S. and Leep, E. M. 1974, Ap. J., 193, 113.
Gatley, I., Becklin, E. E., Matthews, K., Neugebauer, G., Penston, M. V. and Scoville, N. 1974, Ap. J., 191, L 121.
Grasdalen, G. L. 1974, Ap. J., 193, 373.
Harper, D. A. 1974, Ap. J., 192, 557.
Harvey, P. M., Gatley, I., Werner, M. W., Elias, J. H., Evans, N. J., Zuckerman, B., Morris, G., Sato, T. and Litvak, M. M. 1974, Ap. J., 189, L 87.
Kuiper, T. B. H., Zuckerman, B., Kuiper E. N. R. and Kakar, R. K. 1975, in preparation.
Liszt, H. S., Wilson, R. W., Penzias, A. A., Jefferts, K. B., Wannier, P. G. and Solomon, P. M. 1974, Ap. J., 190, 557.
Perinotto, M. and Patriarchi, P. 1974, paper presented at Symposium "HII Regions and the Galactic Center", Frascati, Italy.
Phillips, T. G., Jefferts, K. B., Wannier, P. G., and Ade, P. A. R. 1974, Ap. J., 191, L 31.
Schiffer, F. H. and Mathis, J. S. 1974, Ap. J., 194, 597.
Strom, S. E., Grasdalen, G. L., Strom, K. M. 1974, Ap. J., 191, 111.
Strom, S. E., Strom, K. M. and Grasdalen, G. L. 1975, Ann. Rev. Astron. Ap., 13, in press.
Traub, W. A., Carleton, N. P., and Hegyi, D. J. 1974, Ap. J., 190, L 81.
Zuckerman, B. 1973, Ap. J., 183, 863.
Zuckerman, B. and Evans, N. J. 1974, Ap. J., 192, L 149.
Zuckerman, B. and Palmer, P. 1974, Ann. Rev. Astron. Ap., 12, 279.
Zuckerman, B. and Palmer P. 1975, Ap. J. (in press).

# A NEW RADIO MAP OF THE CORE OF ORION A

S. F. GULL and A. H. M. MARTIN*

Mullard Radio Astronomy Observatory, Cavendish Laboratory,
Madingley Road, Cambridge CB3 0HE, England

Although the Orion Neubla is the closest bright HII region, its
radio continuum structure is only poorly known. It lies at $\delta = -5^{\circ}$,
too far south and too close to the celestial equator to be properly
mapped by E/W Earth-rotation synthesis telescopes. We have used both
E/W and N/S baselines to overcome this problem and in this paper we
describe the resulting map.

The E/W observations were made with the Cambridge One-Mile telescope and the N/S data were obtained with the Owens Valley interferometer, both at a frequency of 5 GHz. Conventional Fourier inversion
produced separate maps, which were afterwards added and cleaned
(Högbom 1974). The resulting map is shown as Fig. 1; it has an equivalent half-power width of 7 arc sec in right ascension and 20 arc sec in
declination.

The brightness distribution is very complicated, with numerous
peaks and hollows. None of these peaks is unresolved, so that any
separation into individual components would seem to be very artificial.
Such peaks as do occur do not correspond in detail with those found by
Webster and Altenhoff (1970) in the only previous high-resolution
observations. We attribute this lack of agreement to their very
asymmetrical beam.

*Present address: Institute of Astronomy, Madingley Road,
Cambridge CB3 0HA, England.

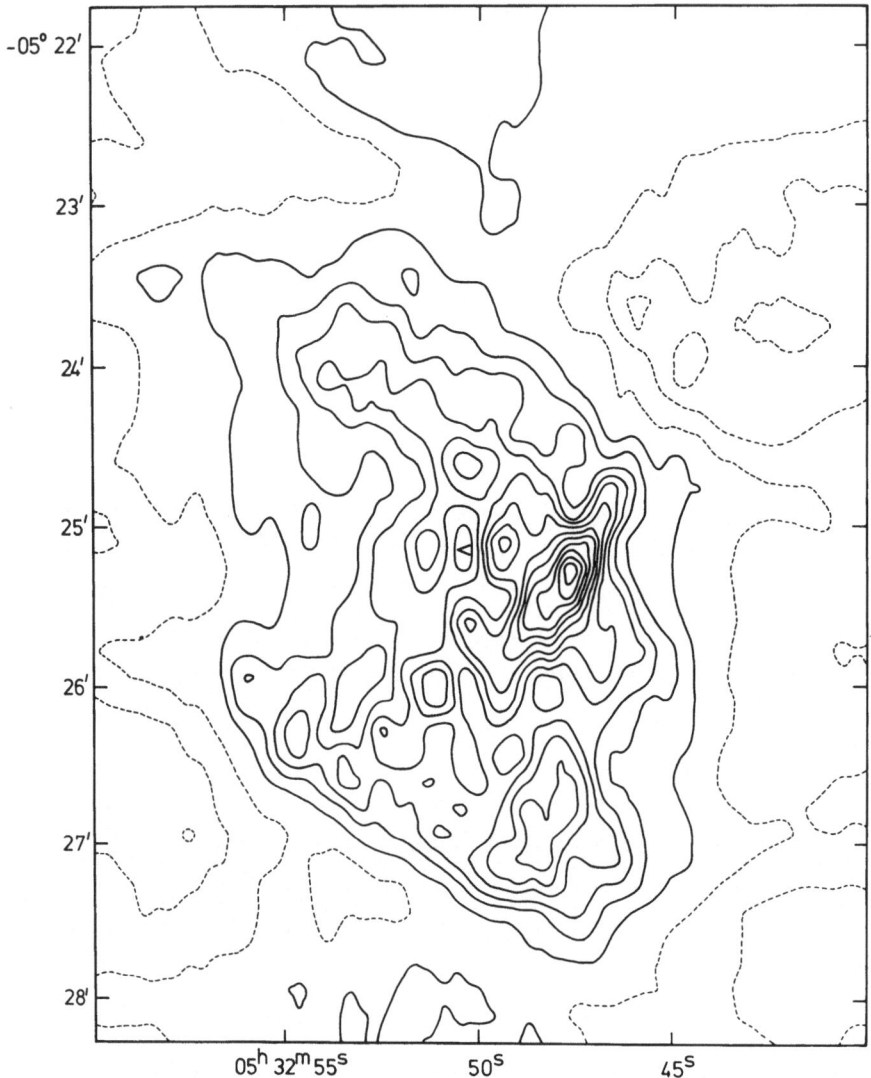

Fig. 1   Brightness distribution of Orion A at 5 GHz.  This map is the
        result of 400 beamshape subtractions.  The contour interval
        is 60 K and the zero and negative contours are shown dotted.
        Local minima are shown thus: <.

The radio structure revealed by this map is closely similar to that seen in optical photographs of the region. Most striking is the correspondence between the optical bar marking the ionization front to the SE of the Trapezium (Meaburn 1975) and the sharp-edged ridge occupying the same position on the radio map. To the NW the radio map shows a similar, but broader edge. Agreement with the optical is here less close, but we note that this edge lies parallel to the line of $H_2O$ maser sources (Landecker and Hills 1975).

To the East the most prominent feature of the optical photograph is the well-known dark bay. It is interesting that this dark bay also shows up in the radio contours, implying that it is due to a real deficiency of ionized gas in this region rather than to obscuration by foreground dust. There are some minor but significant differences between the radio and optical pictures. In the SW corner a broad radio peak occupies an almost blank region on the optical photograph, and the same is true of the extension of the NW boundary to the North. In general, however, the close agreement between the radio and optical indicates a rather uniform obscuration across the face of the nebula.

One of us (SFG) is grateful to the Caltech Radio Astronomy Group for their generous assistance and hospitality.

REFERENCES

Högbom, J. A., 1974, Astr. Astrophys. Suppl., 15, 417.
Landecker, T. L. and Hills, R. E., 1975. In preparation.
Meaburn, J., 1975. This Symposium.
Webster, W. J. and Altenhoff, W. J., 1970, Astrophys. Lett., 5, 233.

# M 17

D. Lemke

Max-Planck-Institut für Astronomie
D-6900  Heidelberg

## ABSTRACT

This paper briefly reviews the results achieved until the end of 1974 by optical, radio, infrared and molecular line observations.

The Omega nebula (M 17, W 38, NGC 6618) exhibits many similarities to the Orion nebula M 42. Both have comparable brightness in most spectral regions, but taking into account the larger distance of M 17, 2 kpc compared to 0.5 for Orion, then M 17 is by far the larger and more luminous HII region. Nevertheless, it is much less investigated. That might be due to the facts that
- M 17 is not as easy accessible because of its low position in the summer sky,
- many other interesting objects are in the neighbourhood
- and, until recently no strange objects like Kleinmann-Low or Becklin-Neugebauer in Orion have been discovered in M 17 which would certainly have stimulated a much more extensive investigation.

## 1. Optical Observations

When reviewing what we know about M 17 today we will follow more or less the historical way and start with optical studies. These have contributed to the problems of the exciting stars and the determination of the distance and the absorption towards the nebula.

Dickel (1966) determined from a $H_\alpha$ and $H_\beta$ surface photometry a visual absorption of 3.6 mag for the brightest parts of M 17, see fig. 2. After radio maps got available in the 1960's the radio maxima were found to be 4 arc min to the west of the $H_\alpha$ peak. Again Dickel (1968) and Ishida and Kawajiri (1968) determined the absorption for these regions from the ratio of the emission measures determined from the different radio maps and $H_\alpha$. The result was: there is a gradient from the E to the W with $\sim 4^m$ absorption at the $H_\alpha$ peak to $7^m$ at the southern radio-peak (fig. 2).

Due to this heavy obscuration the search for the exciting stars in the visible region is difficult, an O5 star for instance, close to the radio/infrared peak will be fainter than $17^m$ at V. Schulte has made in 1955 objective prism photographs which are shown in fig. 3. He found two early type stars, identified here by their large ultraviolet extensions. Only the star Schulte No. 1 was spectroscopically classified by Abt to be an O5 star. Puzzling is the bright star BD-16°4816 which is in arc sec distance only to the brightest radio/infrared peak. Repeated spectroscopic and photometric studies confirmed that it is a K 5 star, probably a supergiant, in the foreground. There is another BD star - 16°4818 which was classified as OB +, but again no detailed informations are available yet.

In summary, only Schulte No. 1 seems to be important for the excitation of M 17. Possible contributions will come from Schulte No. 2, BD-16°4818 and a star found by Kleinmann (1973) in the infrared. When discussing the infrared observations later we will briefly return to the problem of the exciting stars, since both the infrared and the radio data show that these stars known so far can not supply sufficient energy.

Distances for M 17 have been derived from stars close to M 17 by Sharpless (1953) and others. They range from 0.7 to 2.8 kpc. This large scatter is due to the fact that probably most stars used are well outside the nebula. The kinematic distance of M 17 obtained from optical lines with Fabry-Perot-spectroscopy by Georgelin (1970) is 2.4 kpc.

## 2. Radio Observations

A larger number of radio-maps in the cm region of the free-free emission from M 17 has been published in the 1960's.

As an example we will look on Zisk's (1966) 2 cm map (fig. 4). There are two main components: a bright southern source (G 15.0-0.7) and a somewhat fainter and more extended source 3 arc min to the north (G 15.1 - 0.7). For comparison with other maps one should keep in mind the position angles and the extension of the northern source towards the bright visible part of the nebula in the east. All other maps made in the cm region look very similar or show only one large peak when made at longer wavelengths and with larger beams (Schraml and Mezger 1968, Johnston and Hobbs 1973, Gordon and Williams 1971).

The southern peak has been resolved by interferometric means at 11 cm by Webster, Altenhoff and Wink (1971). It revealed one strong and two fainter components within 30 arc sec distance from the main source. The extended G 15.1 -0.7 component could not be detected by the interferometer.

Fig. 5 shows the 3.4 mm map made by Montgomery et al. in 1971. There are two sources, A and C, close to the 2 cm peaks. And there is a third source B which coincides with the $H_\alpha$ maximum. This source will also show up on one of the infrared maps to be discussed later.

All these radio continuum observations are summarized in a spectrum by Johnston and Hobbs (1973), which is shown in fig. 6. This spectrum clearly demonstrates the two emission processes: thermal emission from the dust in the infrared and free-free emission of the gas in the radio region. All the radio points define the line which one expects for radiation from ionized hydrogen, the flux density is of the order of several hundred f.u.

Further information was gained by radio recombination line observations. Reifenstein et al. (1970) determined the Doppler shift of the hydrogen $109\alpha$ line to be $\sim 17$ km/sec. From that they derived a kinematic distance of $\sim 2$ kpc for M 17. In the following table the most important parameters of M 17 derived from radio observations are summarized:

Distance $\qquad$ $D \sim 2.2$ kpc

Radial velocity $\qquad$ $V_{LSR} \sim 17.2$ km/s

Linear diameter of the two main sources $\quad 2\,R \sim 3$ pc

Electron density $\qquad$ $N_e \sim 540$ cm$^{-3}$

Mass (both sources together) $\qquad$ $M_{HII} \sim 550$ M$_\odot$

Emission measure $\qquad$ $E \sim 1.6 \cdot 10^6$ cm$^{-6}$ pc

Electron temperature $\qquad$ $T_e \sim 6400$ K

Free free flux density $\qquad$ $S_\nu \sim 500$ f.u.

Recently Gull and Balick (1974) have studied M 17 in the 76$\alpha$ transitions of H, He and C. They found that the best fit to the observed hydrogen line profile can be made by two Gaussian lines corresponding to two different velocity fields separated by 17 km/sec over most of M 17. A 9 km/sec (blue) component seems to be coincident with the optically obscured region. A 26 km/sec (red) component extends towards the bright H$_\alpha$ region. Using also the CII emission (18 km/sec) which arises in neutral regions close to an HII/HI interface and which was found here in the obscured south-west part of the nebula Gull and Balick proposed a model for M 17: it may be understood as some sort of Zuckerman's (1973) Orion model, rotated by 90$^\circ$. Flow of ionized gas from the background and the foreground may be interpreted as arising from the sides of a "bowl" shaped neutral complex with the bowl bottom in the southwest. This model is a useful step towards a better understanding of the structure of M 17, nevertheless it is obvious from the many new data that M 17 has a more complex structure than the Orion nebula.

## 3. Infrared Observations

In 1970 Low, Aumann and Harper discovered with their Lear jet telescope the large far infrared emission from HII regions and found M 17 to be the second brightest after the Orion nebula. From the flux ratio in their two passbands (45 and 60 to 750$\mu$) they determined the dust temperature to be 85 K.

In the following years M 17 was observed from the ground at several infrared wavelengths with higher spatial resolution. Fig. 7 shows a 10$\mu$map made by Kleinmann (1971). In addition to the two bright peaks which are coincident with the radio peaks he found two stars bright in the near ir. One is the already known BD-16$^{\circ}$4816, the K 5 star. The other could be, according to Herbig, an O or very early B star and therefore may be important for the energy balance of M 17. Both stars are marked on fig. 1.

Fig. 8 shows M 17 at 21$\mu$ mapped by Lemke and Low (1972). At both wavelengths, 10$\mu$ and 21$\mu$, M 17 is optically thin. Again, the two main sources are coincident with the radio and the 10$\mu$ peaks. The similarity of the structure of M 17 in the infrared and in the radio region demonstrates clearly the same distribution of gas and dust inside the ionized region. The 21$\mu$ map reveals a third infrared source E, close to the H$\alpha$ maximum. Its position is also close to the 3 mm-source B found by Montgomery et al. (1971). This source seems also to be indicated by a distortion of the isolines on the 2 cm maps. But due to the lack of observations at other wavelengths we do not yet understand the nature of this object.

In fig. 9 all available infrared data are plotted as a thermal spectrum. Obviously the usual black body curve does not fit all the points of M 17(black dots).But two black body curves, refering to two components of the dust with different temperatures, will do the job. From this two component model one can calculate the masses of dust involved, making the usual assumptions about the size and the emissivity of the interstellar grains. The results are 4 and $10^{-3}$ M$_{\odot}$ for the 75 K and the 240 K dust respectively. When comparing these dust masses with the gas mass derived from the radio observations the normal ~1 % dust content follows for M 17.

According to Harper and Rieke (1975) the two dust components
in M 17 may be produced in the following way: if the HII
region is formed out of a neutral cloud containing icy dust
grains,then these will partly loose their ice and partly
retain it somewhat longer, therefore emitting preferentially
in the middle and the far infrared respectively.

Little can be said so far about the temperature distribution of the dust throughout the nebula because of the different beam sizes used for the infrared maps. Wright (1973)
has proposed a model with central dust depletion for M 17:
ultraviolet radiation from the central star longwards of
912 $\overset{o}{A}$ should heat a circumstellar shell of undepleted dust.
This outer shell of colder dust should be visible in the far
infrared on high resolution maps ($\sim$ 1 arc min). On Harper
and Rieke's (1975) new 100$\mu$map (FWHM = 2.3 arc min) there
is no indication for such a shell.

We will now briefly return to the problem of the energy
balance and the exciting stars. The total infrared luminosity of M 17 was found to be $L_{M\ 17} \sim 5 \cdot 10^6 L_\odot$.
When comparing this value with the total luminosity of the
only exciting star definitly known so far, Schulte No. 1,
than this star can account for only 10 to 20 % of the luminosity of M 17 (Johnson 1973). All other stars mentioned
earlier can not solve this discrepancy because their spectral type is later and they may be well outside the nebula.
There is also the problem that none of the stars, including
Schulte No. 1, is coincident with one of the infrared or
radio peaks. So we are still left over with the task to
find the other exciting stars in M 17, not only to solve
the energy crisis but also for a better understanding of
the spatial structure of M 17. Progress has achieved recently by two groups: Harper and Rieke (1975) discovered near
the southern infrared/radio peak a very compact source
bright at 10$\mu$ which may be coincident or close to the exciting source. Beetz et al. (1975) found 1 arc min north

of the southern peak a cluster of early type stars which may contribute to the excitation.

## 4. Molecular Line Observations

Last year Lada, Dickinson and Penfield (1974) reported the discovery of a molecular source near M 17. Its location is shown in fig. 10 on a contour map of $^{12}C^{16}O$ apparent brightness temperature. The CO peak is located 2 arc min to the south-west of the southern radio/infrared peak of the HII region. It is coincident with a region of probably high absorption as suggested by a look on photographic plates and by Dickel's (1968) study which shows an increase in visible absorption towards the west. Also the isotope $^{13}C^{16}O$ was mapped over this south-west source as shown in fig. 11. There are indicated also the positions of

- a strong time-varying $H_2O$ maser found in 1973 by Johnston, Sloanaker and Bologna (the southernmost cross),
- the position of the 3.4 mm continuum peak A which was shown on the map of Montgomery et al. (1971) earlier (the northernmost cross),
- and the position of a small ( < 10 arc sec) bright infra-red source found subsequently by Kleinmann and Wright (1973).

Also the isotope $^{12}C^{18}O$ was detected at two positions near the CO peak M 17 SW. From a calculated colummn density of $10^{19} cm^{-2}$ at the CO peak the authors estimate the total mass of CO in this object to be at least 80 $M_\odot$ , if it is located at the same distance as M 17. Using the solar C to H abundancy the total mass of the cloud was derived to be about 6000 $M_\odot$, which is several times more than the mass of the HII region M 17. Such a cloud could be stabilized, that  is, prevented from collapse, only by a high kinetic temperature ($\sim$ 800 K) or by rotation. But the observations

show no indication for either. So it was concluded that the
object is probably undergoing collapse.

The discovery of M 17 SW was greatly welcomed by all
astronomers who like to compare M 17 with M 42. Both molecular clouds are associated with pointlike infrared sources: (i) the 600 K Becklin-Neugebauer object in Orion A,
- and (ii) the 200 K Kleinmann-Wright source in M 17 SW.
These two compact infrared sources might be similar condensations at different stages of their evolution. In addition, strong time-varying $H_2O$ masers are located near the
infrared sources in both objects. However, there seems to
be no extended far infrared emission in M 17 SW like that
from the Kleinmann-Low nebula in Orion. Although other molecules have been found in M 17 SW, for instance SO, HCN,
$H_2CO$, it still has to be shown that it is as rich in molecular line radiation as Orion. Recent observations of formaldehyde towards M 17 by Lada and Chaisson (1975) exhibit
a very complex kinematic and spatial structure of this obscured region west of the nebula. Nevertheless their results
support the concept that bright stars and HII regions are
formed at the edges of molecular clouds of high density
and large mass. Besides of M 42 and probably NGC 2024 this
seems to be established now also for M 17.

REFERENCES


Beetz, M., Elsässer, H., Poulakos, C., Weinberger, R., 1975,
      to be published.
Dickel, H.R., 1966, A.J. 71, 852.
Dickel, H.R., 1968, Ap.J. 152, 651.
Georgelin, Y.P., Georgelin, Y.M., 1970, Astr. & Ap. 6, 349.
Gordon, M.A., Williams, T.B., 1971, Astr. & Ap. 12, 120.
Gull, T.R., Balik, B., 1974, Ap.J. 192, 63.
Harper, D.A., Low, F.J., 1971, Ap. J. 165, 9.
Ishida, K., Kawajiri, N., 1968, Publ.Astr.Soc. Japan 20,95.
Johnson, H.M., 1973, Ap.J. 182, 497.
Johnston, K.J., Sloanaker, R.M., Bologna, J.M., 1973,
      Ap.J. 182, 67.
Johnston, K.J., Hobbs, R.W., 1973, Astr.J. 78, 235.

Kleinmann, D.E., 1973, Ap.Lett. 13, 49.
Kleinmann, D.E., Wright, E., 1973, Ap.J. 185, L 131.
Lada, C., Dickinson, D.F., Penfield, H., 1974, Ap.J. 189, L 35.
Lada, C., Chaisson, E., 1975, Ap.J. 195, 367.
Lemke, D., Low, F.J., 1972, Ap.J. 177, L 53.
Low, F.J., Aumann, H.H., 1970, Ap.J. 162, L 79.
Montgomery, J.W., Epstein, E.E., Oliver, J.P., Dworetsky, M.
        M., Forgarty, W.G., 1971, Ap.J. 167, 77.
Reifenstein III, E.C., Wilson, T.L., Burke, B.F., Mezger, P.
        G., Altenhoff, W.J., 1970, Astr. & Ap. 4, 357.
Rieke, G.H., Harper, D.A., 1975, to be published.
Schraml, J., Mezger, P.G., 1969, Ap.J. 156, 269.
Schulte, D.H., 1955, Ap.J. 123, 250.
Sharpless, S., 1953, Ap.J. 118, 362.
Webster, W.J., Altenhoff, W.J., Wink J.E., 1971, Astr. J. 76,
        677.
Wright, E., 1973, Ap.J. 185, 569.
Zisk, S.H., 1966, Science 153, 1107.
Zuckerman, B., 1973, Ap.J. 183, 863.

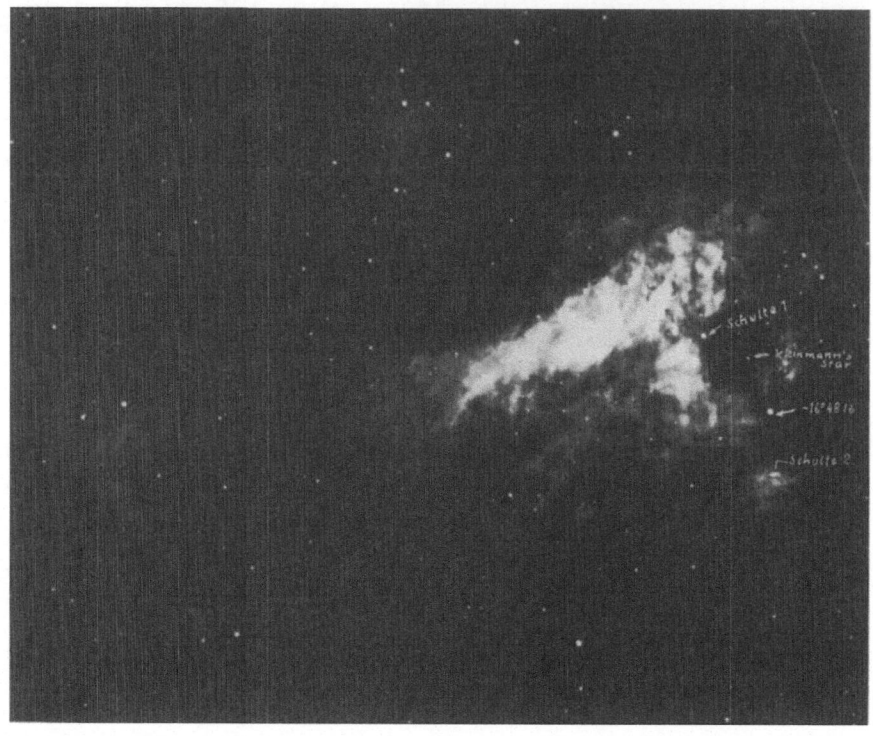

Fig. 1   Photograph of M 17, made by Dr.G.H. Herbig with
the 120-inch telescope at Lick Observatory.
Scale: 6.5 mm/ arc min.

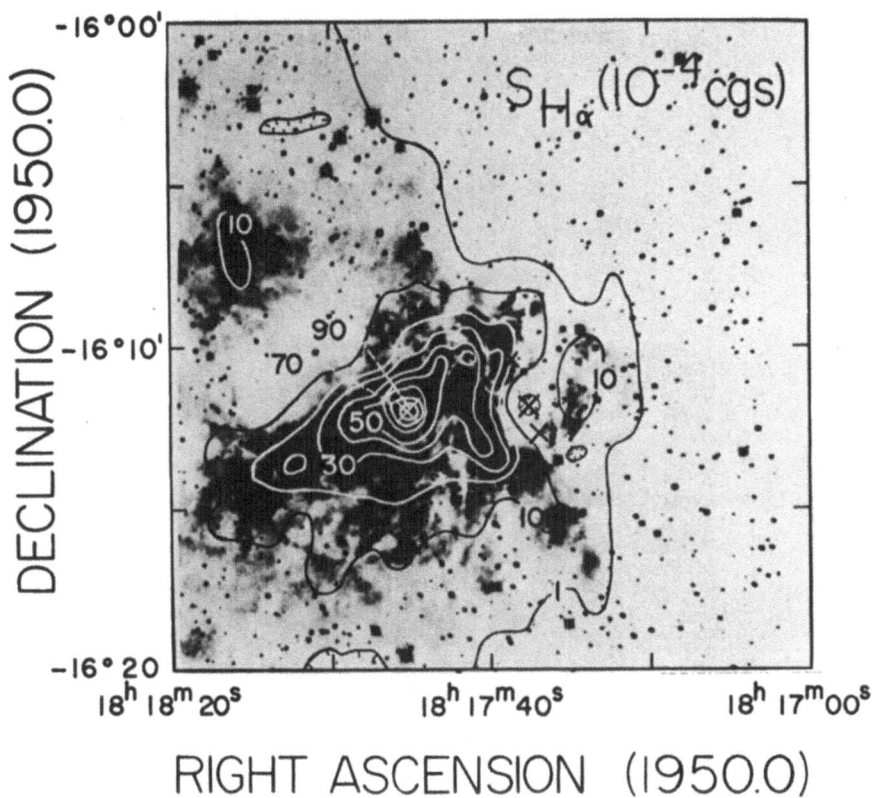

Fig. 2     Surface brightness contours in $H_\alpha$ . The three
crosses in the west refer to the two radio
peaks and an optical "hole". Visual absorption
increases towards the west from $A_V \sim 3.6^m$ at the
$H_\alpha$ maximum to $A_V \sim 7^m$ at the southern radio peak.
(Dickel, H.R., 1968).

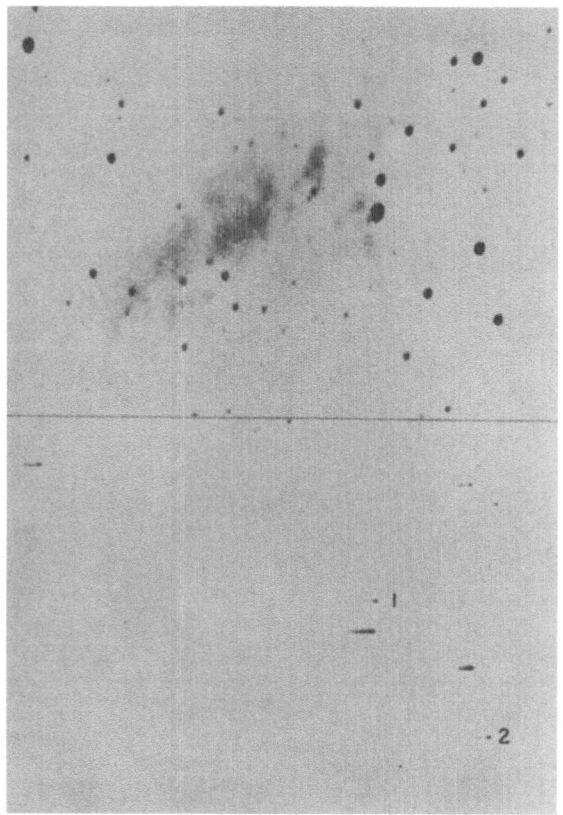

Fig. 3    Objective prism plate of M 17. Stars No. 1 and
          2 are early type stars, they probably contribute
          to the excitation of the nebula. (Schulte, D.H.,
          1965).

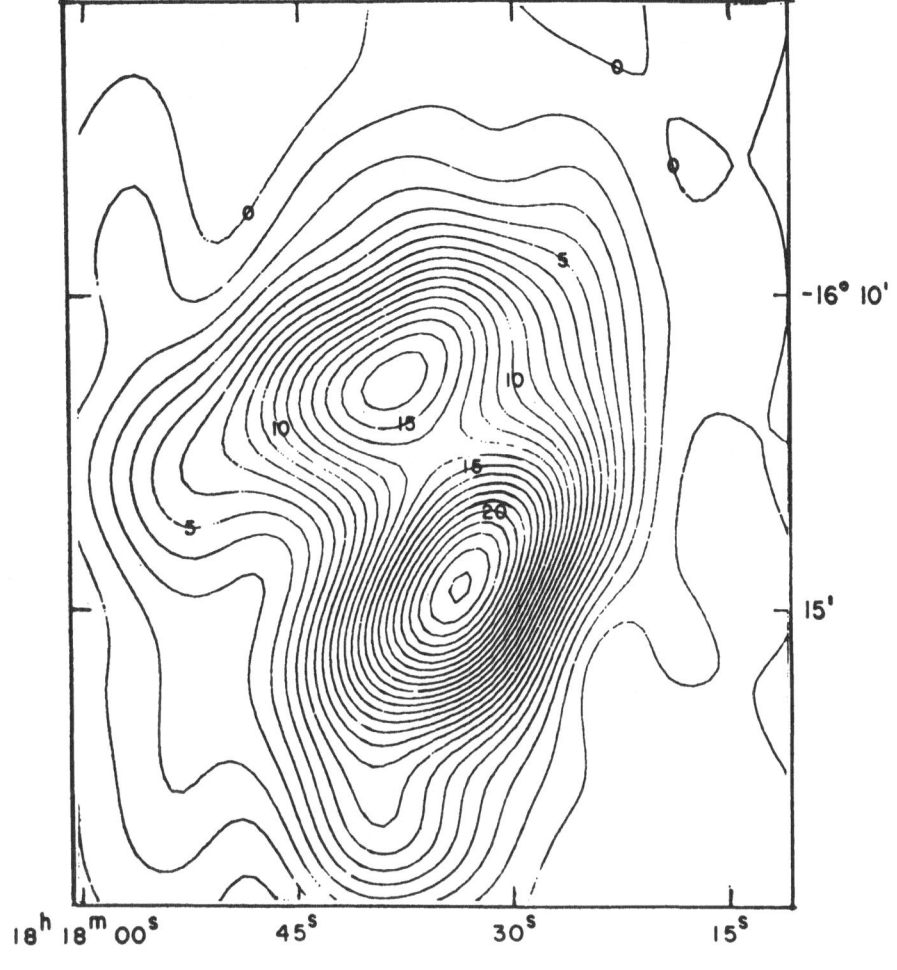

Fig. 4    Radio continuum map of M 17 at 2 cm wavelength.
         (Zisk, S.H., 1966).

386

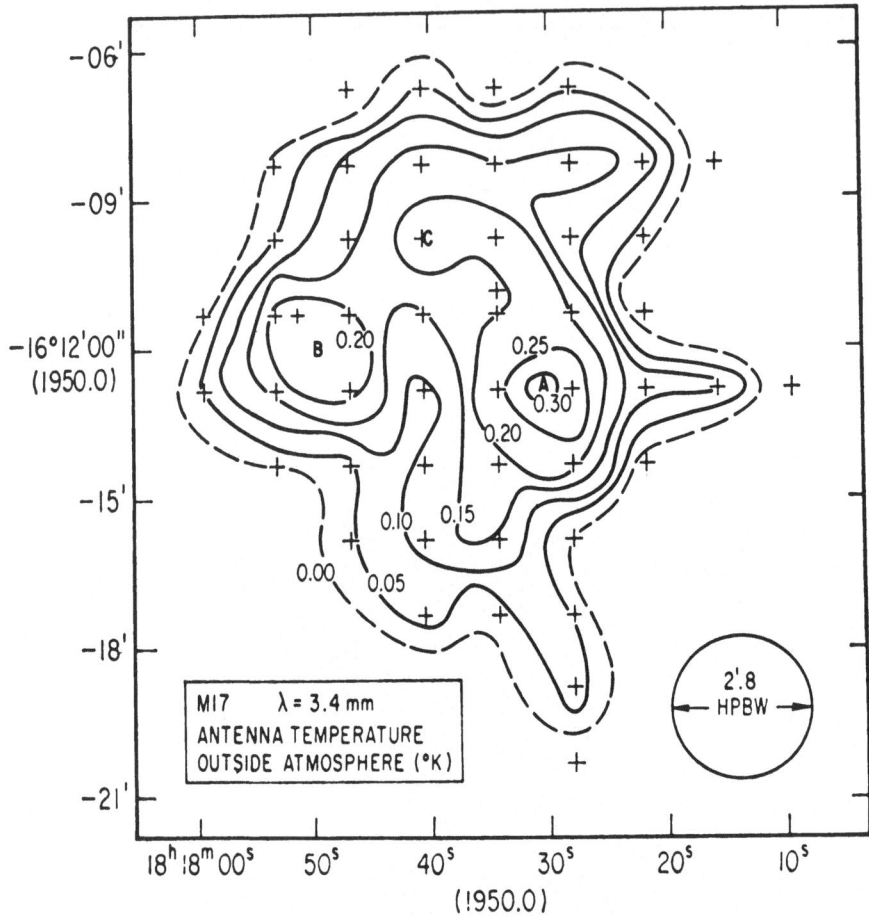

Fig. 5    Contour map of M 17 at 3.4 mm wavelength. Peaks
         A and C agree in position with the two principal
         components in the other radio and infrared maps.
         Peak B may be coincident with the 21μ source E.
         (Montgomery, J.W., et al. 1971).

387

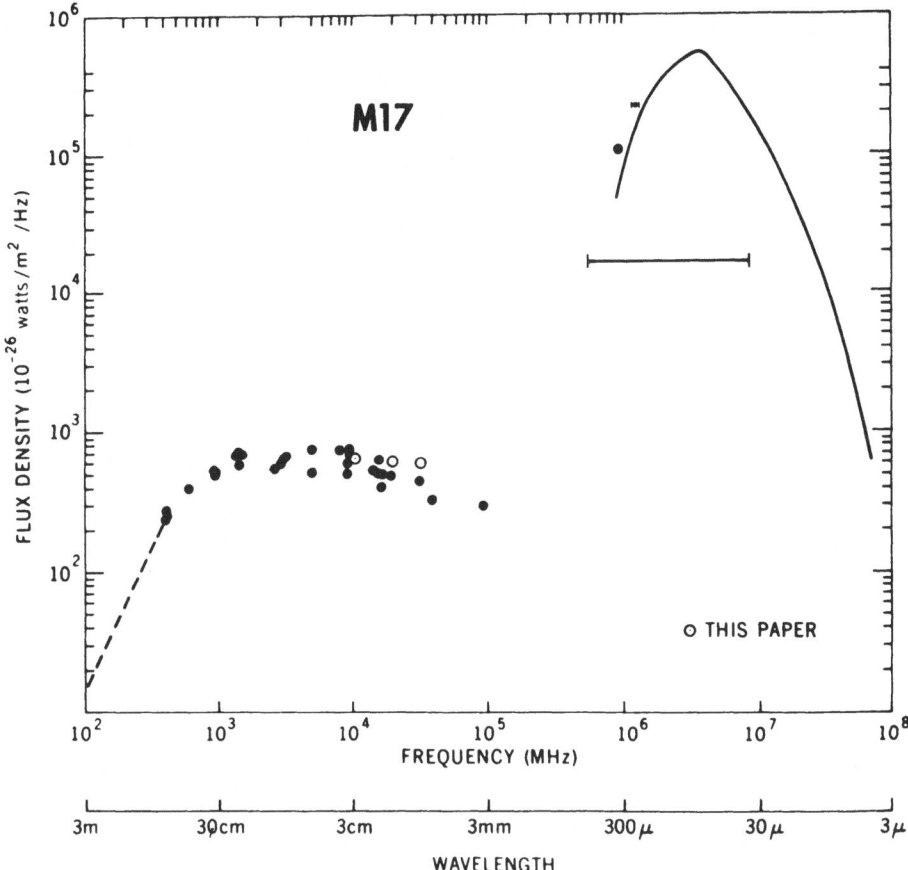

Fig. 6    Spectrum of M 17. At radio wavelengths the
          radiation is due to free-free emission of the
          ionized hydrogen gas, at infrared wavelenghts
          the radiation is emitted by optically thin dust
          mixed with the gas. (Johnston, K.J., Hobbs, R.W.,
          1973).

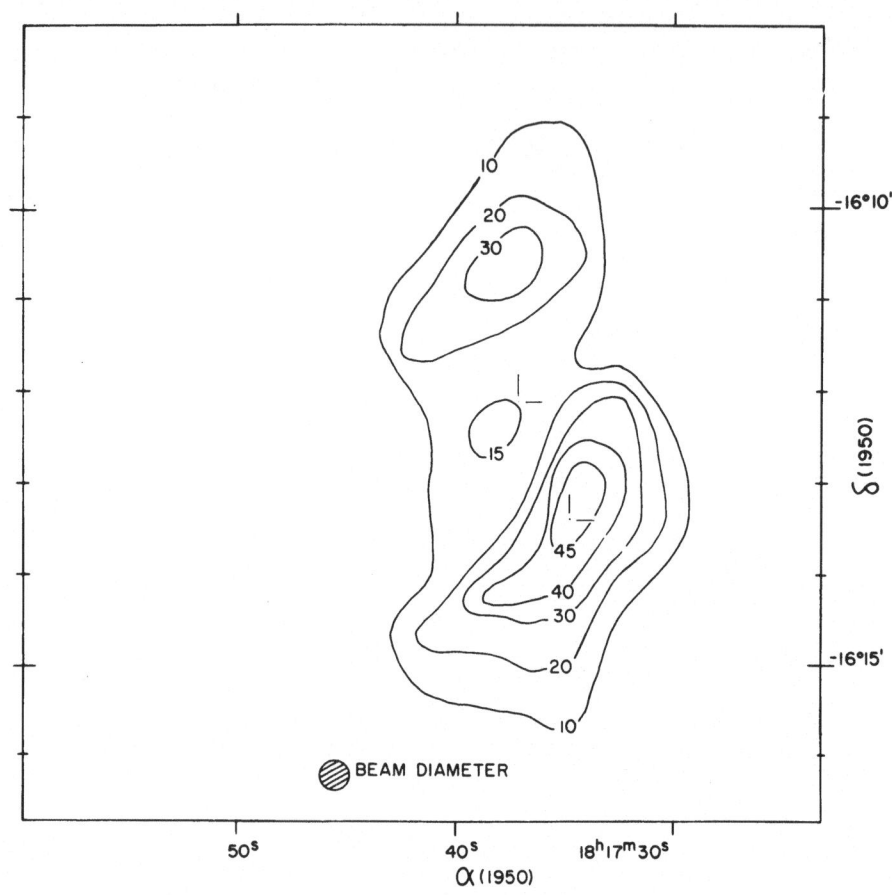

Fig. 7    Isophotal map of M 17 at $10\mu$ in units $4.2 \cdot 10^{-18}$
W $m^{-2}Hz^{-1}$ $sr^{-1}$. Two stars bright at $2.2\mu$ are
indicated by corner markings, the southernmost
is BD-16°4816. (Kleinmann, D.E., 1973).

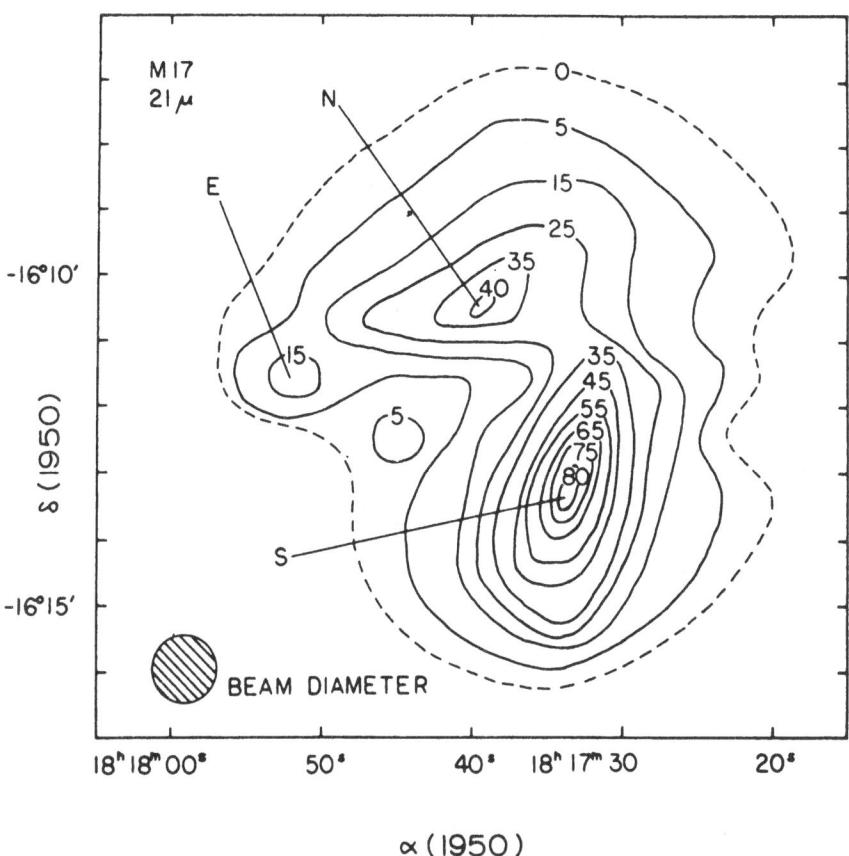

Fig. 8    Isophotal map of M 17 at $21\mu$ in units $\sim 10^{-17}$
W m$^{-2}$ Hz$^{-1}$ sr$^{-1}$. S and N refer to the two principal components also seen in the radio maps,
E indicates a new source. (Lemke, D., Low, F.J.,
1972).

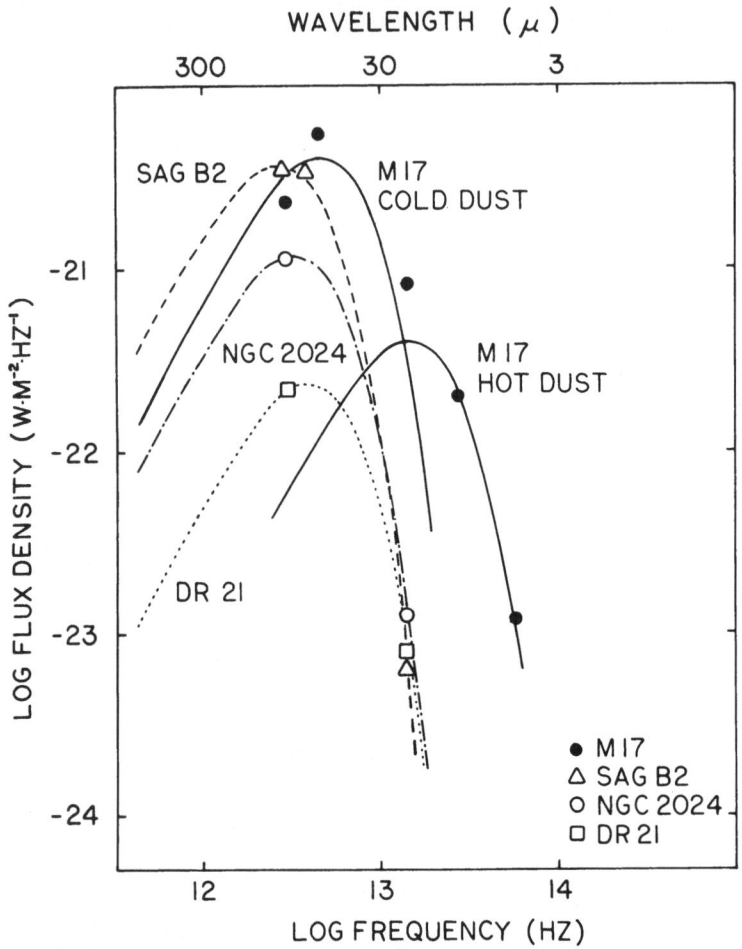

Fig. 9    Thermal spectrum of M 17 (black dots). The two
fitted black body curves refer to two dust components with different temperatures, 240 K and 75 K.
(Lemke, D., Low, F.J., 1972).

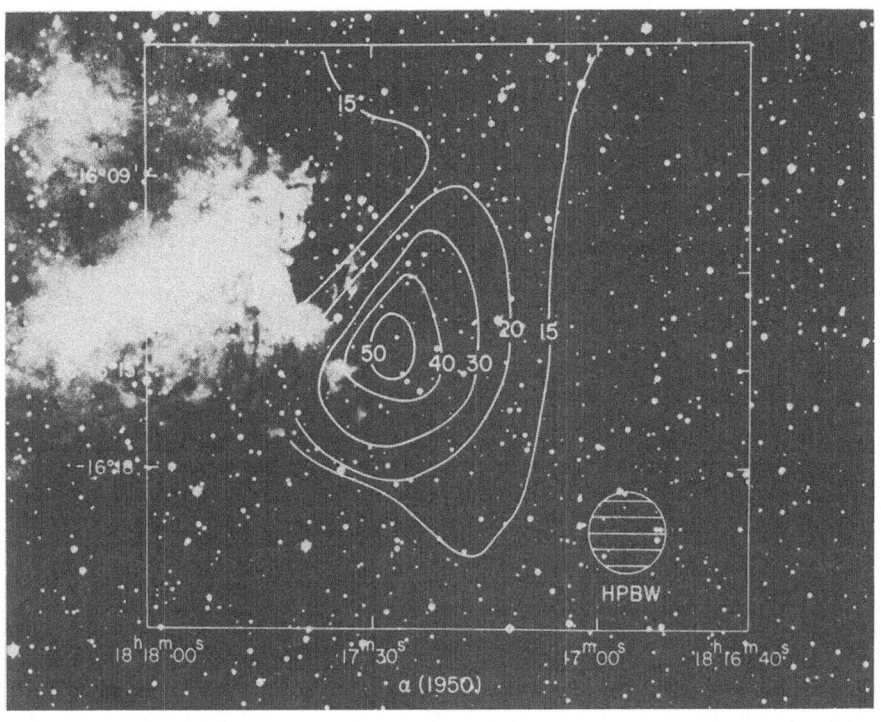

Fig. 10 Contour map of $^{12}C^{16}O$ apparent brightness temperature. This new molecular source is named M 17
SW. (Lada, C., et al. 1974).

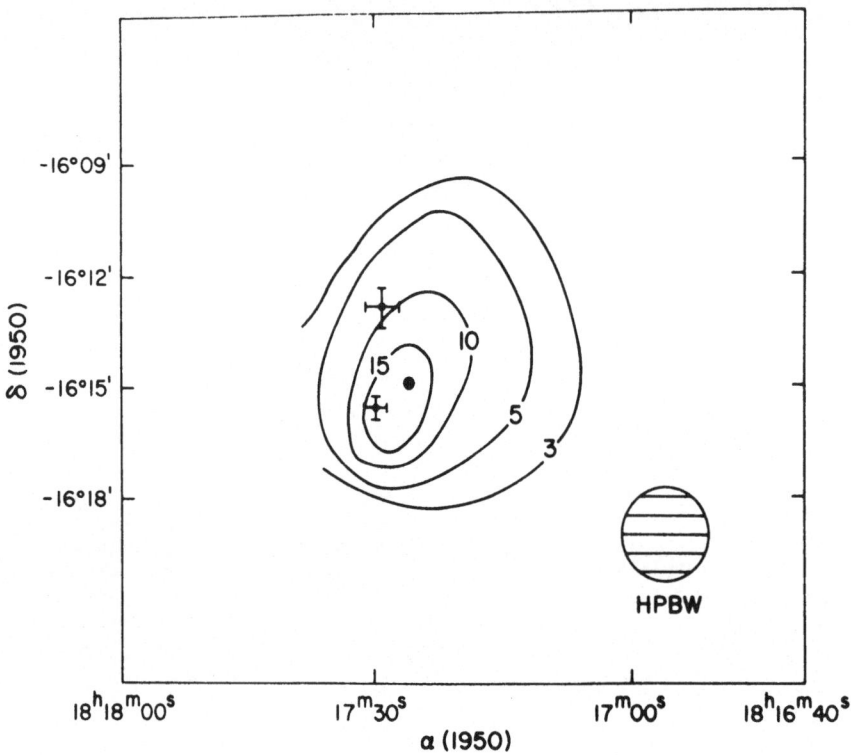

Fig. 11    Contour map of $^{13}C^{16}O$ emission in M 17 SW. The
            northernmost cross indicates the 3.4 mm source
            A(Montgomery, J.W., et al. 1971), the southern-
            most cross indicates the $H_2O$ maser position
            (Johnston, K.J., et al. 1973). The filled-in
            circle represents the infrared source of Klein-
            mann and Wright, 1973. (Lada, C., et al. 1974).

# DR21 AND ASSOCIATED SOURCES - A REVIEW

STELLA HARRIS

Mullard Radio Astronomy Observatory,
Cavendish Laboratory, Cambridge, England

## ABSTRACT

DR21 is one of the radio continuum sources contained within the larger source W75. It has been mapped over a wide range of radio frequencies and angular resolutions, besides being studied in the infrared and in a variety of molecular lines, and is also a region of OH and $H_2O$ maser activity. With high resolution observations the region is found to contain a whole complex of different objects, whose properties may be interpreted as representing the various stages in an early evolutionary sequence.

INTRODUCTION

W75 was one of the sources in the Westerhout (1958) survey detected in the search for discrete sources along the galactic ridge (Fig. 1(a)). The survey was conducted at a frequency of 1.39 GHz, using the 25-m telescope at Dwingelo. Later higher resolution surveys of the central Cygnus X region (Downes and Rinehart 1966; Pike and Drake 1964) showed that this very intense region in fact contained more than twenty discrete sources within a diameter of about five degrees, and that W75 itself was resolved into two principal sources, thereafter designated DR21 and DR23 (Fig.1(b)).

This paper is concerned specifically with observations of DR21, which have subsequently been mapped with varying degrees of resolution over a wide range of frequencies. In 1967 it was observed at 408 and 1407 MHz with the Cambridge One-Mile telescope by Ryle and Downes, who found it to be a thermal source of small angular diameter; its thermal nature being later confirmed by the detection of recombination line emission (Mezger et al. 1967). At the time it had the highest emission measure yet observed in a galactic source, and was probably the first object to be labelled a 'compact HII region'.

RADIO CONTINUUM OBSERVATIONS

Subsequent observations at higher frequencies and correspondingly higher resolutions (Webster and Altenhoff 1970; Wynn-Williams 1971) showed the source to consist of two main components - a very compact object to the north, and a more extended southern one, both being surrounded by a single low-brightness 'outer' envelope.

In 1973 it was observed by the Cambridge 5 km telescope at 5 GHz (Harris 1973) with an angular resolution of 2 arc sec, showing the southern source to be further subdivided into three very compact objects, again with a less dense emission region, the 'inner' envelope, surrounding these three (Fig. 2). All four compact objects were now resolved by the telescope beam, having angular sizes ranging between 4 and 7 arc sec, corresponding to linear sizes of about 0.1 pc at a distance of 3 kpc. Similar structure has since been observed by B. Balick at a frequency of 8 GHz (private communication).

The distance to DR21 is in fact very uncertain. There is no
optically visible feature, and the radial velocities obtained from
recombination lines and 21 cm absorption profiles show strong evidence
for substantial local motions and do not supply a consistent value for
the kinematic distance. However, 3 kpc is a generally adopted value,
and is further supported by the recent 1667 MHz OH line absorption
observations of Paschenko (1973).

With regard to the spectrum, there is a fairly large number of good
flux density measurements for this source. This has made it possible to
obtain a reasonable estimate of the electron temperature, using the high
resolution contours to construct theoretical thermal spectra for various
values of electron temperature, and comparing these with the observational results (Harris 1973). The method gives an electron temperature
of 8000 ± 2000 K, which is in good agreement with the result obtained
by the more analytical method of Salem and Seaton (1974). Adopting an
electron temperature of 8000 K, other parameters have been calculated,
and give the sort of values (electron density ~ $3 \times 10^4$ cm$^{-3}$; emission
measure ~ $5 \times 10^7$ pc cm$^{-6}$) which have now come to be expected in these
ultra-compact HII regions.

## MASER SOURCES AND ASSOCIATED CONTINUUM

The W75 region also contains two OH sources, one of which is known
to have some further associated continuum. The layout of the various
sources is shown in Fig. 3. The first of the OH sources to be discovered was the more southern one, W75(S)OH, which lies about 3 arc min
north of DR21, and was detected in the 1665 MHz line by Palmer and
Zuckerman (1967). Zuckerman et al (1969) observed it at intervals over
the period 1965/68, during which time its flux density increased from
about 50 to 100 f.u.[*] (peaking towards the end of 1966) and then fell
off again towards 1968, since when there appears to have been no further
significant variation (Lekht 1974).

In 1969 Sullivan detected an $H_2O$ source coincident with the OH
position, and monitored it for variability between February 1969 and
March 1970 (Sullivan 1973). It exhibited a similar mode of behaviour,
though on a much more dramatic scale, falling from a peak density of

[*] 1 f.u. = $10^{-26}$ W m$^{-2}$ Hz$^{-1}$

about 5000 f.u. in February 1969, down to a barely detectable level by the end of the year, and has been found still to be a very weak source (~ 100 f.u.) in observations made by R. E. Hills in 1973 (private communication). No continuum radiation has been detected at this position, to a limit of 15 mfu at a frequency of 5 GHz (Harris 1974).

About 15 arc min to the north of W75(S)OH there is a second OH source, W75(N)OH, which also radiates most strongly in the 1665 MHz line (Rydbeck et al. 1969; Zuckerman et al. 1969). This source, however, does have related continuum emission. A nearby continuum source was first detected by Wynn-Williams (1971) at a frequency of 1407 MHz, and this has subsequently been mapped with the 5 km telescope at 5 GHz (Harris 1974), showing it to be a low-brightness structure extending over a diameter of about 16 arc secs (Fig. 4). Its flux densities at 1.4 and 5 GHz are comparable (about 0.1 f.u.), indicating that the source is optically thin at both frequencies.

The OH position measured by Hardebeck (1972) lies somewhat to the southwest of this continuum source and has recently been found by Habing et al. (1974) to coincide with a smaller 19 mfu continuum component having an angular size of $< 4$ arc sec (Fig. 5). There is also an $H_2O$ source in the region (Johnston et al. 1973) but its position is not sufficiently well known to enable it to be associated with either the OH or the main continuum source.

INFRARED OBSERVATIONS

There is also known to be infrared emission from the W75 region. This has been investigated over a wide range of frequencies (Harper and Low 1971; Hoffmann et al. 1971; Lemke and Low 1972; Rieke et al. 1973; Sibille et al. 1974; Wynn-Williams et al. 1974), but in most cases there was not sufficient resolution to be able to associate the emission with any one particular feature. Ground-based observations at $350 \mu$ by Rieke et al. (1973) with a 63" beam showed two distinct sources in the DR21 part of the region. One was coincident with the continuum source, and the other with the W75(S)OH source, the latter being marginally the stronger of the two. Similar structure has been observed by Elias et al. (1975) at a frequency of 1 mm (Fig. 6). Detailed observations over the wavelength range $1.65 \mu \leqslant \lambda \leqslant 20 \mu$ have been carried out by Wynn-

Williams et al. (1974), who searched for emission both in the DR21/
W75(S)OH region and in the region of the W75(N) sources.  DR21 itself
was detected at 20µ, but despite being such a bright radio source, its
emission at this wavelength was not sufficiently strong to allow a
proper mapping.  While its overall extent was similar to that of the
radio continuum, there was evidence that the detailed distributions
differed quite significantly, which is unusual for a compact HII region.
At shorter wavelengths the source was even fainter, and already at 5µ
fell below the value expected from an extrapolation of the free-free
continuum.  The observations indicated a visual absorption of about $50^m$.
At 10µ most of the flux  came from the northern part of DR21, which
therefore appeared to have the hotter energy distribution.  The authors
also detected three 20µ sources in the region of W75(S)OH - one of which
was coincident with the OH source, while the remaining two (W75 IRS1,2)
had no other associated features (Fig. 7).

In W75N the peak of the 20µ emission is coincident with the radio
peak (see Fig. 4), and both have an angular extent of about 20 arc sec.
Hence the distributions appear to be in good qualitative agreement,
though the 20µ emission was again too weak to allow a full mapping.  A
2µ source was detected about 5 arc sec SW of the 20µ peak which the
authors believed could possibly be stellar.


OTHER MOLECULAR LINES

Finally, the region has also been intensively studied in various
different molecular lines (Kutner et al. 1973; Mayer et al. 1973; Morris
et al. 1973; Morris et al. 1974; Snyder and Buhl 1971; Thaddeus et al.
1972).  In particular, Morris et al (1974) have observed it simultaneously in millimeter-wavelength lines of HCN and CS, and have drawn some
interesting conclusions.  Fig. 8 shows their map in the HCN line, which
is seen to extend over the larger part of the entire W75 region, with
prominent and comparable maxima corresponding to the positions of DR21,
W75(S)OH, and W75(N).  The authors suggest that DR21 and W75(S)OH might
serve as prototypes for two classes of far ir/molecular line source,
representing distinct stages in the stellar evolution process - the Class
I sources being those which do not have associated continuum emission,
and the Class II sources those which do.  Their suggestion is that most

of the Class I objects are still at the collapsing protostar stage, in which case the infrared radiation derives from a conversion of the gravitational energy; whereas the Class II objects already contain a massive main-sequence star which has begun to ionise the surrounding gas. Such an object would have a high intrinsic infrared luminosity, which could be further enhanced by such processes as thermal emission from heated dust mixed in with or surrounding the ionised gas. If this interpretation is correct, and if the 0.1 pc components in DR21 are indeed ionised by individual O-stars, then these will have been formed about $10^4$ years previous to the W75(S)OH protostar source. Certainly according to this interpretation one would expect to find higher molecular densities associated with the Class I type sources, which is borne out by the fact that the majority of the less common molecular lines observed in W75 have tended to be found in association with W75(S)OH rather than with the other components. W75(N) continuum is probably a weak case of a Class II object (its ionisation could be produced by a BO star). The W75(N)OH source might possibly be at an intermediate stage - the ionisation process having only just begun.

Thus it appears that the various observations of W75 may well be giving us the picture of a vast molecular cloud containing a whole range of early evolutionary stages. It will be particularly interesting to obtain even higher resolutions in the far infrared and molecular line observations, in order to locate more precisely the concentrations of gas and dust.

## ACKNOWLEDGMENTS

I am grateful to all those authors mentioned in the text who have generously permitted me to reproduce their material, either published or unpublished. I also acknowledge receipt of a Science Research Council studentship.

## REFERENCES

Downes, D. and Rinehart, R. 1966, Ap. J., 144, 937.
Elias, J. H., Gezari, D. Y., Hauser, M. G., Werner, M. W., and Westbrook, W. E. 1975, Ap. J. (to be published).
Habing, H. J., Goss, W. M., Matthews, H. E., and Winnberg, A. 1974, Astr. and Ap., 35, 1.
Hardebeck, E. G. 1972, Ap. J., 172, 583.

Harper, D. A. and Low, F. J. 1971, Ap. J. (Letters), 165, L9.

Harris, S. 1973, M.N.R.A.S., 162, 5P.

Harris, S. 1974, M.N.R.A.S., 166, 29P.

Hoffmann, W.F., Frederick, C. L., and Emery, R. J. 1971, Ap. J. (Letters), 170, L89.

Johnston, K. J., Sloanaker, R. M., and Bologna, J. M. 1973, Ap. J., 182, 67.

Kutner, M. L., Thaddeus, P., Penzias, A. A., Wilson, R. W., and Jefferts, K. B. 1973, Ap. J. (Letters), 183, L27.

Lekht, E. E. 1974, Astr. Zh., 51, 341.

Lemke, D. and Low, F. J. 1972, Ap. J. (Letters), 177, L53.

Mayer, C.H., Waak, J.A., Cheung, A.C., and Chui, M.F. 1973, Ap. J., 182, L65.

Mezger, P.G., Altenhoff, W., Schraml, J., Burke, B. F., Reifenstein III, E. C., and Wilson, T. L. 1967, Ap. J. (Letters), 150, L157.

Morris, M., Zuckerman, B., Palmer, P., and Turner, B. E. 1973, Ap. J., 186, 501.

Morris, M., Palmer, P., Turner, B. E., and Zuckerman, B. 1974, Ap. J., 191, 349.

Palmer, P. and Zuckerman, B. 1967, Ap. J., 148, 727.

Paschenko, M. I. 1973, Astr. Cirk., No. 785.

Pike, E. M. and Drake, F. D. 1964, Ap. J., 139, 545.

Rieke, G. H., Harper, D. A., Low, F. J. and Armstrong, K. R. 1973, Ap. J. (Letters), 183, L67.

Rydbeck, O. E. H., Ellder, J., and Kollberg, E. 1969, Ap. J. (Letters), 156, L141.

Ryle, M. and Downes, D. 1967, Ap. J. (Letters), 148, L17.

Salem, M. and Seaton, M. J. 1974, M.N.R.A.S., 167, 493.

Sibille, F., Bergeat, J., and Lunel, M. 1974, Astr. and Ap., 30, 181.

Snyder, L. E. and Buhl, D. 1971, Ap. J. (Letters), 163, L47.

Sullivan, W. T. 1973, Ap. J. Suppl. No. 222.

Thaddeus, P., Kutner, M. L., Penzias, A. A., Wilson, R. W., and Jefferts, K. B. 1972, Ap. J. (Letters), 176, L73.

Webster, W. J. and Altenhoff, W. J. 1970, Astr. J., 75, 896.

Westerhout, G. 1958, Bull. Astr. Netherlands, 14, 215.

Wynn-Williams, C.G. 1971, M.N.R.A.S., 151, 397.

Wynn-Williams, C.G., Becklin, E. E. and Neugebauer, G. 1974, Ap. J., 187, 473.

Zuckerman, B., Ball, J. A., Dickinson, D. F., and Penfield, H. 1969, Ap. Letters, 3, 97.

Zuckerman, B., Yen, J. L., Gottlieb, C. A. and Palmer, P. 1972, Ap. J., 177, 59.

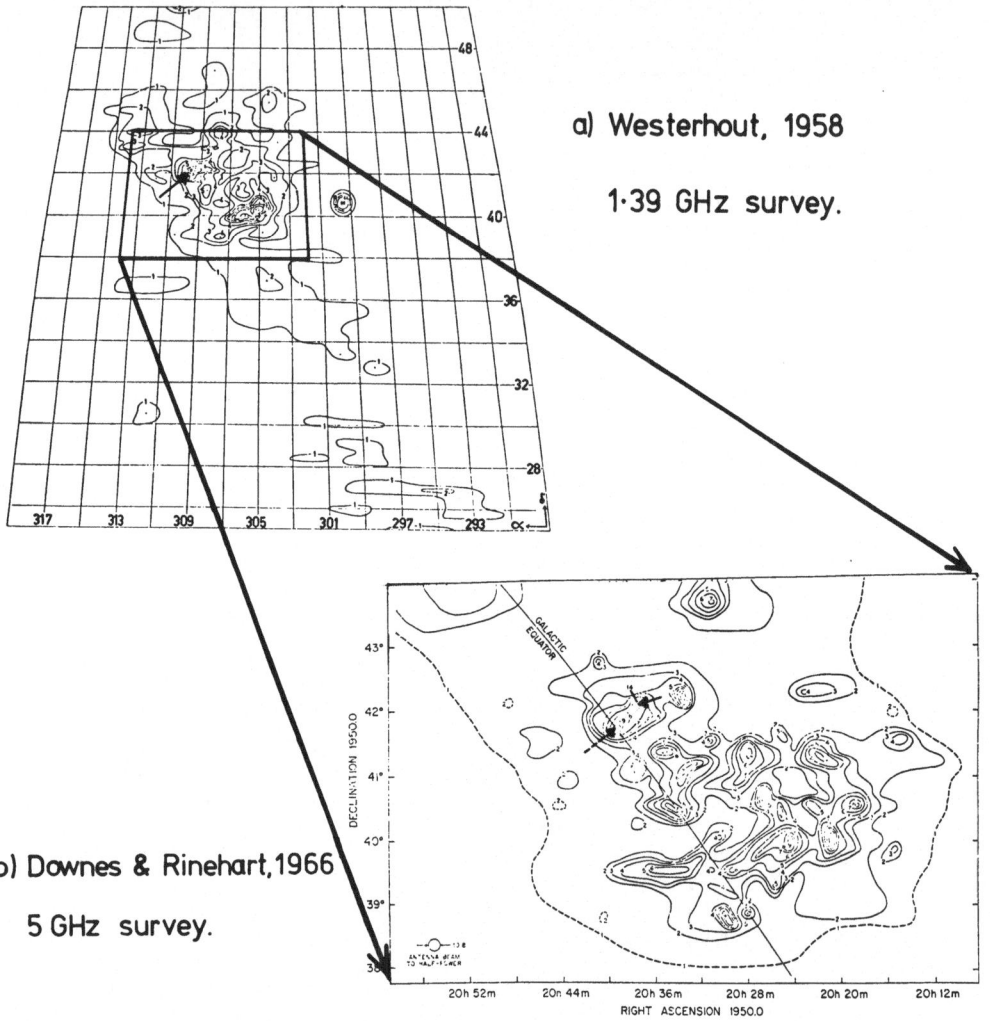

a) Westerhout, 1958

1·39 GHz survey.

b) Downes & Rinehart, 1966

5 GHz survey.

Fig. 1a.  A part of the Westerhout (1958) survey of the Galactic Ridge.
The arrow marks the position of the source W75.

Fig. 1b.  The Downes and Rinehart (1966) survey of the Cygnus X region.
W75 is here resolved into two main sources, DR21 (solid arrow)
and DR23 (dashed arrow).

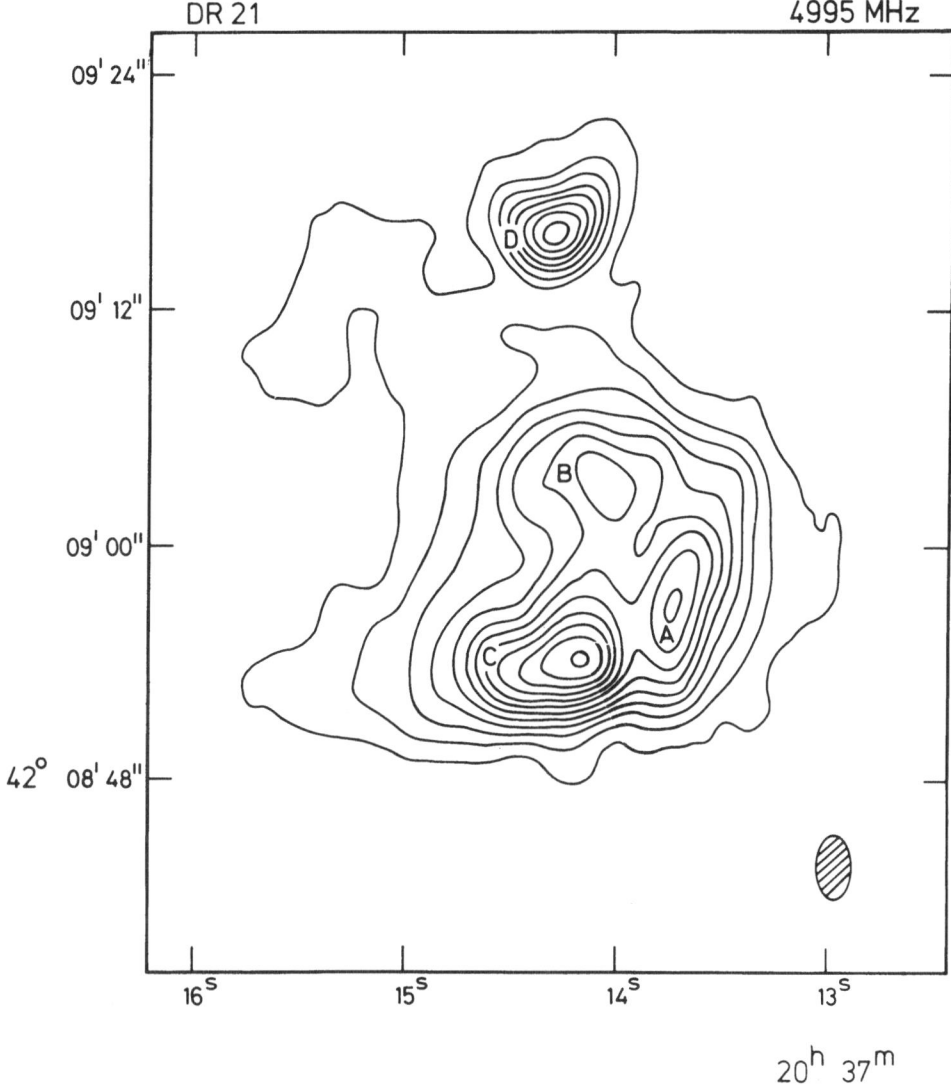

Fig. 2.   Contour map of DR21 at 5 GHz obtained with the Cambridge 5 km
synthesis telescope at a resolution of 2 arc sec.  The contour
interval is 392 K and the coordinates are for 1950.0.
(Reproduced from M.N.R.A.S., 162, 5P, by kind permission of
the editors).

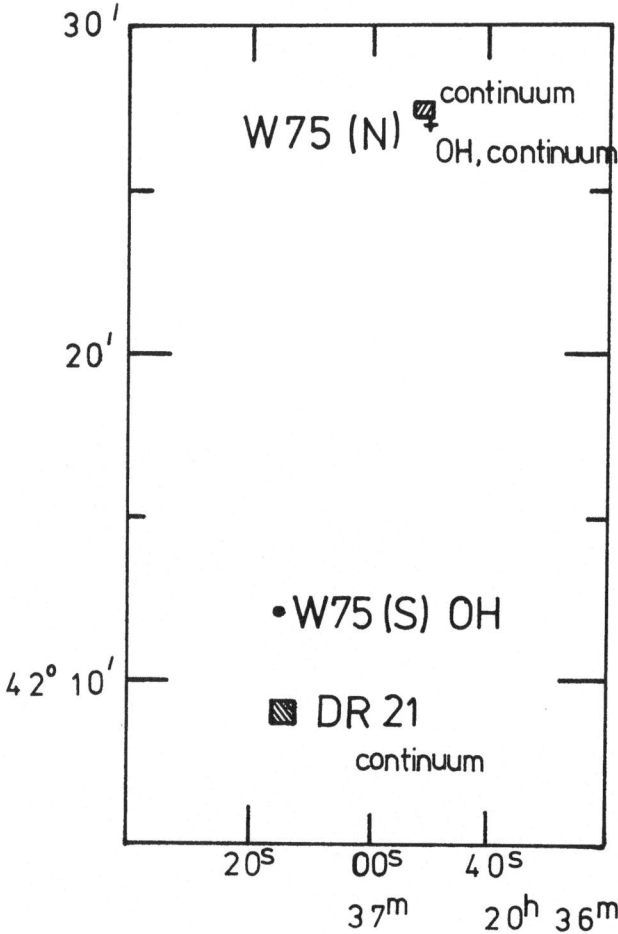

Fig. 3.   Diagram showing the location of the OH and continuum sources in W75.

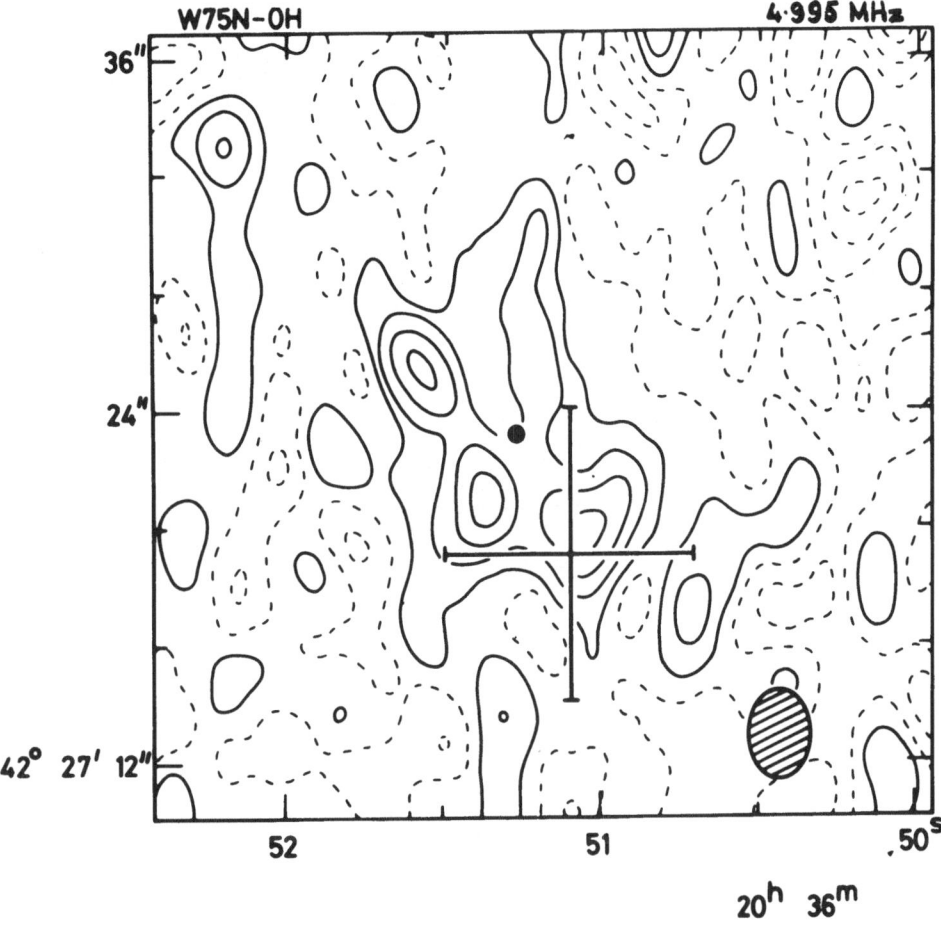

Fig. 4.  Contour map of W75N at 5 GHz obtained with the Cambridge 5 km
synthesis telescope at a resolution of 2 arc sec.  The
coordinates are for 1950.0.  The contour interval is 41 K,
the first positive contour being at +1/2 of this interval.
Negative contours are shown dotted.  The cross marks the
peak position of the 20 μ infrared radiation (Wynn-Williams
et al, 1974).  (Figure reproduced from M.N.R.A.S., <u>166</u>, 29P,
by kind permission of the editors).

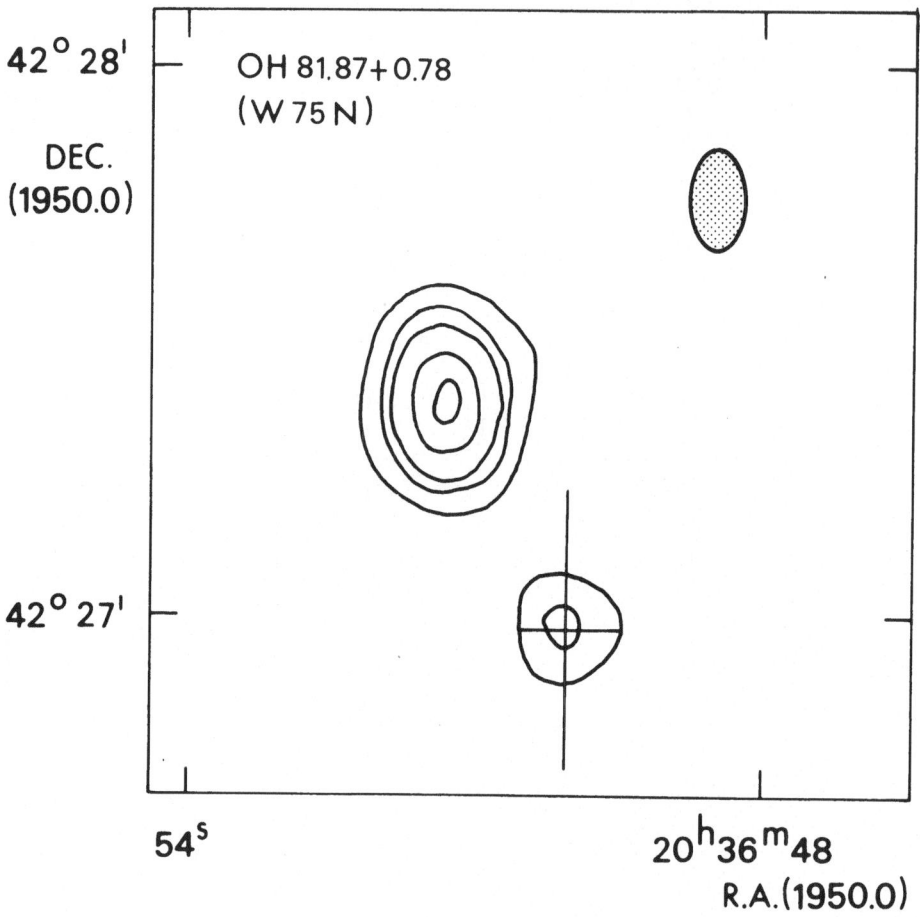

Fig. 5.  Contour map of the W75N and W75(N)OH continuum sources at 5 GHz
obtained with the Westerbork Synthesis Radio Telescope at a
resolution of 7 arc sec, (Habing et al 1974).  Contour values
are 1,3,5,10 and 15 units, where 1 contour unit is 3.6 K in
brightness temperature.  The position of the OH maser, as
measured by Hardebeck (1972), is indicated by a cross.
(Figure kindly supplied by the authors, and reproduced by
permission of the editors of Astronomy and Astrophysics ).

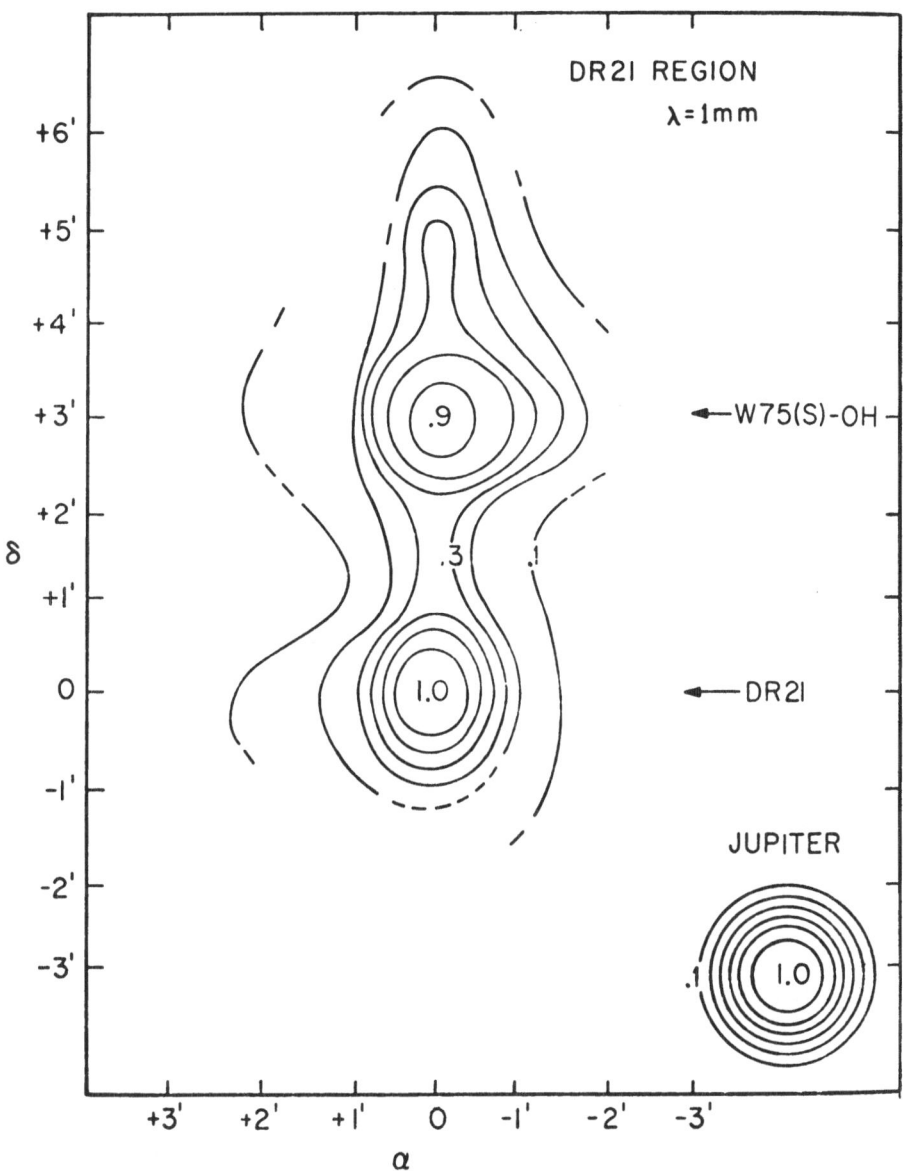

Fig. 6.   Contour map of the DR21/W75(S)OH sources obtained with the
200-inch Telescope at a wavelength of 1 mm, (Elias et al 1975).
The relative heights of the contours are as marked, and the
disc in the lower right-hand corner gives the telescope beam
area as measured on Jupiter.

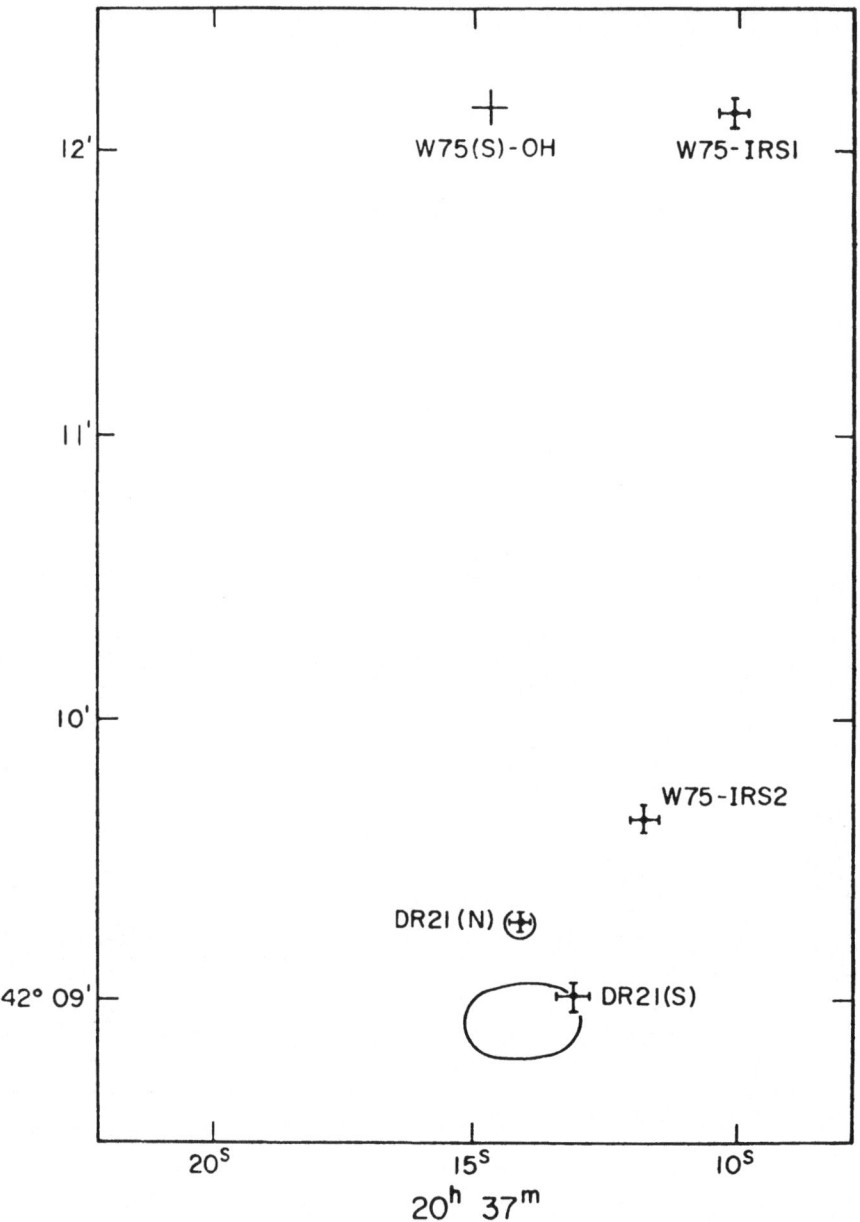

Fig. 7. Infrared sources in the vicinity of the DR21 and the W75(S)OH
source, (Wynn-Williams et al 1974).
(Taken from Ap.J., 187, 473, by kind permission of the authors
and publisher. Copyright, University of Chicago Press).

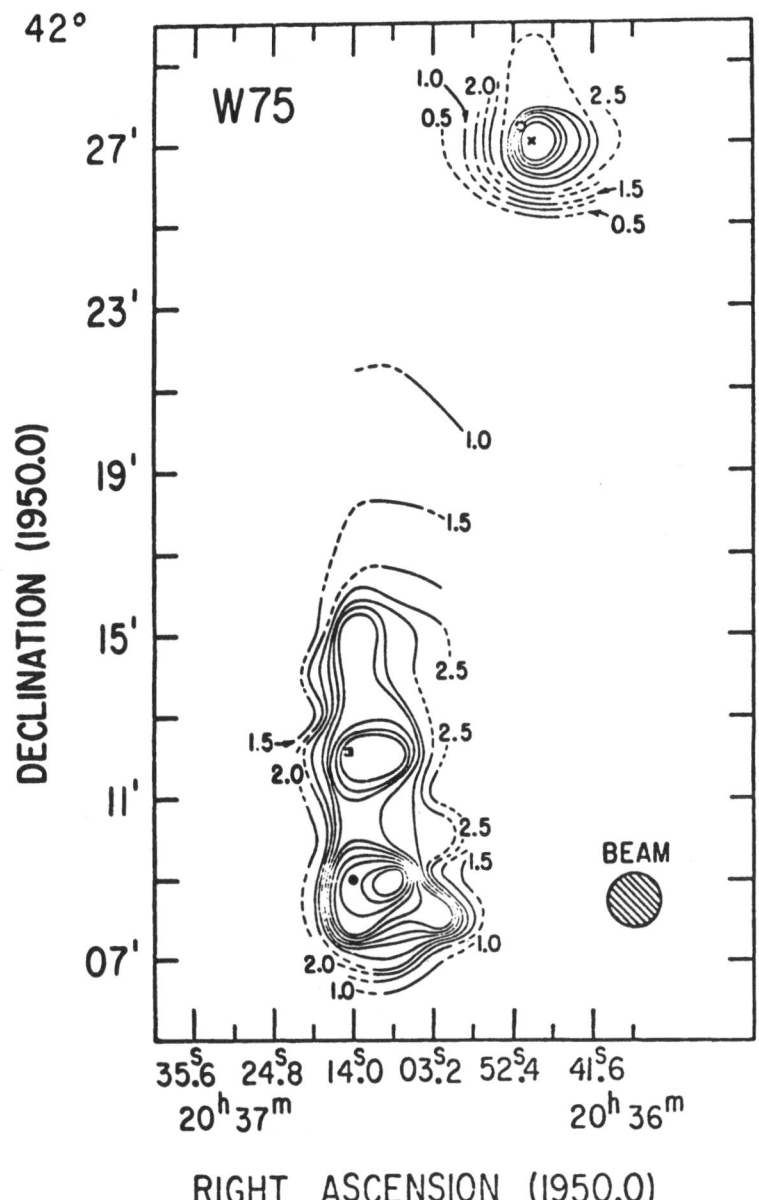

Fig. 8. Contours of equivalent width of HCN J =1→0 emission in the W75 region (Morris et al 1974).
(Taken from Ap.J., 191, 349, by kind permission of the authors and publisher. Copyright, University of Chicago Press).

# THE GIANT H II REGION W3

P. G. MEZGER and J. E. WINK

Max-Planck-Institut für Radioastronomie,
Bonn, FRG

## ABSTRACT

Radio- and IR-observations of W3 are summarized and interpreted and the origin of the thermal IR-radiation is discussed.

## I. INTRODUCTION

W3 is a giant H II region, located in the Perseus arm, at a distance
of about 3 kpc from the sun and 12 kpc from the galactic center. It is
associated (but not identical) with the optically visible H II region
IC 1795. East of IC 1795 are located the H II regions IC 1805 (= W4)
and IC 1848 (= W5) which, like IC 1795, are extended regions of strong
Hα-emission. In the radio range, all three H II regions show extended
thermal emission of weak to intermediate intensity. They are included
in the early λ21cm continuum survey by Westerhout (1958). W-numbers refer to this catalogue.

The λ2cm survey by Schraml and Mezger (1969), made with an angular
resolution of 2', showed that the dominant radio source falls west of
the region of strong Hα-emission referred to as IC 1795 and thus must
be hidden by dust clouds. Mezger and Höglund (1967) first measured the
radial velocity of W3 from observations of its H109α recombination line
emission. Comparison with the radial velocity of IC 1795, obtained from
its Hα-emission, showed that W3 and IC 1795 are approximately at the
same distance. A similar conclusion holds for IC 1805 and IC 1848. It
gradually became clear that here one observes an OB-association whose
subgroups and their associated H II regions are at different evolutionary stages. IC 1848 and 1805 are probably the oldest and most evolved
H II regions, in which the OB-stars have completely ionized and sufficiently dispersed the surrounding gas, so that the stars become visible. In IC 1795 one observes strong Hα-emission from the ionized gas
but the OB-stars are still obscured by dust. W3 represents the earliest
stage in this evolutionary sequence. The gas density is still so high
that the OB-stars have only ionized the inner part of the surrounding
cloud, which is sufficiently opaque to block most of the Hα-emission
associated with the strong radio emission. Much support was given to
this picture of an evolutionary sequence when Mezger et al. (1967) discovered the compact H II region G133.9+1.1, which is associated with
one of the most powerful OH maser sources. This source, which is now
often referred to as W3(OH), is one of the youngest H II regions known
in our Galaxy.

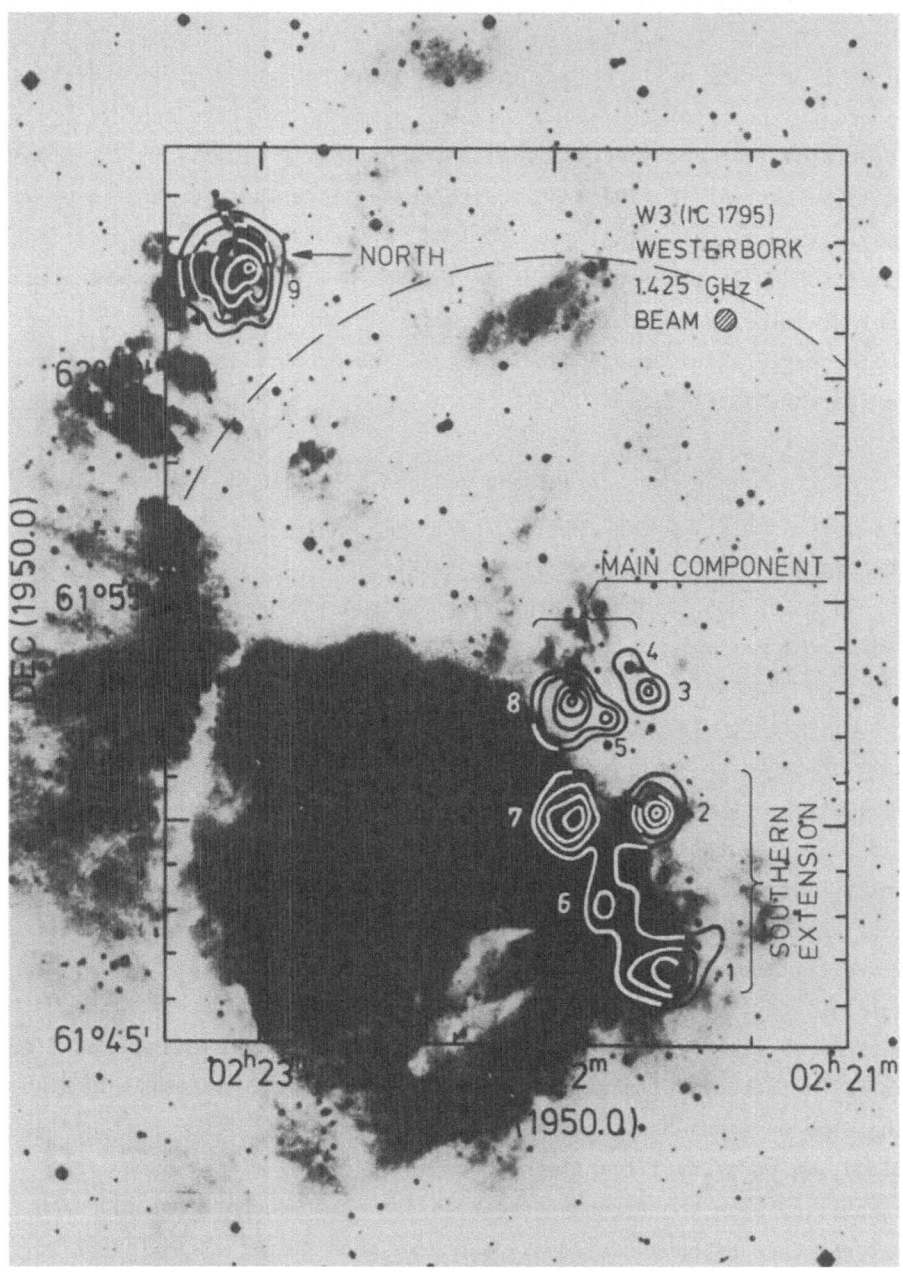

Fig. 1  W3, Main component, Southern extension and Northern component,
observed at λ21cm with the Westerbork synthesis telescope. Resolution 25'' x 28'' (α x δ) (Sullivan and Downes, 1973). Compact
sources 3, 4, 5 and 8 are part of the Main component; 1, 2, 6
and 7 are part of the Southern extension.

Since then, W3 became one of the best studied H II regions both in the radio and in the IR range. Research was stimulated by various considerations; i) one observes a number of O-stars and compact H II regions in various but very early evolutionary stages; ii) in W3, star formation may be initiated by compression of gas in a density wave; iii) the proximity of this giant H II region to the sun; iv) easy observability from the northern hemisphere.


## II. AN OVERVIEW OF EARLIER WORK


Mezger and Wink (1974; hereafter referred to as Paper I), in a recent review paper on compact H II regions, have compiled and interpreted observational data on W3. In this section, we briefly summarize the conclusions from this paper, together with some of the pertinent observational results which are needed for a better understanding of new results presented in the following section.

Figure I, taken from Sullivan and Downes (1973), is the superposition of a contour map of W3 (angular resolution 25" x 28" ($\alpha$ x $\delta$) on a Palomar red print. The numbers by the sub-components refer to Sullivan and Downes, while "North, Main component and Southern Extension" refer to the notation used in Paper I and in this paper. Not visible on this overlay is W3(OH), located at a distance of about 17' southeast of the center of the Main component.


When observed with high angular resolution, both in the IR and at radio wavelengths, W3 Main component breaks up into a number of individual sources, shown in Figure 2. The IR-maps at $\lambda 2.2\mu$ and $\lambda 20\mu$ are from Wynn-Williams et al. (1972), the aperture synthesis radio maps at $\lambda 3.7$cm and $\lambda 11$cm are from Wink et al.(1975).IRS 2 appears to be a heavily

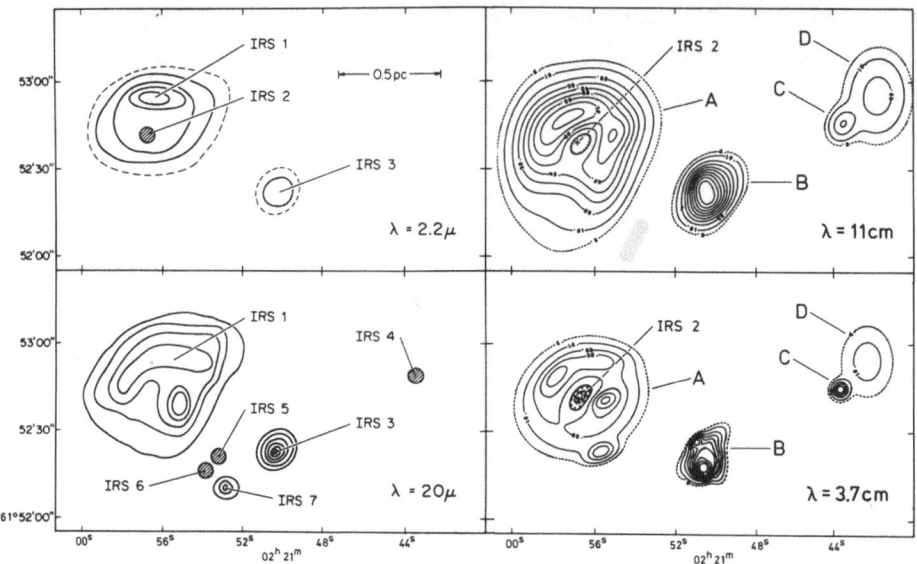

Fig. 2  IR- and radio map of W3, Main component. The IR-maps are from
Wynn-Williams et al. (1972) at λ2.2μ and λ20μ. The radio maps
are from Wink et al. (1975) at λ3.7cm and λ11cm.

obscured O-star; a second star-like object close to IRS 2 has recently
been detected in the near IR by Beetz et al. (1974). There are several
features to be noted: i) the similarity of the brightness distribution
of W3(A) at λ2.2μ and in the radio range with the λ20μ map. The former
maps represent free-free and free-bound emission from the gas while the
λ20μ map represents the thermal emission from dust grains which are
heated by the photon absorption; ii) there are no radio sources with
$S_{8.1GHz} \gtrsim 0.07$ f.u. associated with IRS 5, 6, 7. [*] The radio source
W3(D), on the other hand, shows no IR emission with $S_{20μ} \gtrsim 100$ f.u. IRS
5, which has a higher color temperature than the other IR sources and
no radio continuum emission, is associated with an $H_2O$ maser source.

The radio source W3(OH) has first been resolved by Wink et al. (1973).
It has an angular diameter of 1.7" and an electron density of  $2 \cdot 10^5$
$cm^{-3}$, if a spherical homogeneous density distribution is assumed. It is
surrounded by OH maser sources which may form a ring around the compact
H II region.

[*] Harten (Westerbork, priv. comm.) and Harris and Wynn-Williams (Cambridge, priv. comm.) recently found some weak radio emission from the
region of IRS 5,6,7.

Not much is known about W3 North. The radio observation of this source with the highest angular resolution is that by Sullivan and Downes (1973). It is not known as an IR source. Its $He^+$-abundance is similar to that of W3 Main component, $\simeq 0.07$.

By comparing the flux density derived from single-dish high resolution observations with the sum of the flux densities of the individual components, derived from aperture synthesis observations, we concluded that the ionized gas in W3 Main component is all contained in compact H II regions with well defined Strömgren spheres. This implies that the H II regions are ionization-bounded and still embedded in dense, neutral gas. In fact, Wynn-Williams et.al. (1972), by comparing observed flux densities in the near IR with flux densities expected from the extrapo- lated free-free radio spectrum, could estimate a visual extinction of $14^m$ in front of W3(A) and a considerably higher extinction in the direc- tion of W3(OH). Most of this extinction must occur in the immediate vicinity of the H II region.

Since no optical photons escape the dust envelope around this H II region, all are eventually absorbed by dust grains and reemitted in the IR. Hence, the total IR-luminosity, $L_{IR}$, should equal the stellar luminosity, $L_*$. Applying this argument and using stellar atmospheres, we could show that single MS stars can account for the observed IR- and radio flux densities of the compact H II regions and IR-sources ob- served in W3.

A similar comparison of the sum of the radio flux densities from com- pact components in the southern extension (seen in Fig. 1) with the in- tegrated flux density shows that only 14 % of the total radio flux density comes from the compact components. These sources are also not seen in the IR.

We interpret these observations in the following way: O-stars reach the Main Sequence (MS) embedded in a cocoon of dust and gas. In its early stages, all photons emitted by the star are absorbed by dust and the cocoon is seen as a relatively hot IR source, with no radio con- tinuum source associated. IRS 5 may represent this stage. Next, the star forms a small compact H II region. Most of the photons $\lambda > 912$ A escape the H II region and are absorbed in the dense neutral gas and dust surrounding the H II region. The temperature of the dust grains is lower than in the preceeding cocoon, and the IR-spectrum peaks in the far IR, typically around $\lambda 100\mu$. W3(OH) may represent this stage. The ionized gas expands and more and more of the surrounding neutral gas is ionized. W3(A) probably represents a more evolved stage than W3(OH). Eventually, all of the neutral shell is ionized and Lyman continuum (Lyc) photons escape into the low-density gas which surrounds the condensations out of which the O-stars have formed. An example of this stage may be the Southern Extension of W3. Note that here the radio emission is associated with strong H$\alpha$-emission (Figure 1). IC 1795, 1805 and 1848 represent later stages in this evolutionary sequence of H II regions and O-star subgroups.

In this picture, O-stars do not form out of very massive molecular clouds. It rather appears that, in large clouds of neutral gas of medium average density, condensations of some 100 $M_\odot$ form, which contract and form O-stars in their center. The gas between these condensations has a relatively low density. From the total mass of ionized hydrogen in a more evolved compact H II region like W3(A) plus the mass contained in the surrounding neutral shell we estimate that perhaps one half of the mass of a contracting condensation is transformed into stars of spectral type O and later. If O-stars were formed inside of very massive, dense clouds, it would be hard to see how they could disperse the surrounding gas and dust during their lifetime and thus become visible at all. The picture suggested here helps to overcome this difficulty.

Only some stages of this evolutionary sequence of O-stars and associated IR-sources and compact H II regions have been worked out in detail. We refer to the review papers in this volume by Kahn, by Kippenhahn and by Krügel.

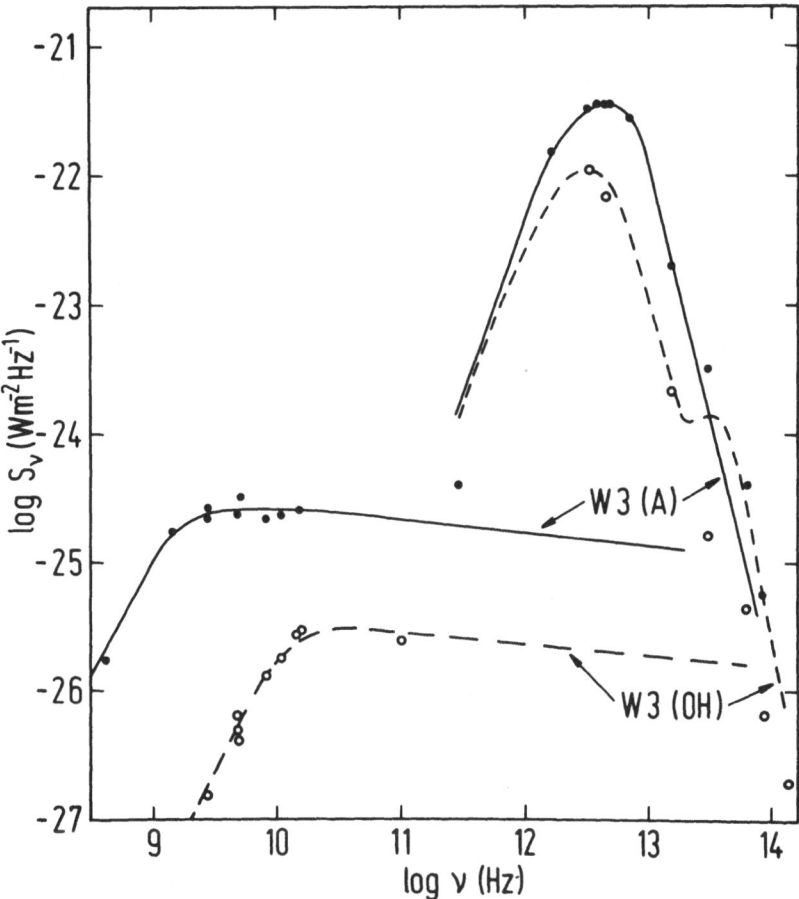

Fig. 3 Combined radio and IR spectra of W3(A) and W3(OH). Observational
data from various observers; lines represent model fits
by Krügel and Mezger (1975).

III. <u>THE ORIGIN OF THE IR-RADIATION</u>

Figure 3 shows the combined radio- and IR-spectra of two of the compact H II regions in the W3 complex, W3(A)/IRS1 and W3(OH)/IRS8. These spectra have been compiled from various sources by Krügel and Mezger (1975; hereafter referred to as Paper II). IR-observations with high angular resolution (i.e. some arcsec) are only available at wavelengths $\lambda \leq 20\mu$ (see Fig, 2), at which the main contribution comes from hot dust inside the H II region, which is heated by absorption of L$\alpha$- and Lyc-radiation. The peak of the IR-emission, however, occurs at wavelengths shortward of $\lambda 100\mu$, corresponding to a blackbody temperature of $\sim 70$ K. It is still a controversial question whether this cool dust is located in the outer part of the H II region or rather in a dense shell of dust and neutral gas surrounding the H II region.

A straightforward answer to this question could be given if the apparent sizes of both radio and far IR-sources were known. However, the angular resolution of balloon-borne far IR-telescopes was insufficient to resolve any of the compact IR-sources in the W3 complex. Only recently did Fazio et al. (1974) map W3 at $\lambda 69\mu$ with sufficient angular resolution (60'' and 30'', respectively) to resolve or at least to obtain meaningful upper limits of the size of three IR-sources. The $\lambda 69\mu$ map by Fazio et al. is shown in the lower part of Figure 4. Apart from W3 Main component and OH a third IR source was found which we here designate "East."

The upper part of the diagram shows a $\lambda 6$cm map of the W3 complex, observed by Smith, Altenhoff and Schraml with the MPIfR 100-m telescope (angular resolution 2.6'). The counterpart of W3 East is clearly seen in the radio map; the flux density of this source is $S_5 \sim 0.6$ f.u. It appears to be optically thin at $\lambda 6$cm. Fazio et al. also found some $\lambda 69\mu$ emission immediately to the east of W3(A). This may be the counterpart of the plateau seen in the radio map.

Some observed or inferred parameters of the three sources are compiled in Table 1. $N_c$ is the intrinsic Lyc-photon flux inferred from $L_*$ and stellar model atmospheres (Paper I). $N_c'$ is the number of Lyc-photons

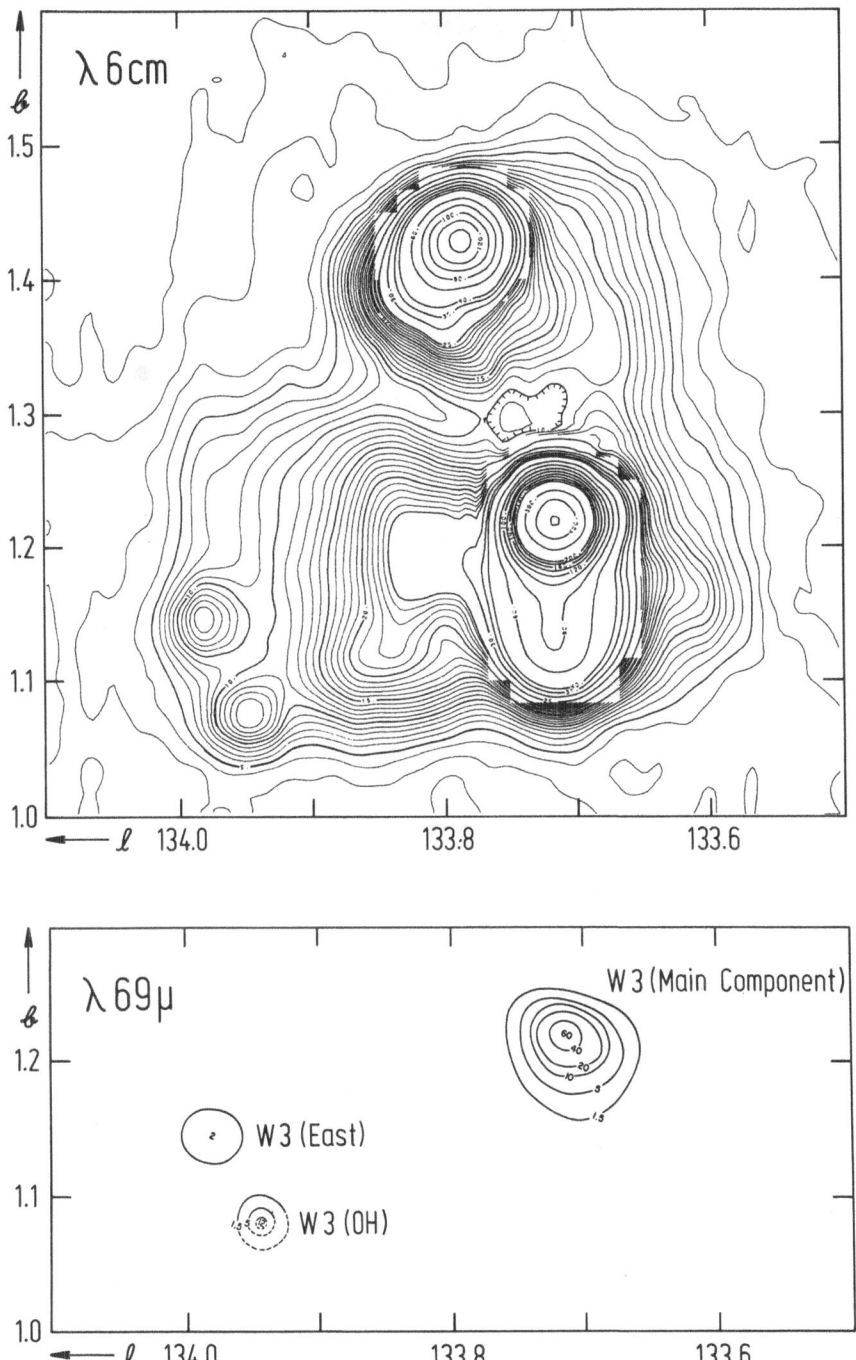

Fig. 4  λ6cm map of the W3 region observed by Smith, Altenhoff and
Schraml with the MPIfR 100-m telescope. Angular resolution 2.6'.
λ69μ map of the W3 region by Fazio et al. (1975) observed with
angular resolutions of 30" and 60", respectively. Not shown in
the diagram is a region of weak IR emission to the east of Main
component.

TABLE 1

| Source Name | $\ell$ | $\flat$ | $L_{IR} - L_*$ | $N_c$ | $N'_c$ | $N'_c/N_c$ | R | $\Theta_{Radio}$ | $D\Theta_{Radio}$ | $\dfrac{\Theta_{20\mu}}{\Theta_{Radio}}$ | $\dfrac{\Theta_{69\mu}}{\Theta_{Radio}}$ |
|---|---|---|---|---|---|---|---|---|---|---|---|
| | Degree | | erg · s$^{-1}$ | s$^{-1}$ | | | | arc sec | cm | | |
| W3(A)/IRS1 | 133.716 | 1.220 | 3.3 E 39 | 5.0 E 49 | 1.6 E 49 | 0.32 | 0.65 | 40 | 1.8 E 18 | 1 | <3 |
| W3(OH)/IRS8 | 133.945 | 1.072 | 7.1 E 38 | 1.0 E 49 | 3.2 E 48 | 0.32 | | 1.6 | 7.1 E 16 | <2 | ≲38 |
| W3(East) | 133.987 | 1.147 | 1.3 E 38 | 4 E 47 | 5 E 47 | ~1.0 | | ≲60 | ≲2.7 E 18 | | ≲2.4 |

absorbed by the gas. All but the parameters for W3 East are taken from Paper II. $L_{IR}$ of W3 East is estimated from the ratio of its λ69μ flux density to that of W3(OH) (mean value) as given by Fazio et al., multiplied with the luminosity of W3(OH).

At least in the case of W3(OH), the λ69μ source is much more extended than the radio source. This is an indication that in this source most of the far IR-radiation originates outside the compact H II region. However, O-stars emit 50 % of their energy at λ < 912 A in the Lyc-region. The fact that, for W3(A) and W3(OH), the number $N'_c$ of Lyc-photons absorbed by the gas is considerably smaller than the estimated stellar Lyc-photon flux $N_c$ (Table 1) is a first (but not a unique) indication that in some cases large fractions of the total stellar radiation are absorbed by the gas inside the H II region. A better indicator for absorption of Lyc-photons inside the H II region is the ratio R of its $He^+$- to $H^+$-Strömgren sphere, weighted with the square of the proton density. If the ratio γ of He-photons $N_{He}$(504 > λ/A > 228) to H-photons, $N_H$(912 > λ/A > 504), is less than 0.20, then R < 1. On the other hand, all O-stars have γ ≳ 0.20 and hence we would expect R = 1 for all but a few small H II regions which are ionized by B-stars. However, since the absorption cross section of dust for He-photons is about six times that for H-photons, the stellar Lyc-radiation field gets depleted of He-photons (see the review paper by L.F. Smith, this volume). As a result, R can become smaller than unity even if γ > 0.20. What counts is the ratio $N'_{He}/N'_H$ = γ' of He-to H-photons available for ionization of the gas. γ' is a function of the optical absorption depths for He- and H-photons inside the H II region.

$$\gamma' = \gamma e^{-(\tau_{He} - \tau_H)} \qquad (1)$$

This and subsequent formulae are from Mezger, Smith and Churchwell (1974). Hence $R(\gamma') = R(\tau_{He}, \tau_H)$, once $\gamma$ is fixed by the spectral type of the ionizing star. Similarly, the ratio $N'_c / N_c$, discussed above, can be expressed as a function of $\tau_H$, $\tau_{He}$.

$$\frac{N'_c}{N_c} = f_{net} = \frac{e^{-\tau_H} + \gamma e^{-\tau_{He}}}{1 + \gamma} \qquad (2)$$

The total IR-luminosity of the H II region is equal to the amount of stellar radiation absorbed by the dust inside and outside the H II region. With

$$f_{eff} = \frac{e^{-\tau_H} + \gamma \varepsilon e^{-\tau_{He}}}{1 + \gamma \varepsilon} \qquad (3)$$

and $\varepsilon = \langle E_{He} \rangle / \langle E_H \rangle \simeq 1.85$ the ratio of the average energies of He- and H-photons, $L_{IR}$ can be written as a function of various optical absorption depths, $\tau_{He}$, $\tau_H$, $\tau_u$, $\tau_o$ *)

$$L_{IR} = \underbrace{N_\alpha h\nu_\alpha}_{I} + \underbrace{(1 - f_{eff}) N_c \langle E_{Lyc} \rangle}_{II} \qquad (4)$$

$$+ \underbrace{\left[ f_{eff} N_c \langle E_{Lyc} \rangle - N_\alpha h\nu_\alpha \right] (1 - e^{-\tau_o})}_{III}$$

$$+ \underbrace{L_{\lambda > 912} (1 - e^{-\tau_u})}_{IV}$$

---

*) This expression for $L_{IR}$ differs from that given by Mezger et al., 1974, by the addition of term III.

The physical meaning of these four terms is:

I. Contribution of trapped L$\alpha$-photons; $N_\alpha \simeq N'_c$

II. Heating of dust inside the H II region by direct absorption of Lyc-photons; $\langle E_{Lyc} \rangle$ is the average energy of Lyc-photons.

III. Contribution of Lyc-photons absorbed by the gas inside the H II region. Most of this radiation is thought to escape from the H II region in the form of forbidden lines of C, N, O and Ne. $\tau_o = \tau'_o + \tau''_o$ is the sum of absorption depths for visual photons inside $(\tau'_o)$ and outside $(\tau''_o)$ of the H II region.

IV. Contribution of the UV-radiation $\lambda > 912$ A. $\tau_u = \tau'_u + \tau''_u$ is the sum of absorption depths for UV-photons inside $(\tau'_u)$ and outside $(\tau''_u)$ of the H II region. Eq. (1) and hence $R(\gamma')$, eqs. (2) and (3) can be evaluated as a function of $\tau_H$ if the absorption cross section of dust grains for various wavelength ranges are known. With x the number of dust grains per H-atom we adopt the following values:

$$a_o = x\sigma_{He}/x\sigma_H = 5$$
$$b_1 = x\sigma_u/x\sigma_H = 1$$
$$b_2 = x\sigma_o/x\sigma_H = 1$$

We assume the stellar parameters given by Mezger et al. (1974) for weighted average values for the Salpeter original luminosity function. Further, we assume $\tau_{He} = a_o \tau_H R^{1/3}$ valid for a spherical H II region of constant electron density. The resulting functions $L_{IR}/L_*$, R and $N'_c/N_c$ as a function of $\tau_H$, the absorption depth for H-photons, are shown in Fig. 5.

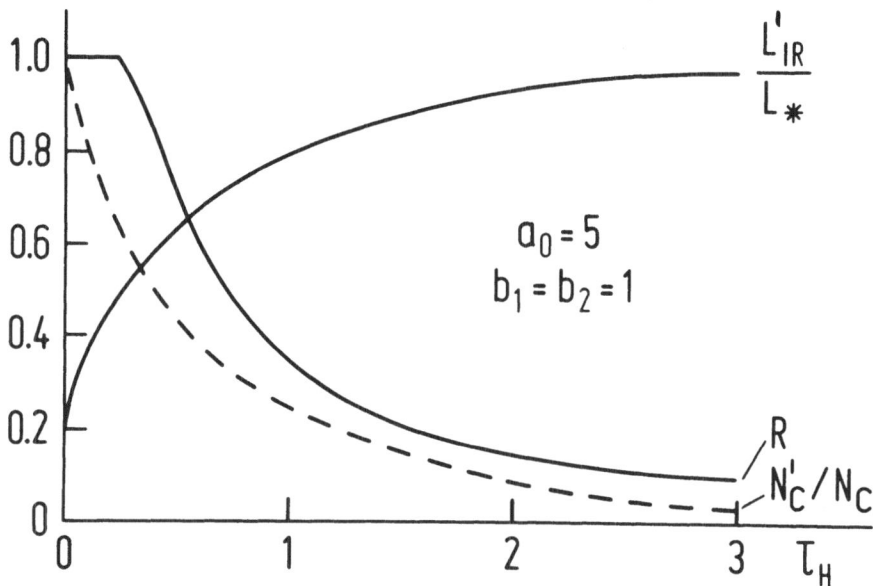

Fig. 5   Diagram showing as a function of the absorption depth $\tau_H$ for H-
photons inside the H II region: The fraction $L'_{IR}/L_*$ of the total
stellar luminosity radiated in the IR by dust   inside the H II
region. The ratio R of $He^+$- to $H^+$-Strömgren spheres, weighted
with the square of the proton density. The fraction $N'_c/N_c$ of
stellar Lyc-photons absorbed by the gas in the H II region.

Note that $L'_{IR} = L_{IR}$ ($\tau''_o = 0$, $\tau''_u = 0$), i.e. $L'_{IR}$ is the energy contribution by only those photons which are absorbed by dust inside the H II
region. Therefore $L'_{IR}/L_*$ is that fraction of the total stellar luminosity
$L_*$ which is radiated in the IR from inside the H II region. Consider now
an H II region in a dense dust shell, i.e. $\tau''_u$, $\tau''_o \rightarrow \infty$. Eq. (4) then
yields immediately $L_{IR} = L_*$. Thus ($1 - L'_{IR}/L_*$) is that fraction of the
total stellar luminosity that either escapes into interstellar space
or is absorbed and reradiated in the far IR by the dust surrounding the
H II region. For $\tau_H \ll 1$, absorption of $L\alpha$-photons is the only source
of heating of the dust inside the H II region. For the stellar parameters adopted here, this amounts to 0.21 $L_*$. With increasing $\tau_H$, R
and $N'_c/N_c$ decrease as the result of absorption of Lyc-photons by dust
inside the H II region. For R = 0.65, the value estimated from observations for W3(A), 65 % of the  total IR radiation should be emitted

by the H II region. These considerations, which hold for any compact H II region, we now apply to the three compact H II regions of W3, whose observed and inferred parameters are listed in Table 1. [*]

W3(A): Both R and $N_c'/N_c$ are less than unity. This implies that most of the IR radiation originates in the H II region proper. This conclusion is also supported by the fact that $\Theta_{69\mu} \simeq \Theta_{Radio}$. The surface brightness distribution of both dust and ionized gas (Fig. 2) shows a ringlike structure, indicating that ionized gas and dust form a shell rather than a uniform spherical H II region.

W3(OH): Here the observational results appear to be contradictory. $N_c'/N_c = 0.32$ suggests that most of the IR-radiation originates inside the H II region. $\Theta_{69\mu} \simeq 40 \, \Theta_{Radio}$ indicates that most of the IR-radiation originates in the neutral gas surrounding the very compact H II region. R is not known, since no recombination lines have yet been observed from this source. Support for the latter conclusion - i.e. IR-emission primarily from the surrounding neutral gas - is obtained from the IR-spectrum Fig. 3. The peak of the IR-spectrum is primarily determined by the mean distance of the absorbing dust from the ionizing star. The similarity of the spectra of W3(A) and W3(OH) indicates that this mean distance must be similar in the two sources. This means that the dust responsible for the IR-emission from W3(OH) must form a shell around the compact H II region whose diameter is similar to that of the shell of ionized gas and dust in W3(A). If this is correct, practically no Lyc-photons are absorbed in the compact H II region. This implies that the H II region is dust-depleted and $N_c$ - estimated from $L_* = L_{IR}$ for a ZAMS-star - is an overestimate, since the star has not yet reached the MS.

W3 East: Not much information is available for this source. From the parameters in Table 1, we conclude that this source is ionized by a B-star, that most of its Lyc-photons are absorbed by the gas and that a

---

[*] Note, however, that Figure 5 holds for stellar parameters which are average values weighted by the Salpeter original luminosity function while in individual model fits (such as Paper II) one uses stellar parameters pertaining to single OB-stars.

large fraction of its IR-radiation originates in the neutral gas surrounding the H II region.

Similar, but quantitative conclusions pertaining to the absorption
characteristics of dust and the distribution of dust and ionized and
neutral gas have been reached by Krügel and Mezger (1975), who constructed models and fitted them to the observed integral parameters and
the IR- and radio spectra. The curves in Fig. 3 result from this model
fitting, which is described in more detail in the review paper by Krügel
(this volume). It appears that W3(A) and W3(OH) represent two different
stages in the evolution of O-stars and their compact H II regions. The
review paper by Krügel will also deal with these implications.

## REFERENCES

Beetz, M., Elsässer, H., Weinberger, R. 1974, Astron. & Astrophys. 34
    335
Fazio, G.G., Kleinmann, D.E., Noyes, R.W., Wright, E.L., Zeilik II, M.
    Low, F.J. 1974, Proc. ESRO Symp. "H II Regions and the Galactic
    Center" (A.M.F. Moorwood, ed.) page 79
Krügel, E., Mezger, P.G. 1975, paper submitted to Astron. & Astrophys.
    (Paper II)
Mezger, P.G., Höglund, B. 1967, Ap. J. 147, 490
Mezger, P.G., Altenhoff, W., Schraml, J., Burke, B.F., Reifenstein, G.C.,
    Wilson, T.L. 1967, Ap. J. 150, L157
Mezger, P.G., Wink, J.E. 1974, Inv. rev. paper, 2nd European Meeting in
    Astron., Trieste, to be published in "Memorie della Sociêta Astro-
    nomica Italiana (Paper I)
Mezger, P.G., Smith, L.F., Churchwell, E. 1974, Astron. & Astrophys. 32,
    269
Schraml, J., Mezger, P.G. 1969, Ap. J. 156, 269
Sullivan III, W.T., Downes, D. 1973, Astron. & Astrophys. 29, 369
Westerhout, G. 1958, BAN 488, 215
Wink, J.E., Altenhoff, W.J., Webster, W.J. 1973, Astron. & Astrophys.
    22, 251
Wink, J.E., Altenhoff, W.J., Webster, W.J. 1975, Astron. & Astrophys.
    38, 109
Wynn-Williams, C.G., Becklin, E.E., Neugebauer, G. 1972, MNRAS 160, 1

# THE W 49 REGION

T.L. Wilson

Max-Planck-Institut für Radioastronomie,
Bonn, FRG

## ABSTRACT

The radio source W 49 consists of: (1) an H II region, W 49A, at galactic longitude $\ell = 43.2^{\rm O}$ and galactic latitude $b = 0.0^{\rm O}$, and (2) a supernova remnant W 49B, at $\ell = 43.3^{\rm O}$, $b = -0.2^{\rm O}$. Mezger, Schraml and Terzian (1967) have summarized the observations of these sources made before 1967. This review emphasizes data taken since that time.

## 1. Distance to the sources from the sun

The W 49 region is not observable optically, because of extinction by dust. The velocity of the H II region, W 49A, measured from radio recombination lines is $\sim 9$ km s$^{-1}$ (see e.g. Gordon and Wallace, 1971). H I absorption line velocities (Akabane and Kerr, 1965) extend to $\sim 70$ km s$^{-1}$, the maximum velocity for this direction from H I emission results. Therefore, the source is at the far kinematic distance. This distance from the sun, calculated from the Schmidt (1965) Model, is 14 kpc. In contrast to W 49A, the distance to W 49B, the SNR, is not well established. From the H I observations Kazès and Rieu (1970) and Radhakrishnan et al. (1972) favor a distance of $\sim 10$ kpc to the SNR, while Sato, Akabane and Kerr (1967) conclude the SNR is $\sim 1$ kpc closer to the sun than W 49A. Mezger et al. (1967) assumed that W 49A and W 49B were the same distance from the sun, and Pastchenko and Slysh (1973) from their OH data also favor the same distance for the two sources. Although radio recombination lines have been detected toward W 49B, these lines are thought to be formed along the line of sight to the SNR; we will discuss these spectral lines in Section 4A. Therefore, we do not know the velocity of the SNR itself, and without this velocity, no definite distance estimate is possible. Furthermore, the absorption line velocity, which would place W 49B in the same general region of space as W 49A, would lie in the range from $\sim 0$ to $\sim 20$ km s$^{-1}$. However, gas close to the sun could also have such a velocity, and therefore the detection of absorption features in the range 0 to 20 km s$^{-1}$ might not lead to an unambiguous distance. It is highly probable that the SNR is more than 10 kpc from the sun, but one can, at present, only postulate, as have Mezger et al. (1967), that the H II region and SNR are the same distance from the sun and have formed from the same interstellar cloud.

## 2. Continuum observations of the W 49 region

Figure 1(a) contains a map of the W 49 region made with the 1 Mile Telescope at 0.4 GHz and Fig. 1(b), a map made with the 100-m telescope at 10.7 GHz. In both maps, the smallest separation between the western

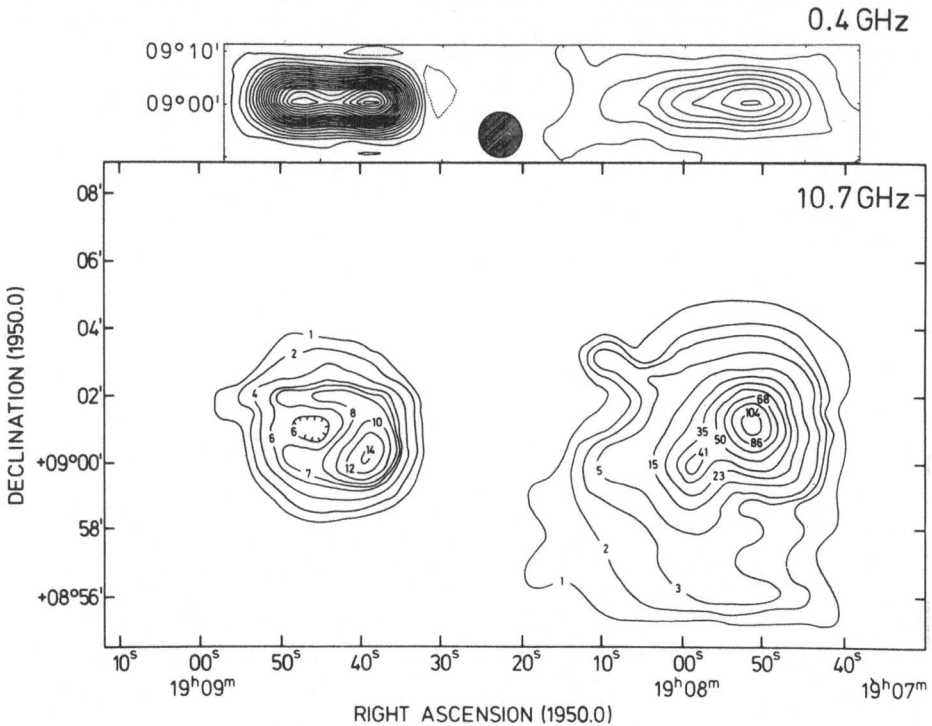

Figure 1(a). A continuum map of the W 49 region made with the Cambridge 1 Mile Telescope (Wynn-Williams, 1969). The Declination axis has been compressed by a factor of 6.4 to give a circular beamshape. The brightness temperature contour interval is 126 K. Figure 1(b) is a map made with the 100-m telescope (Downes and Wilson, 1974). The angular resolution of both maps is the same in R.A., ∿ 1.3'. The brightness temperature interval of the 10.7 GHz map is 0.31 K.

source, the H II region W 49A, and the eastern source, the SNR W 49B is 4', while the distance from the center of W 49A to the center of W 49B is 12'. At a distance of 14 kpc, 1' equals 4.1 pc.

A.   The SNR W 49B

With an angular resolution of 1.3' the SNR W 49B has a partial shell-like structure at 10.7 GHz (Fig. 1(b)), and a double-peak shape at 0.4 GHz (Fig. 1(a)). From observations at 0.4 and 1.4 GHz, Wynn-Williams (1969) concluded that the spectral index $\alpha$, where $S_\nu \sim \nu^{-\alpha}$, is 0.45 in

the western part and 0.6 in the eastern part of the SNR. Baker et al.
(1975), from a comparison of their data at 15 GHz with Wynn-Williams'
(1969) data at 1.4 GHz, conclude that the difference in the spectral
indices of the eastern and western parts of the SNR must be small between 1.4 and 15 GHz. Aizu and Tabara (1967) have estimated that the age
of the SNR is $\sim$ 3000 years. Estimates based on either the uniform
expansion or "blast wave" model, assuming typical SNR and interstellar
medium conditions yield ages between 3000 and 6000 years.


B.  The H II region W 49A

Figure 1 shows that there is substructure in the H II region. Mezger
et al. (1967) had fitted the free-free continuum radio spectrum of
W 49A with  a model consisting of 2 H II regions.  These were pictured
as a high density, high emission measure, small apparent diameter region,
called A2, embedded in an H II region, A1, which had lower density,
lower emission measure and larger apparent diameter.  The A2 region itself might consist of a number of condensations which would have about
the same emission measure, $10^8$ pc cm$^{-6}$.  This emission measure was
derived from the turnover frequency of the radio continuum assuming
thermal emission from a spherical, uniform density sphere.  Shaver and
Goss (1970) fitted a single smooth curve to essentially the same continuum flux density data used by Mezger et al.  Shaver and Goss interpreted
this as meaning that W 49A consisted of a few components, each having a
different emission measure.  More fundamentally, the continuum radio
spectrum of an H II region might be affected by density gradients in the
plasma, which would change the value of the turnover frequency and,
hence, the emission measure of the source which is derived from the turnover frequency (for example, see the discussion of Panagia and Felli
(1975)).  Direct determinations of the source structure, for example in
Fig. 2(a), show at least 6 high density H II regions on a more extended
continuum background.  Wink et al. (1975) have shown that the condensations have a spread in turnover frequencies between $\sim$ 2 GHz and $\sim$ 5 GHz.

The total integrated flux density of the 5 GHz map, in Fig. 2(a), is
61 per cent of the flux density observed with a single telescope.  The

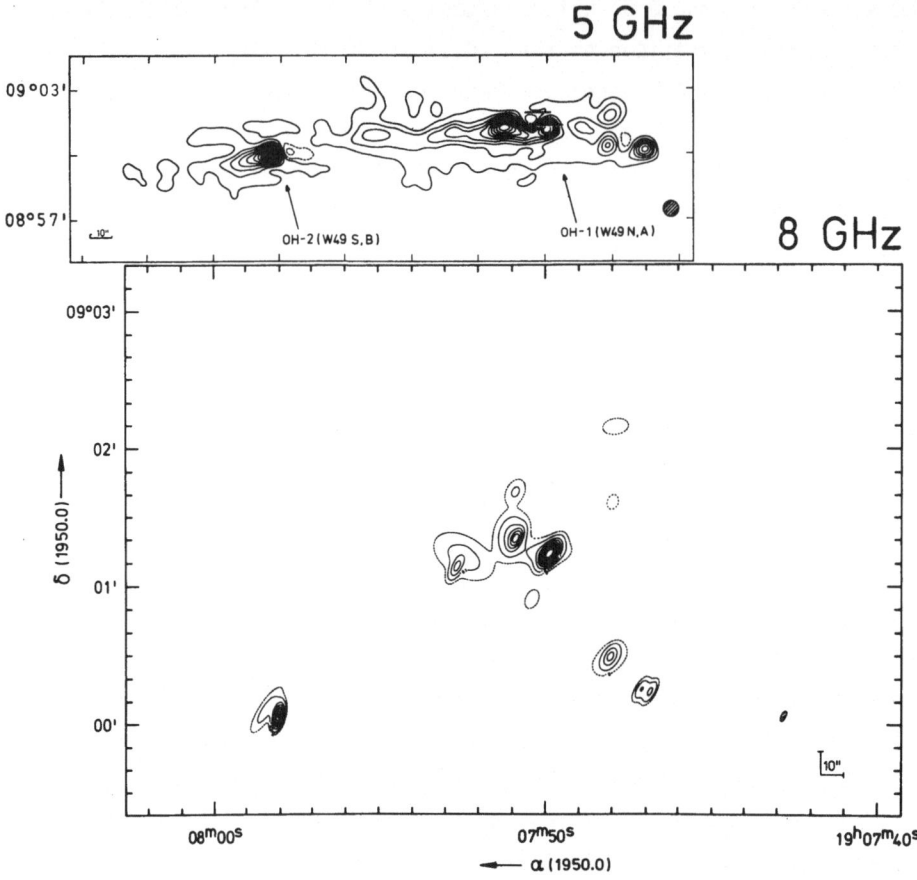

Figure 2(a). A continuum map of W 49A made with the Cambridge 1 Mile Telescope (Wynn-Williams, 1971). The angular resolution is 6.5" x 42" ($\alpha$ x $\delta$), and the brightness temperature contour interval is 40 K.
Figure 2(b). The result of a best fit of Gaussians to a continuum map of W 49A made with the NRAO interferometer (Wink et al., 1975). The map shown was obtained by convolving the model fit solution with a 6.5" x 2" Gaussian, which represents the interferometer beam.

small condensations contain 34 per cent of the single dish flux. As usual, the observations of H II regions made with higher angular resolution give sources with higher electron densities and lower masses. Wink et al. (1975) and Schraml and Mezger (1969) have found that most of the mass of W 49A is contained in a source underlying the small condensations. The gaussian full width to half power of this source is 2' x 1' ($\alpha$ x $\delta$). If this were a uniform sphere at a distance of 14 kpc, the mass of ionized gas would be 2700 $M_\odot$, and this source would have a spherical size (which is 1.47 times the gaussian size) of $\sim$ 12 pc x 6 pc. It is highly likely that this region is made up of a number of small H II regions which are not resolved into individual sources by the beam.

In Table 1, we compare the parameters of the condensations in W 49 measured at 8 GHz with the parameters of the core of the Orion nebula, measured at 15 GHz. The distance to W 49A is $\sim$ 28 times larger than the distance to Orion, so that the linear size corresponding to the beam width is $\sim$ 0.3 pc in both cases. Structure is definitely present in Orion A on a scale finer than 2', so we would similarly expect structure in W 49A on a scale smaller than $\sim$ 3", the best current angular resolution for continuum observations.

A COMPARISON OF THE DERIVED PHYSICAL PARAMETERS OF THE CONDENSATIONS IN W49 AND ORION A

|  | W49 Condensations [x] | Orion A [y] |
|---|---|---|
| Angular resolution of radio telescope used | 2" x 6.5" | 2' |
| Linear size of nebula | 0.5 pc | 0.6 pc |
| $\langle N_e^2 \rangle^{1/2}$ | $5.8 \times 10^3 \ cm^{-3}$ | $2.2 \times 10^3 \ cm^{-3}$ |
| Emission Measure at center of nebula | $3.6 \times 10^7 \ cm^{-6} pc$ | $3.2 \times 10^6 \ cm^{-6} pc$ |
| Excitation Parameter | 86 | 55 |
| Mass | 10 Solar masses | 7 Solar masses |

x  Wink, Altenhoff and Webster, 1975
y  Schraml and Mezger, 1969

## 3.  Maser Emission Centers in W 49A

The review article by Litvak (1974) contains a discussion of the physical conditions in OH and $H_2O$ maser centers.

## A.  The ground state lines of OH

Spectra of the 4 circularly polarized ground state OH lines have been taken with a single telescope by Palmer and Zuckerman (1967), Ball and Meeks (1968) and Weaver et al. (1968). The single dish spectra are complex: For example, Palmer and Zuckerman report 19 features at 1.667 GHz, 22 at 1.665 GHz, 3 at 1.612 GHz and 2 at 1.720 GHz. Rogers et al. (1967) and Raimond and Eliasson (1969) showed that most of the main-line OH emission comes from the source named OH-1 (also referred to as W 49N, W 49(A) or $A_{51}$), shown in Fig. 2(a). Some emission, mostly between 15 and 18 km s$^{-1}$, comes from the source named OH-2, W 49S, W 49(B), or $A_{59}$. Harvey et al. (1974) have detected a third emission center, 37.4" north and 6.4" east of OH-1. This position is indicated in Fig. 2(a). This new OH center has a single velocity feature at 19 km s$^{-1}$. Zuckerman et al. (1969) reported time variations in the 1.720 GHz line at $\sim$ 13 km s$^{-1}$; this variability has been confirmed by Robinson et al. (1970). The 1.720 GHz OH emission arises at position OH-2 (Hardebeck, 1971). Interferometer observations of the OH centers, with angular resolutions of 0.04" and 0.01", show that the sizes of individual regions are $\sim$ 0.05" (Moran et al., 1968). Harvey et al. (1974) report some clustering of the individual OH sources on a scale < 0.3".

B. Excited state lines of OH

Zuckerman et al. (1968) reported the detection of the masering F=1$\rightarrow$0 transition of the $^2\pi_{\frac{1}{2}}$, J=$\frac{1}{2}$ state of OH at 4.766 GHz, in W 49A. The velocity of this line is $\sim$ 8 km s$^{-1}$. Later observations (Zuckerman and Palmer, 1970) showed that the linear polarization of this line is < 30 per cent, and that this line arises within 3' of the continuum peak of W 49A. It is generally assumed that the $^2\pi_{\frac{1}{2}}$, J=$\frac{1}{2}$ lines are unpolarized. Maser OH emission from one other state, the $^2\pi_{3/2}$, J=5/2, F=3$\rightarrow$3 at 6.035 GHz was detected by Zuckerman et al. (1972). The $^2\pi_{3/2}$, J=5/2 lines generally exhibit high circular polarization, but the antenna temperature of the lines in W 49, $\sim$ 0.4 K, was too low to allow a polarization measurement. The radial velocities of the 5 lines are 9.2, 10.4, 12.2, 12.7 and 17.5 km s$^{-1}$. Zuckerman et al. commented that the velocity of these lines appears to be anticorrelated with the velocity of the 18 cm ground state features, and that there is a hint of Zeeman splitting in the spectra. Such a splitting, 0.65 km s$^{-1}$, would imply a magnetic field of $\sim$ 12 milligauss. "Pseudo-thermal" OH emission was first detected in Sgr B2 by Gardner et al. (1971). "Pseudo-thermal" OH line emission appears to arise in an extended region of space and the lines themselves have widths much greater than the masing features. Rickard et al. (1974) preliminarily report the detection of such line emission in W 49A.


## C.  $H_2O$ observations

The 3 centers of $H_2O$ line emission coincide, to the map scale of Fig. 2(a), with the centers of OH line emission.  Each of these centers has a number of subcomponents.  An extensive series of measurements with a single antenna of W 49N is given by Sullivan (1973).  W 49S was first detected by Johnston et al. (1973).  The third $H_2O$ center was detected by Hills and Landecker (priv. comm.) using the 100-m telescope.  The $H_2O$ emission from W 49N shows:  (i) strong time variations on a period of months, (ii) many line emission features covering a velocity range from $\sim$ -150 km s$^{-1}$ to $\sim$ +200 km s$^{-1}$ with the strongest features between $\sim$ -80 and $\sim$ +25 km s$^{-1}$, (iii) very large flux densities in the individual features, ranging up to $\sim$ 80,000 flux units, (iv) unpolarized $H_2O$ emission.  High angular resolution interferometer observations of the $H_2O$ centers in W 49N, using very long baselines, were reported by Burke et al. (1970), Johnston et al. (1971), Moran et al. (1973) and Knowles et al. (1974).  Moran et al. report that the velocity features between -16 and +12 km s$^{-1}$ are distributed over 1" x 1";  the sources can be grouped into 3 regions.  Assuming a uniform disk model, the size of the individual $H_2O$ centers is < 0.0005" and the size of the $H_2O$ feature at -1.8 km s$^{-1}$ is < 0.0003".  Knowles et al. point out that the low level $H_2O$ emission between 15 and 42 km s$^{-1}$ is partially resolved and therefore has a size of $\sim$ 0.003".  A direct comparison of the location of the $H_2O$ and OH centers using an 18-cm and 1.35-cm simultaneous VLB (Mader et al., 1974) shows that the centers are <u>not</u> in the same position and most probably do not have the same source of excitation.


## 4.  <u>Large Scale Dynamics of the Ionized Gas</u>


## A.  The SNR W 49B

Cesarsky and Cesarsky (1973) have detected an H166$\alpha$ recombination line, and Downes and Wilson (1974) have measured an H134$\alpha$ line toward W 49B. These lines have a velocity of $\sim$ 60 km s$^{-1}$ and most probably arise from a region along the line of sight to the SNR, since an H recombination line with a similar velocity and width has been detected about one degree away, at $\ell = 44^o$, by Gordon and Cato (1972).

## B.  The H II Region W 49A

Gordon and Wallace (1971) have mapped W 49A in the H109$\alpha$ and H137$\beta$ lines.  The angular resolution used in the mapping was 6', which is comparable to the total source size (see Fig. 1).  Gaussian fits to the H109$\alpha$ spectra show a velocity gradient from 8 to 10 km s$^{-1}$, increasing with R.A.  This velocity trend, 2 km s$^{-1}$, is small compared with the line width, about 25 - 30 km s$^{-1}$ and hence depends critically on a symmetrical line shape.  It is generally recognized that the H109$\alpha$ line is not in LTE.  Departures from the LTE ratio of the H137$\beta$ to the H109$\alpha$ line, 0.276, indicate non-LTE conditions in the recombination line formation.  Gordon and Wallace find the largest departure from LTE at the center of the nebula.  Their attempt to solve for the electron density, size and number of small, high density H II regions making up the dense centers in W 49A, following the analysis of Hjellming and Gordon (1971), is not correct, since this procedure neglects the influence of the lower density, outer parts of the nebula, where a small departure from LTE will cause a substantial change in the line radiation from inner parts of the H II region (see the discussion by Brocklehurst and Seaton (1972)).

Pankonin, Parrish and Terzian (1974) have measured the H221$\alpha$ line at 0.6 GHz and the H247$\alpha$/H248$\alpha$ lines at 0.4 GHz in W 49A with an angular resolution of $\sim$ 10'.  The velocities of the lines are close to the velocity of the H109$\alpha$ line, so the gas is associated with the H II region itself.  The line width from both lines is less than 15 km s$^{-1}$. If the line broadening were due only to the temperature of the line emitting gas, the kinetic temperature of this gas would be less than 5000 K.  The dense ionized condensations in the H II region observed with interferometers are optically thick at < 1 GHz, so these recombination lines must be emitted in a low density H II region.

The latest published helium recombination line measurements, together with a collection of previous data are given by Churchwell et al. (1974). These authors find that the average ratio of singly ionized helium to ionized hydrogen is $0.060 \pm 0.007$. The uncertainty quoted is the standard error of the mean. The ratio of doubly ionized helium to ionized hydrogen is $< 0.009$. Unpublished measurements (Churchwell, Smith and Mezger, priv. comm.) with an angular resolution of 2.6' at 5 GHz give a slightly higher ratio of singly ionized helium to ionized hydrogen, that is $0.069 \pm 0.005$. The quoted error is one standard deviation. Chaisson and Ball (1971) have reported the detection of the C 85$\alpha$ line (10.527 GHz), at a velocity of $+4.4$ km s$^{-1}$, and an unidentified feature located $+55$ km s$^{-1}$ from the C 85$\alpha$ line. The C-line at $+4.4$ km s$^{-1}$ is most probably associated with W 49 itself. This line has a corrected full width to half power of $5.9 \pm 3.9$ km s$^{-1}$. Chaisson and Ball have shown the unidentified line is due to recombination, since the line is also present in the 94$\alpha$ spectra, at 7.8 GHz. Pankonin, Parrish and Terzian (1973) have also detected the 247$\alpha$ and 248$\alpha$ transition of this feature, and suggest it is the C-line from a source along the line of sight to W 49A. The source of the radiation would have a velocity of $\sim +60$ km s$^{-1}$ and therefore would be located in the Sagittarius arm. Pankonin, Thomasson and Wilson (unpublished) have subsequently detected an H166$\alpha$ line with a velocity of $\sim 60$ km s$^{-1}$, in addition to a C166$\alpha$ feature centered at $\sim 6$ km s$^{-1}$ and having a full width to half power of $\sim 20$ km s$^{-1}$.

## 5. The Molecules in Neutral Gas

### A. CO and CS in W 49A

Wilson et al. (1974) have observed CO emission at 2 positions toward W 49A, and a full map of this source in CO has been made by Liszt and Mufson (1975). Scoville and Solomon (1973) have made CO observations with a resolution of 1'. The individual scans were spaced by 1', in the form of a cross, centered on the position of the continuum peak of W 49A. There are two CO clouds at 3 and 12 km s$^{-1}$, the gaussian full width to half power of the 3 km s$^{-1}$ cloud is 2.5', the gaussian full width to

half power of the 12 km s$^{-1}$ cloud is 4'. Both CO clouds are centered 1'
north of the position of the continuum peak. The CO is optically
thick. If the gas fills the telescope beam, the brightness temperature
is the excitation temperature of the gas. Provided the CO is in LTE, the
excitation temperature equals the kinetic temperature of the gas, and
this is $\sim$ 10 K. The density in the neutral gas required for collisional
excitation of the CO is > 500 cm$^{-3}$. Presumably most of the neutral gas
is in the form of hydrogen molecules, which are unobservable. The J=3-2
and J=2-1 transition of the CS molecule at 98 and 157 GHz require
considerably higher densities for collisional excitation. The detection
of these two lines, reported by Scoville and Solomon, and the J=2-1 line
by Turner et al. (1973) would imply a hydrogen molecule density of
$\sim$ 6 x 10$^5$ cm$^{-3}$. Scoville and Solomon report that the size of this
higher density region is < 2'.

B.   OH and H$_2$CO

Absorption line measurements of OH are limited by the large beamsize
and the mixture of emission and absorption near 15 km s$^{-1}$. Pastchenko
and Slysh (1973) report two 1.72 GHz OH absorption features at 10 and
17 km s$^{-1}$ toward W 49A, but no linewidths are given.

The H$_2$CO line at 4.8 GHz is nearly always seen in absorption.
Observations toward strong sources might allow one to place the H$_2$CO
relative to the ionized gas. Fig. 3 presents the H$_2$CO and CO profiles
for the continuum peak. The angular resolution of the H$_2$CO observations
(made by Bieging, Downes and Wilson with the 100-m telescope) is 2.6'.
Wilson (1970) had failed to detect the H$_2$CO feature at $\sim$ 17 km s$^{-1}$
toward W 49B found by Whiteoak and Gardner (1970); later observations
by Whiteoak and Gardner (1974) have confirmed the presence of this line,
which is also clearly present as an OH absorption (Pastchenko and Slysh,
1973). Figure 4 shows that the H$_2$CO line temperature integrated over
velocity is weak and not obviously concentrated towards the peak of
W 49B.

Scoville and Solomon have concluded from the differences in the
H$_2$CO and CO profiles that the 3 km s$^{-1}$ cloud lies behind the H II region,

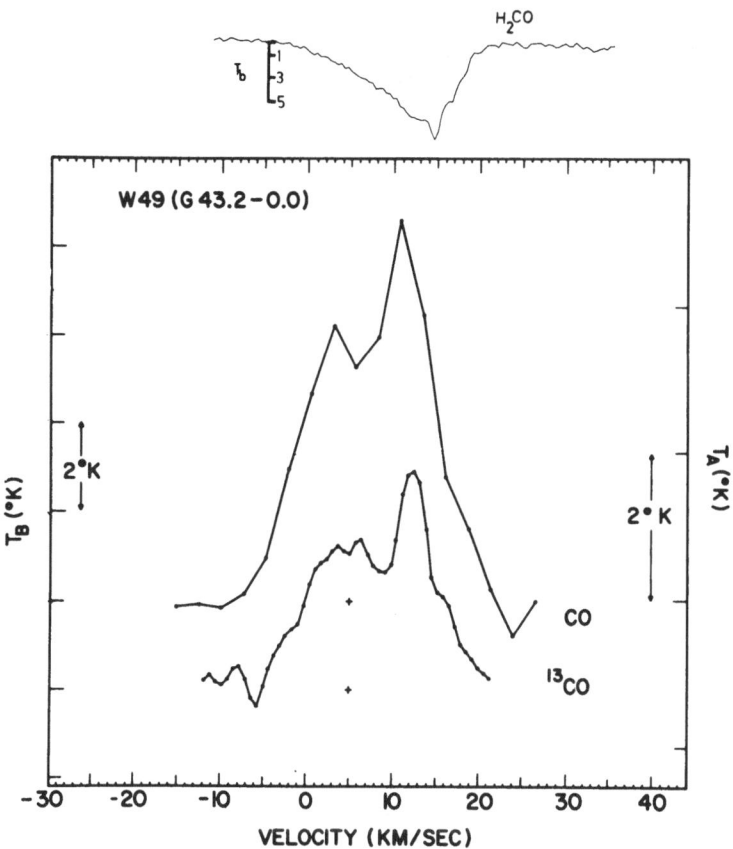

Figure 3. The 4.8 GHz $H_2CO$ (Bieging, Downes and Wilson), and 115 GHz $^{12}CO$ and 110 GHz $^{13}CO$ lines (Scoville and Solomon 1973) measured toward the peak of W 49A.

and absorbs only the 2.7 K cosmic background. The $H_2CO$ mapping of Scoville and Solomon (1973) was made with a 6.6' beam and therefore is severely limited by beam smearing. A preliminary $H_2CO$ map made with the 100-m telescope is given in Fig. 4. This map indicates the size of the $H_2CO$ cloud measured with a 2.6' beam is < 1/2 the size measured with a 6.6' beam. Scoville and Solomon reported a velocity gradient of ~ 2 km s$^{-1}$, the higher velocities at larger R.A.'s. The Effelsberg results show a more complex pattern, but we find that the spectral lines with larger widths have lower average radial velocities. We agree with Scoville and Solomon that the center of the $H_2CO$ cloud lies NW of

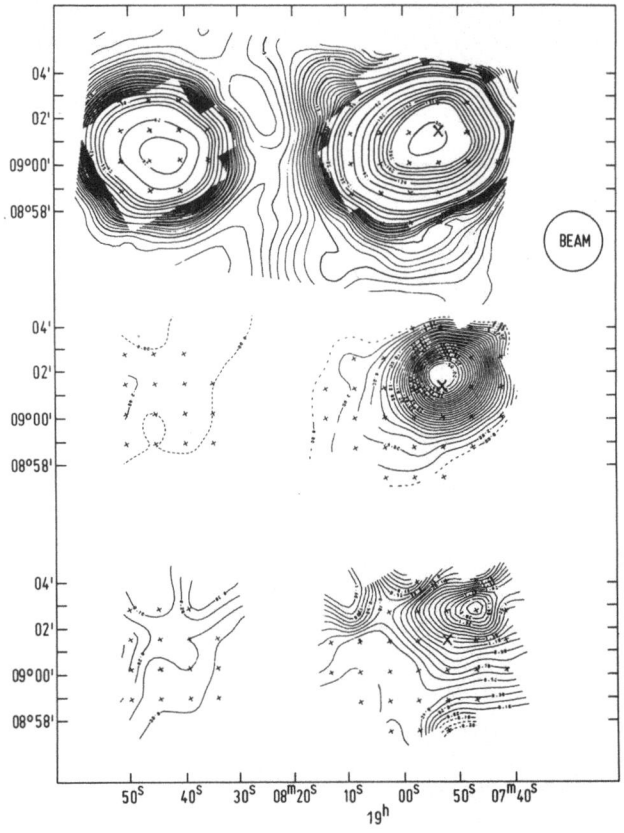

Figure 4. The continuum, integrated $H_2CO$ line, and integrated $H_2CO$ opacity maps for W 49. The integrated $H_2CO$ line and opacity maps cover the velocity range 0 to 30 km s$^{-1}$. The crosses show the positions where line data were taken, and the large cross in the maps marks the position of the nominal peak. The contour unit of the continuum map is 0.13 K, $T_a$. The units of the integrated $H_2CO$ line map are degrees K · km s$^{-1}$, and the units of the integrated opacity map are km s$^{-1}$. The beam size is 2.6'.

the continuum peak. Fomalont and Weliachew (1974) have observed the $H_2CO$ with an interferometer. The antenna spacings ranged from 100 to 600 ft. and the results were analyzed using model fitting. This $H_2CO$ map shows opacity increasing to the NE rather than the NW, from single dish results. The highest column density reported by Fomalont and Weliachew (1973) is about the same as the maximum column density obtained with the 100-m telescope.

## C.  CH, HCN and U93.2 toward W 49A

Rydbeck et al. (1975) report the detection of the three $^2\pi_{\frac{1}{2}}$, $J = \frac{1}{2}$ lines of CH, near 3.3 GHz.  The lowest frequency line in the triplet, the F=0→1 transition, at 3.26 GHz is strongest near $\sim$ 15 km s$^{-1}$.  This line is made up of at least 2 velocity components, at $\sim$ 15 and $\sim$ 5 km s$^{-1}$. In addition, Rydbeck et al. have detected CH lines at 41 and 62 km s$^{-1}$. At those velocities, the CH line ratios are closer to LTE.

Snyder and Buhl (1971) have detected an HCN line, which has a velocity of 6 km s$^{-1}$, and a width of 18 km s$^{-1}$.  Turner (1974) has detected a transition from an unidentified molecule at 93.2 GHz. Following the suggestion of Green et al. (1974), Thaddeus and Turner (1975) have confirmed that this spectral line is due to the molecular ion $NH_2^+$.

## 6.  Infra-red Observations

The review article by Wynn-Williams and Becklin (1974) contains a discussion of radio and infra-red observations.

## A.  High angular resolution measurements

Becklin et al. (1973) have measured the IR flux densities at 1.65, 2.2, 4.8, 10 and 20 $\mu$ for the 2 main OH and $H_2O$ maser sources in W 49, W 49N (OH-1) and W 49S (OH-2).  All these data were taken with beam sizes less than 7".  Rieke et al. (1973) have taken flux densities of these regions at 350 $\mu$ with an angular resolution of 63".  Becklin et al. draw the following conclusions from a comparison of the IR and radio data:  (i) for W 49N (OH-1), the $H_2O$ maser cannot be pumped by isotropic continuum IR, unless there is considerable IR absorption, since the flux density from the $H_2O$ maser far exceeds the IR flux density;  (ii) for both W 49N and W 49S the interstellar extinction is at least 50 mag, and much of this absorption may be associated with the H II regions them- selves, and (iii) there is no relation between the peak flux density of the OH, $H_2O$ and IR.

## B.  Low angular resolution measurements

The far IR measurements of Harper and Low (1971), covering the range 45 - 750 μ, give an IR luminosity of $1.9 \times 10^7$ $L_\theta$.  Harper and Low also pointed out that the IR luminosity is proportional to the radio continuum flux density.

Using the numerical evaluation of the integrated Salpeter Luminosity Function given in Table 3 of Mezger, Smith and Churchwell (1974), for spectral type O4 to K5, we need a mass of $4.4 \times 10^4$ $M_\theta$ stars to give the observed radio flux density.  The total luminosity of these stars is $2 \times 10^7$ $L_\theta$, roughly the observed IR luminosity.  If the distribution of stars extends from O4 to O9, the mass of stars would be reduced by $\sim 10$, but the ratio of total luminosity to radio flux density is only 0.8 times lower.  The mass of the stars was calculated assuming that none of the Lyman continuum photons were absorbed inside the H II region. Leibowitz (1973), Petrosian (1973), Jura and Wright (1974) and Mezger, Smith and Churchwell (1974) have investigated the possibility that absorption of photons with $\lambda \leq 912$ Å exists inside H II regions.  Using the approach of Mezger et al. one can relate the number ratio of ionized helium to ionized hydrogen to the amount of absorption by dust.  The main assumptions in the analysis of Mezger, Smith and Churchwell are: (1) the true number density of singly ionized helium to ionized hydrogen is 0.1, (2) the cross sections for absorption of photons for wavelengths shorter than 912 Å given by Mezger et al. are correct, and (3) the number of photons capable of ionizing helium and hydrogen are those in Table 3 of Mezger et al.  If the amount of absorption inside the H II region is correctly given by these calculations, we would require 1.7 times more photons to ionize hydrogen and helium in order to give the observed radio continuum flux.  However, the increased number of photons required to provide the ionization required to give the observed radio flux density, taking into account the absorption inside the H II region, would also increase the IR luminosity by 1.7, so that the observed IR luminosity is too low by a factor of 1.7, which would imply $\sim 60$ per cent of the photons generated in the H II region do not heat the dust.

# References

Aizu, K., Tabara, H., 1967, Prog. Theoret. Phys., 37, 296.

Akabane, K., Kerr, F.J., 1965, Austral. J. Phys., 18, 91.

Ball, J.A., Meeks, M.L., 1968, Astrophys. J., 153, 577.

Baker, J.R., Green, A.J., Landecker, T.L., 1975, Astron. Astrophys., in preparation.

Becklin, E.E., Neugebauer, G., Wynn-Williams, C.G., 1973, Astrophys. Lett., 13, 147.

Brocklehurst, M., Seaton, M.J., 1972, Mon. Not. Roy. Astronom. Soc., 157, 179.

Burke, B.F., Papa, D.C., Papadopoulos, G.D., Schwartz, P.R., Knowles, S.H., Sullivan, W.T., Meeks, M.L., Moran, J.M., 1970, Astrophys. J., 160, L63.

Cesarsky, D.A., Cesarsky, C.J., 1973, Astrophys. J., 183, L143.

Chaisson, E.J., Ball, J.A., 1971, Astrophys. J., 169, 495.

Churchwell, E., Mezger, P.G., Huchtmeier, W., 1974, Astron. Astrophys., 32, 283.

Downes, D., Wilson, T.L., 1974, Astron. Astrophys., 34, 133.

Fomalont, E.B., Weliachew, L., 1973, Astrophys. J., 181, 781.

Gardner, F.F., Ribes, J.C., Sinclair, M.W., 1971, Astrophys. J., 169, L109.

Gordon, M.A., Wallace, D.C., 1971, Astrophys. J., 167, 235.

Gordon, M.A., Cato, T., 1972, Astrophys. J., 176, 587.

Green, S., Montgomery, J.A., Thaddeus, P., 1974, Astrophys. J., 193, L89.

Hardebeck, E.G., 1971, Astrophys. J., 170, 281.

Harper, D.A., Low, F.J., 1971, Astrophys. J., 165, L9.

Harvey, P.J., Booth, R.S., Davies, R.D., Whittet, D.C.B., McLaughlin, M., 1974, Mon. Not. Roy. Astronom. Soc., 169, 545.

Hjellming, R.M., Gordon, M.A., 1971, Astrophys. J., 164, 47.

Johnston, K.J., Knowles, S.H., Sullivan, W.T. III, Moran, J.M., Burke, B.F., Lo, K.Y., Papa, D.C., Papadopoulos, G.D., Schwartz, P.R., Knight, C.A., Shapiro, I.I., 1971, Astrophys. J., 166, L21.

Johnston, K.J., Sloanaker, R.M., Bologna, J.M., 1973, Astrophys. J., 182, 67.

Jura, M., Wright, E.L., 1974, Astrophys. J., 188, 473.

Kazès, I., Rieu, N.-Q., 1970, Astron. Astrophys., 4, 111.

Knowles, S.H., Johnston, K.J., Moran, J.M., Burke, B.F., Lo, K.Y., Papadopoulos, G.D., 1974, Astronom. J., 79, 925.

Leibowitz, E.M., 1973, Astrophys. J., 186, 899.

Liszt, H.S., Mufson, S.L., 1975, in prep.

Litvak, M.M., 1974, in Ann. Rev. of Astron. and Astrophys., vol. 12, ed. G.R. Burbidge, D. Layzer, J.G. Phillips (Ann. Reviews, Inc., Palo Alto), 97.

Mader, G.L., Johnston, K.J., Moran, J.M., Knowles, S.H., Mango, S.A., Schwartz, P.R., 1974, Bull. Am. Astronom. Soc., 6, 442.

Mezger, P.G., Schraml, J., Terzian, Y., 1967, Astrophys. J., 150, 807.

Mezger, P.G., Smith, L.F., Churchwell, E., 1974, Astron. Astrophys., 32, 269.

Moran, J.M., Burke, B.F., Barrett, A.H., Rogers, A.E.E., Carter, J.C., Ball, J.A., Cudaback, D.A., 1968, Astronom. J., 73, S27.

Moran, J.M., Papadopoulos, G.D., Burke, B.F., Lo, K.Y., Schwartz, P.R., Thacker, D.L., Johnston, K.J., Knowles, S.H., Reisz, A.C., Shapiro, I.I., 1973, Astrophys. J., 185, 535.

Palmer, P., Zuckerman, B., 1967, Astrophys. J., 148, 727.

Panagia, N., Felli, M., 1975, Astron. Astrophys., 39, 1.

Pankonin, V., Parrish, A., Terzian, Y., 1973, Astrophys. J., 180, L113.

Pankonin, V., Parrish, A., Terzian, Y., 1974, Astron. Astrophys., 37, 411.

Pankonin, V., 1975, Astron. Astrophys., 38, 445.

Pastchenko, M.I., Slysh, V.I., 1973, Astron. Astrophys., 26, 349.

Petrosian, V., 1973, in Interstellar Dust and Related Topics, IAU Symp. 52, ed. J.M. Greenberg and H.C. van der Hulst (D. Reidel, Dordrecht).

Radhakrishnan, V., Goss, W.M., Murray, J.D., Brooks, J.W., 1972, Astrophys. J. Suppl., 24, 49.

Raimond, E., Eliasson, B., 1969, Astrophys. J., 150, L171.

Rickard, L.J., Zuckerman, B., Palmer, P., 1974, Bull. Am. Astronom. Soc., 6, 340.

Rieke, G.H., Harper, D.A., Low, F.J., Armstrong, K.R., 1973, Astrophys. J., 183, L67.

Robinson, B.J., Goss, W.M., Manchester, R.N., 1970, Austral. J. Phys., 23, 363.

Rogers, A.E.E., Moran, J.M., Crowther, P., Burke, B.F., Meeks, M.L., Ball, J.A., Hyde, G.M., 1967, Astrophys. J., 147, 369.

Rydbeck, O.E.H., Kollberg, E., Hjalmarson, A., Sume, A., Elldér, J., Irvine, W.M., 1975, Research Lab. of Electronics and Onsala Space Observatory Research Report 120.

Sato, F., Akabane, K., Kerr, F.J., 1967, Austral. J. Phys., 20, 197.

Schmidt, M., in Galactic Structure, ed. A. Blaauw and M. Schmidt, (Univ. of Chicago Press, Chicago), 513.

Schraml, J., Mezger, P.G., 1969, Astrophys. J., 156, 269.

Scoville, N.Z., Solomon, P.M., 1973, Astrophys. J., 180, 31.

Shaver, P.A., Goss, W.M., 1970, Austral. J. Phys., Astrophys. Suppl., No. 14, 133.

Snyder, L.E., Buhl, D., 1971, Astrophys. J., 163, L47.

Sullivan, W.T. III, 1973, Astrophys. J. Suppl., 25, 393.

Thaddeus, P., Turner, B.E., 1975, Astrophys. J., in press.

Turner, B.E., Zuckerman, B., Palmer, P., Morris, M., 1973, Astrophys. J., 186, 123.

Turner, B.E., 1974, Astrophys. J., 193, L83.

Weaver, H., Dieter, N.H., Williams, D.R.W., 1968, Astrophys. J. Suppl., 16, 219.

Whiteoak, J.B., Gardner, F.F., 1970, Astrophys. Lett., 5, 5.

Whiteoak, J.B., Gardner, F.F., 1974, Astron. Astrophys., 37, 389.

Wilson, T.L., 1970, Astrophys. Lett., 7, 95.

Wilson, W.J., Schwartz, P.R., Epstein, E.E., Johnson, W.A., Etcheverry, R.D., Mori, T.T., Berry, J.J., Dyson, H.B., 1974, Astrophys. J., 191, 357.

Wink, J., Altenhoff, W., Webster, W.J., 1975, Astron. Astrophys., 38, 109.

Wynn-Williams, C.G., 1969, Mon. Not. Roy. Astronom. Soc., 142, 453.

Wynn-Williams, C.G., Becklin, E.E., 1974, Proc. Astronom. Soc. Pacific, 86, 5.

Zuckerman, B., Palmer, P., Penfield, H., Lilley, A.E., 1968, Astrophys. J., 153, L69.

Zuckerman, B., Ball, J.A., Dickinson, D.F., Penfield, H., 1969, Astrophys. Lett., 3, 97.

Zuckerman, B., Palmer, P., 1970, Astrophys. J., 159, L197.

Zuckerman, B., Yen, J.L., Gottlieb, C.A., Palmer, P., 1972, Astrophys. J., 177, 59.

# THE W 51 SOURCE COMPLEX

J. BIEGING

Max-Planck-Institut für Radioastronomie,
Bonn, FRG

## ABSTRACT

Recent observational results for the giant H II complex W 51 are reviewed. The radio continuum emission is produced by two types of component:  a diffuse, very extended ionized gas with low emission measure;
and a number of compact sources with high emission measures and electron densities. The compact components are associated with near-infrared sources and with massive molecular clouds, the densities of which
range from $n_{H_2} \sim 10^3$ cm$^{-3}$ to $10^5$ cm$^{-3}$. Some of the molecular clouds
in the W51 complex may be part of a gas stream flowing along the
Sagittarius spiral arm.

# I. INTRODUCTION

The W51 complex consists of a number of discrete radio sources superimposed on an extended, low-brightness background. The region has been
mapped with the 100-m telescope as part of the 5 GHz continuum survey
of the galactic plane (Altenhoff et al. 1975). A preliminary version of
the map is shown in Figure 1. At least nine discrete sources are resolved, in addition to a shell-like source which extends some 30' in
galactic latitude. As the 5 GHz map shows, the radio emission can be
divided roughly into 2 regions, conventionally labelled A and B. W51A
consists of the strongest components, G49.5-0.4 and G49.4-0.3, plus
two or three weaker sources and a somewhat extended, low-brightness
background. The W51B sources include G49.2-0.4, G49.1-0.4 and G48-9-0.3,
possibly some weaker components, and as with W51A, a low-level background. The shell-like source extends to meet and, probably, blend with
the W51B sources.

The existence of such a large complex of radio emission is a consequence of the distribution of spiral arms in the Galaxy. At the longitude of W51 ($1=49^{\circ}$) the line of sight intersects the Sagittarius arm
tangentially (Kerr and Westerhout 1965). It is perhaps not always appreciated that this intersection is not simply a grazing of the line of
sight across the outermost edge of the arm. Burton and Shane (1970) have
shown that the arm is inclined away from the circle of constant galactic
radius, and curves out toward the Sun. As a result, we see the arm nearly in cross section, with a total path length through the arm of at
least 5 kpc at $1 = 49^{\circ}$. Since H II regions are generally good tracers
of spiral arms, it is not surprising that we find a large concentration
of thermal sources at this longitude. It is important, then, to bear in
mind the possible projection effects which may be present toward W51.

# II. RADIO CONTINUUM AND INFRARED OBSERVATIONS

## a) Large-scale Structure

The W51 complex has been mapped over a frequency range from 53 MHz to

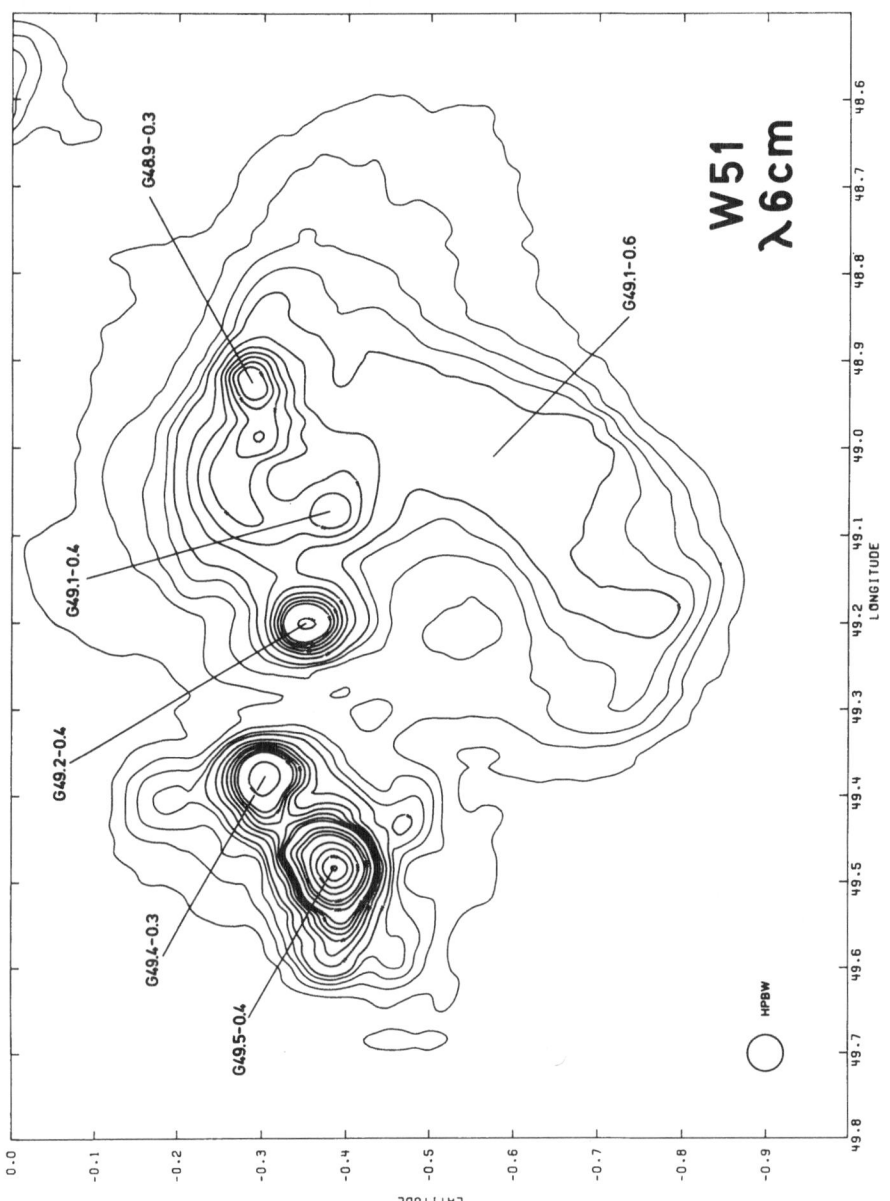

Figure 1  Continuum map of W51 at 5.0 GHz, obtained with the 100-m
telescope (Altenhoff et al. 1975).

15 GHz, with angular resolutions in the range 34' to 2' (Gol'nev et al. 1966, Pauliny-Toth et al. 1966, Kundu and Velusamy 1967, Mezger and Henderson 1967, Wendker and Yang 1968).

The highest resolution map in the decimeter wavelength range is that by Shaver (1969), obtained with the one-mile Molonglo Cross. Shaver has compared his map with nigh frequency maps by Schraml and Mezger (1969) at 15.4 GHz, and by MacLeod and Doherty (1969) at 10.6 GHz, to determine the spectral indices of the various components. He finds that the ratio of brightness temperatures implies that the components in W51A are thermal and optically thick at 408 MHz. The components in W51B also appear to be thermal, except that the underlying source has a non-thermal spectrum, as does the shell source.

Terzian and Balick (1969) have mapped the region at 195, 430 and 611 MHz with the Arecibo telescope, and conclude from the composite spectrum that the low-frequency radiation is a combination of thermal and non-thermal emission. Low-frequency observations of W51 in the range 53 to 318 MHz have recently been published by Parrish (1972). From a model calculation assuming an isothermal nebula, he finds for the W51 complex as a whole that the best-fit electron temperature is 8000 K. Based on the observed thermal turnover in the spectrum of the integrated flux density, and the extended brightness temperature distribution at 197 MHz, Parrish argues that the dominant low-frequency component of W51 is an extended distribution of ionized hydrogen with a half-width of about 40'. However, because the shell-like non-thermal source may contribute a significant fraction of the observed low-frequency brightness distribution, Parrish's size estimates for the extended component may be somewhat overestimated, at least perpendicular to the galactic plane.

The existence of a diffuse, low surface-brightness envelope is also apparent from centimeter wavelength maps of W51. Relatively high-resolution maps with beam sizes from 2' to 6' have been made at 5 GHz (Mezger and Henderson 1967; Gardner and Morimoto 1968), 10.6 GHz(MacLeod and Doherty 1968) and 15.5 GHz (Schraml and Mezger 1969). MacLeod and Doherty, with 2:8 angular resolution at 10.6 GHz, conclude that some 60 per cent of the integrated flux density of the source complex is from the

extended background. They argue that the total flux spectrum implies a
mixture of optically thick and thin components for frequencies below
1 GHz but that all components are optically thin above 1 GHz.

The extended background component could be explained either as a
superposition of evolved H II regions along the line of sight, or as an
extended low-density halo surrounding the bright central thermal sources.
If the spiral arm has been an active site of star formation for a
sufficiently long time, one might expect to see a number of evolved
H II regions as objects of low emission measure. The integrated flux
density from these objects could be large if we see many of them superimposed along the line of sight. MacLeod and Doherty (1968) note that
the W51 extended background is similar to that of the W43 source complex,
which also lies at the tangent point of a spiral arm. On the other hand,
the close positional coincidence of the extended background and the compact sources suggests that the extended component may be an envelope of
ionized gas surrounding the dense central sources. Excitation of the
low-density gas could be provided by Lyman-continuum photons which have
escaped from the central regions.

## B) Compact structures    and associated infrared sources

High resolution interferometric observations of W51A have revealed
the existence of small-scale structure with high brightness temperatures. Martin (1972) has synthesized the G49.5 and G49.4 components
with the Cambridge One-Mile Telescope at 2.7 and 5.0 GHz. His results
for G49.5 are shown in Figure 2. There are eight clearly distinguishable
components with a wide range of intensities. Martin finds that the sum
of the flux densities for the eight components is about 70 per cent of
the total flux density of G49.5-0.4. This discrepancy suggests that the
compact sources are embedded in a diffuse ionized envelope which is
undetected in the high resolution synthesis map. Comparison of the 2.7
and 5.0 GHz maps shows that the compact sources are all optically thin
at 5.0 GHz, but that the brightest ones are becoming optically thick at
2.7 GHz. Assuming $T_e$ = 9000 K, and a distance of 7.3 kpc, Martin derives
physical parameters based on the uniform sphere model of Mezger and
Henderson (1968). The excitation parameter U can be calculated for each

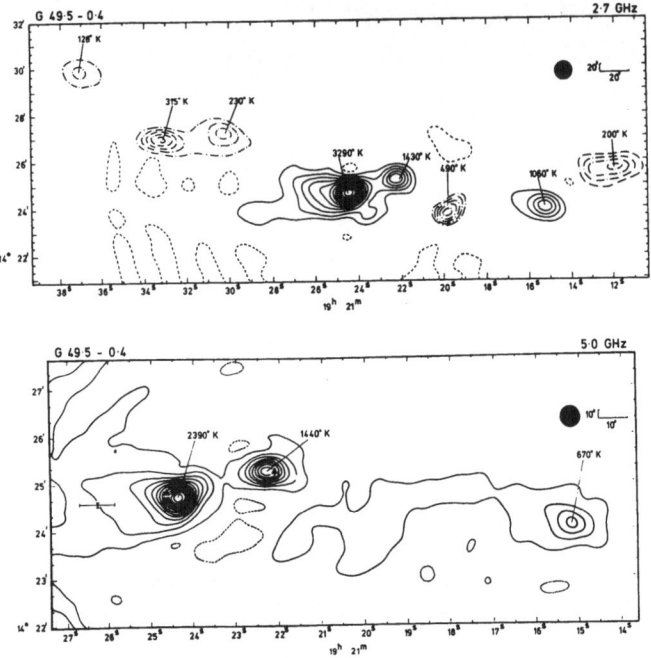

Figure 2  High-resolution synthesis map of G49.5-0.4 at 2.7 and 5.0 GHz. Half-power beamwidth indicated by filled circle (from Martin 1972).

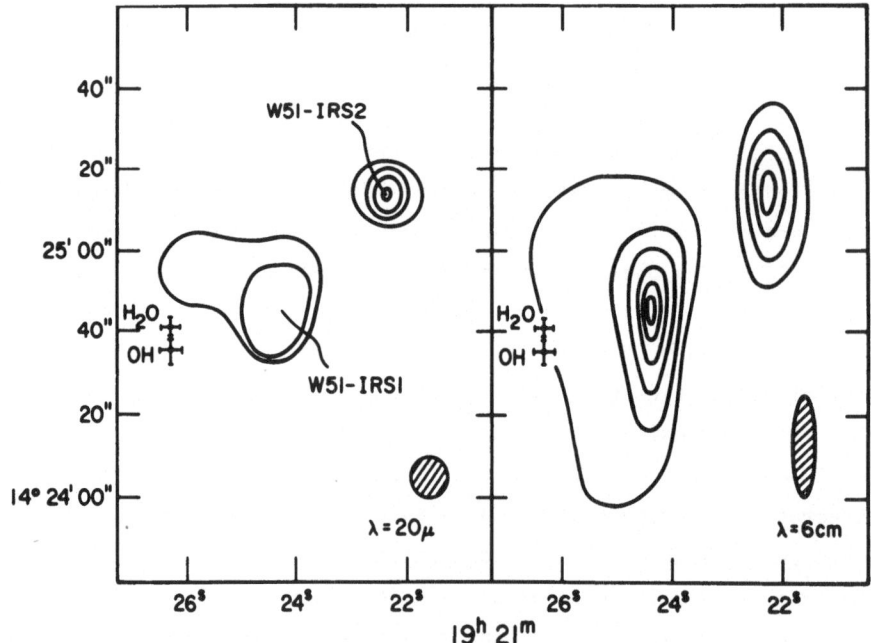

Figure 3  Brightness distributions of the central sources in G49.5-0.4 at λ20μ and λ6cm (Wynn-Williams, Becklin, and Neugebauer 1974a). Crosses indicate positions of the OH and H$_2$O maser sources.

component, given the flux density and distance. Martin's values for the eight components are listed in Table 1.

TABLE 1

Excitation of G49.5-0.4

| Component (Martin 1972) | U (cm$^{-2}$ pc) | $N_c$ (photons s$^{-1}$) | Spectral Type (Panagia 1973) | L/L$_o$ |
|---|---|---|---|---|
| G49.5 a | 57 | 0.8 x 10$^{49}$ | 06 | 2.5 x 10$^5$ |
| b | 113 | 6.1 | 04 | 13 |
| c | 93 | 3.4 | 05 | 6.8 |
| d | 116 | 6.5 | 04 | 13. |
| e | 170 | 21. | 2x04+05 | 32.8 |
| f | 69 | 1.4 | 06 | 2.5 |
| g | 79 | 2.1 | 05 | 6.8 |
| h | 58 | 0.8 | 06 | 2.5 |
| Total | | 4.2 x 10$^{50}$ | | 8.0 x 10$^6$ |

The corresponding flux of Lyman continuum photons, $N_c$, required to produce the observed radio flux is given in column three of Table 1. Note that this value is a lower limit if the radio source is becoming optically thick at 2.7 GHz, which may be true for the brightest components. In column four we list the minimum exciting star required to produce $N_c$ Lyman continuum photons per second, and in column five the total stellar luminosity for the given spectral type, based on the calculations of Panagia (1973) for early-type stars. The required excitation can be produced by a single hot O-star in every case but that of component (e), which needs two or three O4 stars. The total expected stellar luminosity from these exciting stars is 8.0 x 10$^6$ L$_o$.

Further evidence for even smaller-scale structure in G49.5-0.4 has been found by Balick (1972) and by Turner et al. (1974). The G49.5e source appears to consist of several subcomponents, each requiring an exciting star with spectral classes in the range 05 or later. Balick's aperture synthesis at 2.7 and 8.1 GHz indicates an electron density greater than $3 \times 10^4$ cm$^{-3}$, and a diameter of the order of 0.06 pc for these objects.

The excitation for the eight components discovered by Martin (1972) can be compared with the excitation of the whole G49.5-0.4 source. Gardner and Morimoto (1968) measured a total 5 GHz flux density of 109 Jy for the G49.5-0.4 source (with background subtracted). Assuming a distance of 7.3 kpc and $T_e$ = 9000 K, this implies $N_c = 5.4 \times 10^{50}$ photons s$^{-1}$, or 30 per cent more than the calculated Lyman continuum flux from Martin's eight components. Evidently there are additional sources of Lyman continuum photons, probably a large number of cooler stars, which do not produce compact, high emission measure H II regions, but which contribute to the ionization of the extended, less dense gas.

If we assume that the ionizing flux $N_c$ is due to stars in spectral classes 04 through K5 and apply the integrated Salpeter luminosity function as computed by Mezger, Smith and Churchwell (1974), the pre-dicted (minimum) stellar luminosity from the exciting stars is $1.1 \times 10^7$ $L_o$. This value agrees well with the total infrared luminosity measured by Harper and Low (1971). It is possible, then, to account for both the observed radio and infrared luminosities if we assume that all the Lyman continuum photons ionize the gas, and that eventually all the stellar radiation is absorbed by dust and reradiated in the far infra-red. In this picture we do not allow any absorption of Lyman photons by dust within the H II regions. However, recent observations of the H- and He109$\alpha$ recombination lines by Mezger and Smith (1975) indicate that the relative abundance of ionized helium is

$$\frac{N(He^+)}{N(H^+)} \approx 0.08 \pm 0.01 \ (G49.5\text{-}0.4).$$

Their result requires that some (60 $\pm$ 5) per cent of the ionizing pho-tons are absorbed by dust, assuming a normal helium abundance of 10 per cent. In that case, the number of exciting stars must be increased to explain the observed radio emission in the presence of dust absorption. At the same time, sixty per cent of the total stellar luminosity must escape the nebula without heating the associated dust grains, to account for the observed infrared luminosity. In short, there are two possibilities: (a) no ionizing photons are absorbed by dust, but ultimately all stellar radiation heats the surrounding grains; or (b) some 60 per cent of ionizing photons are absorbed by grains, but at the same time, 60 per cent of the total stellar energy escapes without being absorbed by dust. Case (a) requires a helium abundance about 20 per cent below the cosmic value, while (b) permits a normal helium abundance.

The brightest compact components found by Martin have counterparts in the near infrared. A $\lambda 20\mu$ map by Wynn-Williams, Becklin and Neugebauer (1974) is shown in Figure 3. The bright emission components in the infrared exactly correspond to the brightest features on Martin's 5.0 GHz map, part of which is reproduced on the right of Figure 3. Note that the positions of the OH and $H_2O$ maser sources (Wynn-Williams et al. 1974b; Hills et al, 1972) do not coincide with either an infrared or a radio continuum peak. The nearest continuum source is some 30" away, or about 1.0 pc at an assumed distance of 7 kpc. (The difference in position between $H_2O$ and OH sources is only barely significant, given the errors. Their angular separation corresponds to a linear distance of 0.2 pc at 7 kpc.) Wynn-Williams et al. (1974a) found no discrete 20$\mu$ source greater than 25 Jy at the position of the OH/$H_2O$ source.

The near infrared spectra of the two sources show the long wavelength excess characteristics of such emission from compact H II regions (Wynn-Williams et al. 1974a). Longward of about 5$\mu$, the measured flux density exceeds the predicted free-free value, by two orders of magnitude at 20$\mu$. The infrared excess is believed to come from heated dust associated with the compact H II regions. Shortward of 3$\mu$, the observed flux density is less than predicted, indicating the effect of either interstellar or circumstellar extinction. For IRS1, Wynn-Williams et al. (1974a) estimate a visual absorption of $A_v \geq 40^m$, and for IRS2, $A_v \geq 20^m$. [However, a recent re-determination of the visual extinction for these sources by Soifer (private communication) indicates that the true values for $A_v$ may be some $20^m$ greater than calculated by Wynn-Williams et al. (1974a)].

The size of the stronger source, IRS2, shows a wavelength dependence, in the sense that the diameter increases with increasing wavelength. This dependence could be explained by a temperature gradient in the circumnebular dust, such that the dust is cooler toward the outside. It might also be due to spatially separated components (as in Orion A) with different spectral indices, which are unresolved by present observations.

The W51 complex has also been detected in the far infrared. For the band 45 to 750μ, Harper and Low (1971) measured fluxes of 64. x $10^{-14}$ W cm$^{-2}$ for G49.5 and 12 x $10^{-14}$ W cm$^{-2}$ for G49.4 with color temperatures estimated to be 70 K for each component. Assuming a distance of 7.3 kpc for both sources, these fluxes imply IR luminosities of 1.1 x $10^7$ $L_o$ and 2.0 x $10^6$ $L_o$, respectively. At 100μ, Hoffman et al. (1971) measured a flux density of 1.3 x $10^5$ Jy, a factor of 1000 more than the 6 cm continuum flux density. High-resolution observations of G49.5-0.4 at 55 and 100μ by Harvey, Hoffmann, and Campbell (1975) show the presence of a bright, compact far-infrared source whose position lies between the IRS1 and IRS2 peaks. Surrounding this compact source is a more extended, lower-brightness component which may be identical with that observed by Hoffmann et al. (1971) and by Harper and Low (1971). At 350μ, Rieke et al. (1973) made scans for G49.5-0.4 with a 1' beam. Their results show that the source is clearly extended, but without the structure one would expect if the 350μ radiation were produced only in the compact radio components. If the source of the 350μ radiation is heated dust, then that dust must be fairly uniformly distributed throughout the nebula.

III. SPECTROSCOPY OF THE IONIZED GAS

Recombination line observations indicate, in agreement with measurements of the continuum, that there are two types of components in the source complex. Low frequency transitions near 400 MHz have been observed by Gordon et al. (1974), Parrish et al. (1972), Pankonin et al. (1975), and Parrish and Pankonin (1975). These observers find no evidence for pressure broadening at low frequencies; this result sets limits on the density of the ionized gas in which these lines are formed. Parrish et al. (1972) and Pankonin et al. (1975) find that the electron density must be less than 50 $cm^{-3}$, assuming an electron temperature of $10^4$ K. Based on the frequency dependence of the line intensities, they conclude that this component of the ionized gas has an electron density of about 20 $cm^{-3}$, and an emission measure of $10^4$ $cm^{-6}$ pc. They identify this low-density component with the extended background continuum source observed at decimeter and centimeter wavelengths.

The dense, compact components within the source complex are optically thick at low frequencies and as a result do not contribute significantly to the high-quantum number recombination lines. At higher frequencies, however, these dense regions will dominate in forming these lines. Unfortunately, the determination of physical parameters of the ionized gas from recombination line data is complicated both by the source geometry, which is manifestly not simple, and by the probable non-LTE conditions of the gas. An analysis of the H109α line by Wilson et al. (1970) assumed LTE conditions and uniform spherical geometry. Because the angular resolution of their observations was not sufficient to separate the compact subcomponents, the assumption of uniform spherical geometry was invalid. As a result, these authors seriously overestimated the masses and underestimated the excitation parameters of the W51 thermal components.

Hjellming and Davies (1971) have computed average values for electron density, emission measure and diameters, based on a non-LTE analysis. Since, under their assumptions, the averages are weighted by the square of the electron density, the derived emission measures and temperatures apply only to the densest parts of the H II regions. The physical significance of the computed values is therefore unclear, in the absence of knowledge of the precise source geometry. In the case of a complicated thermal source complex like W51, which comprises a wide range of sizes and electron densities in the ionized components, a proper recombination line analysis will require observations with high angular resolution, so that the effects of the source geometry can be calculated.

Wilson et al. (1970) have determined kinematic distances from the measured recombination line velocities of the brightest W51 components, assuming the Schmidt (1965) model for galactic rotation. They conclude that W51 consists of two H II complexes, separated by some 1500 parsecs. The four brightest sources in W51B have an average velocity of 68 km $s^{-1}$, while the two W51A sources have a mean velocity of 56 km $s^{-1}$. Although the W51B velocities are higher than the maximum permitted velocity for $1 = 49°$, there is known to be a "high-velocity stream" of neutral hydrogen at this longitude, which Burton and Shane (1970) place near the tangent point at a distance of 6.5 kpc. Evidently the H II regions of W51B are located within this neutral stream. Since molecular absorption has been observed in the W51A components at velocities well above the recombination line velocities, Wilson et al. (1970) locate the two W51A sources at the far kinematic distance, about 8 kpc from the Sun.

## IV. SPECTROSCOPY OF THE NEUTRAL GAS

Observations of molecular lines in the W51 region indicate a complex picture, with several gas clouds at different velocities and with different physical conditions. Both the formaldehyde absorption and the CO emission spectra of W51A suggest two or more blended components at about 57 and 67 km $s^{-1}$ (Zuckerman et al. 1970; W. Wilson et al. 1974; Penzias et al. 1971). Figure 4 shows CO emission spectra obtained by Liszt (1974) at some of the continuum peaks of W51. The pairs of spectra for the two isotopic species agree well with the velocities of the recombination lines in that the strongest CO emission always occurs at the velocity of the ionized gas. For G49.5-0.4 (W51A), there is a second weaker feature at about 67 km $s^{-1}$, which is also the velocity of the strongest $H_2CO$ absorption. The top two pairs of spectra are for W51B components, and show no emission at velocities below about 60 km $s^{-1}$, supporting the conclusion that W51B is a separate kinematical feature, distinct from the lower-velocity W51A components.

Scoville and Solomon (1973) have mapped parts of W51 in the 6 cm $H_2CO$ absorption and in the 2.6 mm CO emission lines. They find evidence

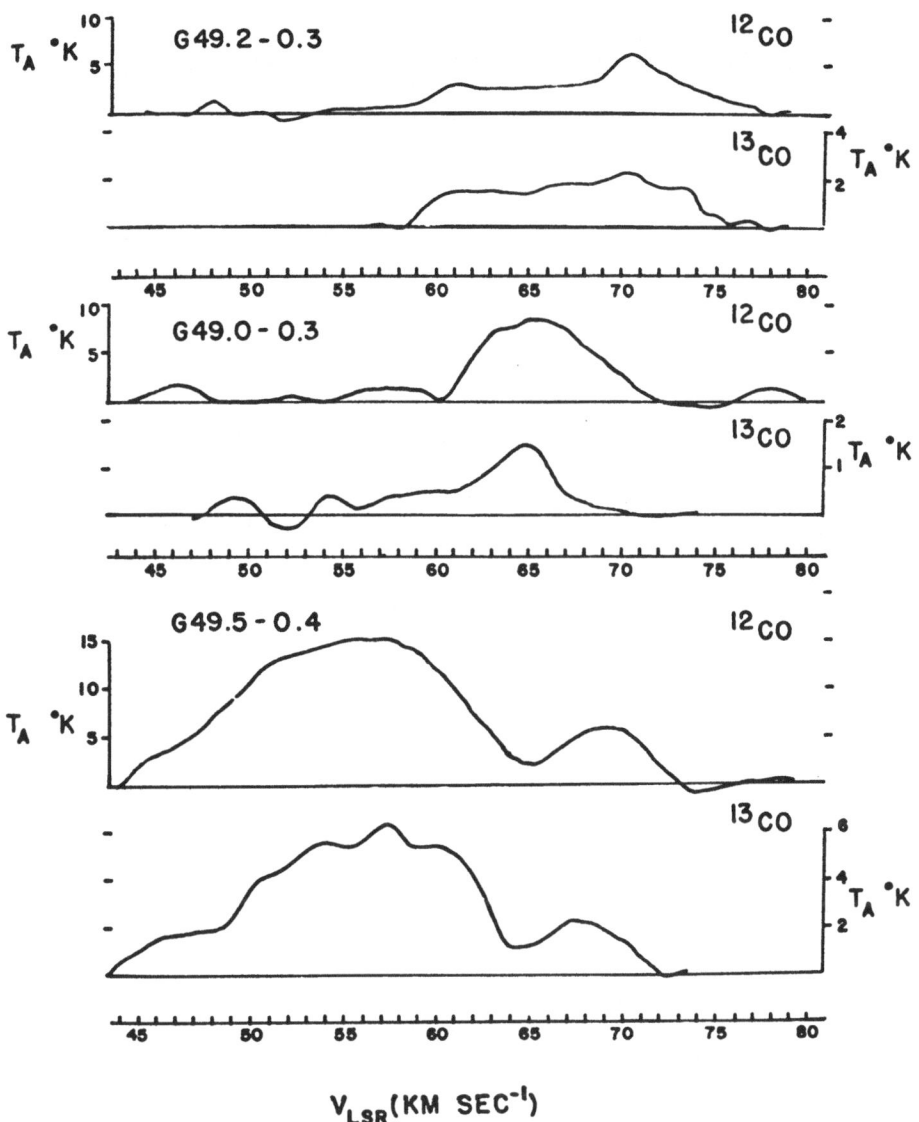

Figure 4   Emission spectra of two isotopic species of carbon monoxide,
in the directions of continuum sources G49.2-0.3, G49.0-0.3,
and G49.5-0.4 (Liszt 1974).

for three distinct clouds: at 50 km s$^{-1}$, a cloud associated with G49.4-0.3; at 57 km s$^{-1}$ a cloud near G49.5-0.4; and at about 65 km s$^{-1}$ a more extended feature which they believe to be associated with both components. Based on the CO intensity, they argue that the 57 km s$^{-1}$ line arises in a cloud with a total mass greater than 10$^5$ solar masses. The relatively weak H$_2$CO absorption feature implies that the continuum source lies in front of the massive neutral cloud.

Fomalont and Weliachew (1973) also find evidence for a separation of H$_2$CO absorption into distinct velocity groups. They conclude from the smooth variation of opacity that there is no evidence for small-scale ($\sim$ 1') clumping of the formaldehyde clouds. Recent 18 cm OH absorption observations by Slysh (1974) indicate that there are four distinct clouds seen in the OH lines, at velocities of 50, 57, 63 and 67 km s$^{-1}$. The strongest continuum source, G49.5, lies behind all four, but G49.4 must be in front of the gas at 57 and 67 km s$^{-1}$.

Bieging, Downes and Wilson (1975) confirm the existence of these kinematic features in their maps of the λ6-cm H$_2$CO line opacity over the W51A region (Figure 5). The thermal sources G49.5-0.4 and G49.4-0.3 each show molecular absorption at about their respective recombination line velocities (58 and 53 km s$^{-1}$), indicating the existence of dense neutral clouds out of which the ionized regions presumably have formed. In addition, there is strong absorption at velocities from 60 to 70 km s$^{-1}$, a range which is forbidden by the Schmidt (1965) model for galactic rotation. We identify this molecular gas with the high-velocity stream first detected by Burton (1966) in neutral hydrogen emission. The absence of H$_2$CO absorption on G49.4-0.3 at 67 km s$^{-1}$, the velocity at which the line is strongest on G49.5-0.4, indicates that the high-velocity stream passes between the two H II regions. The absorption at 63 km s$^{-1}$ may be due to molecular clouds in this gas stream where it recrosses the line of sight closer to the Sun. The relative locations of the main thermal sources and molecular clouds are shown schematically in Figure 6.

The recombination line velocities of the W51B sources indicate that they may also be associated with this high-velocity stream of gas. Lin

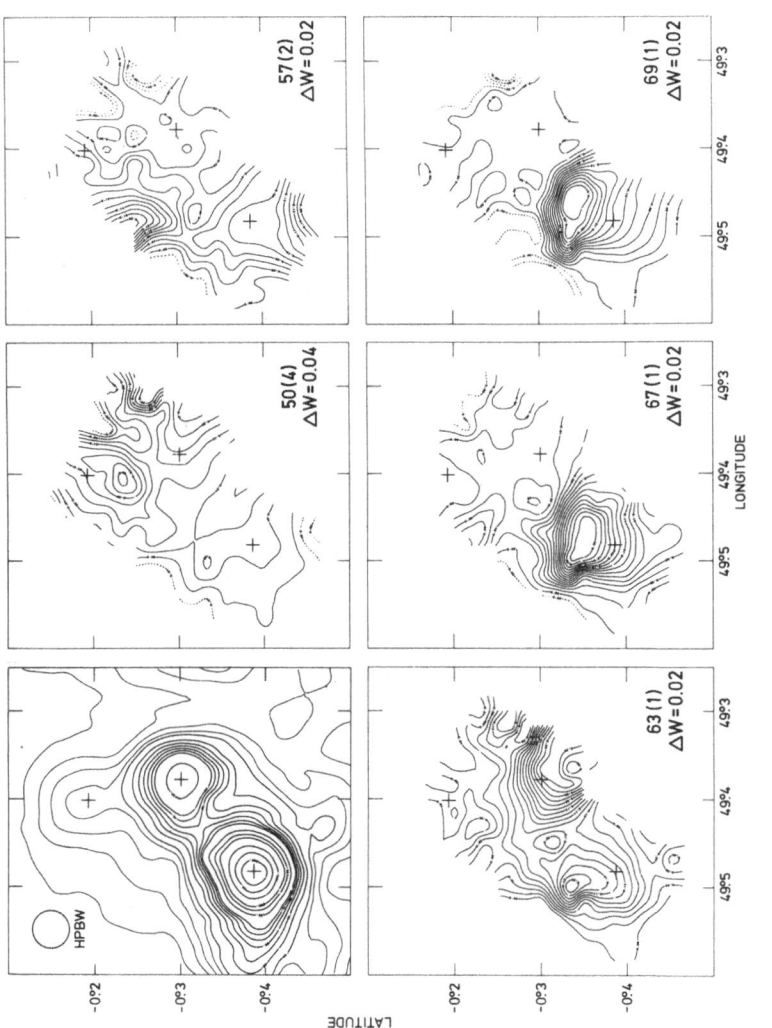

Figure 5   Contours of integrated optical depth in the 4830 MHz line of $H_2CO$, for G49.5-0.4 and G49.4-0.3. Center velocity and width (in parentheses) of the range of integration are given in km s$^{-1}$. Contour interval, $\Delta W$, is in km s$^{-1}$. The 5.0 GHz continuum map is shown in the upper left. Crosses mark positions of the continuum peaks. (Bieging, Downes and Wilson 1975).

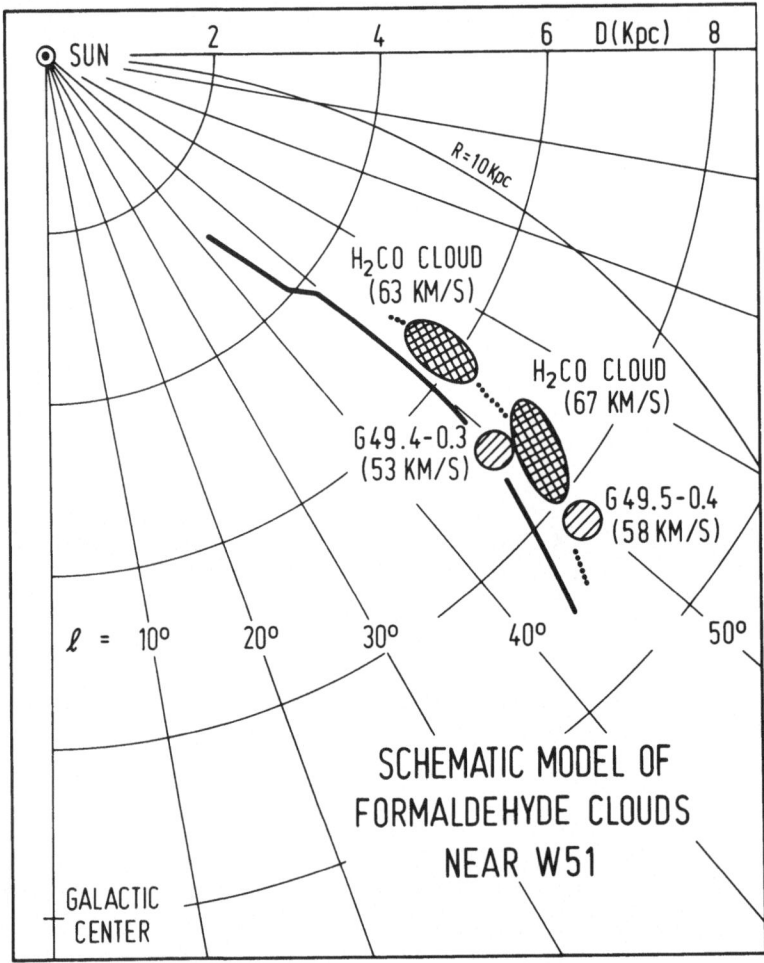

Figure 6   Schematic model for the location of bright H II
           regions and molecular clouds along the Sagittarius
           Spiral arm. The thermal sources G49.5-0.4 and G49.4-
           0.3 are each associated with a dense molecular cloud
           having the same velocity as the H II region. Location
           of Sagittarius arm (heavy line) and high-velocity gas
           stream (dotted line) are from Burton and Shane (1970).

(1970) and Burton and Shane (1970) have suggested that such streaming
motion is associated with the passage of a density wave through the region. If so, one might expect to see distinct differences in the
physical properties of sources within and outside of the streaming gas.
Unofrtunately, spectroscopic observations of W51 have been limited almost exclusively to the A components. It would clearly be of interest
to obtain observations of the W51B sources in as great detail as already exist for W51A.

Emission line observations of carbon monosulfide in G49.5-0.4 show
that the region contains a dense central molecular cloud. Penzias et al.
(1971) concluded from measurements of the J = 3-2 transition near 147
GHz that a density of $n_{H_2} \geq 4 \times 10^5$ cm$^{-3}$ is required to account for the
observed CS line temperature. More recently, however, Liszt and Linke
(1975) have mapped the molecular source in the three lowest rotational
transitions of CS, at 49, 98, and 147 GHz. They argue that, after cor-
recting for the effects of instrumental resolution, the ratio of line
intensities places an upper limit on the molecular hydrogen density of
$n_{H_2} \leq 4 \times 10^4$ cm$^{-3}$, although this limit probably does not exclude a
more condensed core within the cloud. The position and angular extent
of the CS emission agree well with the OH and $H_2O$ maser sources and the
millimeter lines of HCN (Snyder and Buhl 1971) and $H_2S$ (Thaddeus et al.
1972), all of which are believed to be associated with dense molecular
clouds ($n_{H_2} > 10^5$ cm$^{-3}$). The velocities of all these lines are near
58 km s$^{-1}$, which is also that of the recombination line of G49.5-0.4
(Wilson et al. 1970). The agreement in both position and velocity is
strong evidence for a condensed molecular cloud near or within the H II
region.

## V. SUMMARY

Recent observational results for the W51 source complex have yielded
these conclusions:

(1) Low-frequency continuum and recombination line results indicate
that W51 has an extended, diffuse ionized component with an electron
density of about 20 cm$^{-3}$, a temperature of $10^4$ K, and an emission

measure of $10^4$ cm$^{-6}$ pc.

(2) The O-stars required to excite the eight compact sources in
G49.5-0.4 (Martin 1972) can account for 70 per cent of the observed
radio free-free emission. Assuming the Salpeter luminosity function
applies to the stellar content of G49.5-0.4, the minimum luminosity deduced from the total radio free-free emission is just equal to the observed total far infrared luminosity of the nebula. If the observed
ionized helium abundance implies absorption of Lyman-photons by dust
within the H II region, then as much as 40 per cent of the ionizing
flux may be absorbed. At the same time, however, some 40 per cent of
the total stellar luminosity must escape without heating the grains, to
account for the observed infrared luminosity.

(3) The brightest near infrared sources in G49.5-0.4 are associated
with the compact radio components which require the greatest excitation.

(4) The agreement in the velocities of recombination lines and CO
emission lines indicates that the H II regions of W51 are associated
with large molecular clouds. The relative positions of neutral and
ionized gas along the line of sight can be deduced from studies of OH
and $H_2CO$ absorption spectra. The higher-velocity molecular clouds may
be part of a gas stream also deteced in the 21 cm neutral hydrogen line.

(5) Millimeter wavelength studies of high excitation molecular
transitions (CS, $H_2S$, HCN) show that there exists a dense molecular
cloud ($n_{H_2} \sim 10^5$ cm$^{-3}$) with dimensions on the order of 2 pc, centered
near the position of the OH/$H_2O$ maser source in G49.5-0.4.

ACKNOWLDEGEMENTS

I wish to thank W. Altenhoff, P.G. Mezger, V. Pankonin, and L.F.
Smith for providing observational results prior to publication. I am
particularly grateful to E. Churchwell for his critical reading of the
manuscript.

# REFERENCES

Altenhoff, W., Downes, D., Pauls, T., and Schraml, J. 1975, in preparation.

Balick, B. 1972, Ap. J., 176, 353.

Bieging, J.H., Downes, D., and Wilson, T.L. 1975, in preparation.

Burton, W.B. 1966, B.A.N., 18, 247.

Burton, W.B., and Shane, W.W. 1970, in IAU Symposium No. 38, The Spiral Structure of Our Galaxy, ed. W. Becker and G. Contopoulos (New York: Springer Verlag), p. 397.

Fomalont, E.B., and Weliachew, L. 1973, Ap. J., 181, 781.

Gardner, F.F., and Morimoto, M. 1968, Aust. J. Phys., 21, 881.

Gol'nev, V.Ya., Kipovka, N.M., and Pariskii, Yu.N. 1966, Sov. Astron. A.J., 9, 690.

Gordon, K.J., Gordon, C.P., and Lockman, F.J. 1974, Ap. J., 192, 337.

Harper, D.A., and Low, F.J. 1971, Ap. J., 165, L9.

Harvey, P.M., Hoffmann, W.F., and Campbell, M.F. 1975, Ap. J., 196, L31.

Hills, R., Janssen, M.A., Thornton, D.D., and Welch, W.J. 1972, Ap. J. 175, L59.

Hjellming, R.M., and Davies, R.D. 1970, Astron. & Astrophys. 5, 53.

Hoffman, W.F., Frederick, C.L., and Emery, R.J. 1971, Ap. J. 170, L89.

Kerr, F.J., and Westerhout, G. 1965, in Galactic Structure, ed. A. Blaauw and M. Schmidt (Chicago: Univ. of Chicago Press), p. 167.

Kundu, N.R., and Velusamy, T. 1967, Ann. d'Astrophys. 30, 59.

Lin, C.C. 1971, in Highlights of Astronomy (Vol. 2), ed. C. de Jager (Dordrecht: Reidel Publishing Company), p. 88.

Liszt, H.S. 1973, Ph. D. dissertation, Princeton University.

Liszt, H.S., and Linke, R.A. 1975, Ap. J., 196, 709.

MacLeod, J.M., and Doherty, L.H. 1968, Ap. J., 154, 833.

Martin, A.H.M. 1972, MNRAS 157, 31.

Mezger, P.G., and Henderson, A.P. 1967, Ap. J., 147, 471.

Mezger, P.G., Smith, L.F., and Churchwell, E. 1974, Astron. & Astrophys. 32, 269.

Panagia, N. 1973, Astron. J., 78, 929.

Pankonin, V., Parrish, A., and Terzian, Y. 1974, Astron. & Astrophys. 37, 411

Parrish, A. 1972, Ap. J., 174, 33.

Parrish, A., and Pankonin, V. 1975, Ap. J. (in press).

Parrish, A., Pankonin, V., Heiles, C., Rankin, J., and Terzian, Y. 1972, Ap. J., 178, 673

Pauliny-Toth, I.I.K., Wade, C.M., and Heeschen, D.S. 1966, Ap. J. Suppl. 13, 65.

Penzias, A.A., Jefferts, K.B., and Wilson, R.W. 1971, Ap. J., 165, 229.

Penzias, A.A., Solomon, P.M., Wilson, R.W., and Jefferts, K.B. 1971, Ap. J., 186, L53.

Rieke, G.H., Harper, D.A., Low, F.J., and Armstrong, K.R. 1973, Ap. J. 183, L67.

Schmidt, M. 1965, in Galactic Structure, ed. A. Blaauw and M. Schmidt (Chicago: Univ. of Chicago Press), p. 513.

Schraml, J., and Mezger, P.G. 1969, Ap. J., 156, 269.

Scoville, N.Z., and Solomon, P.M. 1973, Ap. J., 180, 31.

Shaver, P.A. 1969, MNRAS, 142, 273.

Slysh, V.I. 1974, Sov. Astron. Journal, 51, 685.

Snyder, L.E., and Buhl, D. 1971, Ap. J., 163, L47.

Terzian, Y., and Balick, B. 1969, Astron. J., 74, 76.

Thaddeus, P., Kutner, M.L., Penzias, A.A., Wilson, R.W., and Jefferts, K.B. 1972, Ap. J., 176, L73.

Turner, B.E., Balick, B., Cudaback, D.D., Heiles, C., and Boyle, R.J. 1974, Ap. J., 194, 279.

Wendker, H., and Yang, K.S. 1968, Astron. J., 73, 61.

Wilson, T.L., Mezger, P.G., Gardner, F.F., and Milne, D.K. 1970, Ap. Lett. 5, 99.

Wilson, W.J., Schwartz, P.R., Epstein, E.E.. Johnson, W.A., Etcheverry, R.D., Mori, T.T., Berry, G.G., and Dyson, H.B. 1974, Ap. J., 191, 357.

Wynn-Williams, C.G., Becklin, E., and Neugebauer, G. 1974a, Ap. J. 187, 473.

Wynn-Williams, C.G., Werner, M.W., and Wilson, W.J. 1974b, Ap. J., 187, 41.

Zuckerman, B., Buhl, D., Palmer, P., and Snyder, L.E. 1970, Ap. J. 160, 485.

# CO/ DUST RATIO IN THE ρ OPHIUCHI COMPLEX

P. ENCRENAZ, E. FALGARONE, R. LUCAS

Département de Radioastronomie, Observatoire de Meudon

Meudon, FRANCE

Maps in the $J = 1 \rightarrow 0$ lines of $C^{12} O^{16}$ and $C^{13} O^{16}$ have been obtained in the ρ Ophiuchi cloud complex, from the Millimeter Wave Observatory. A region approximately 30' x 30' has been fully covered with a half power beam width of 2.3' and a velocity resolution of .65 and .67 km/s respectively. A comparison with a map of visual extinction obtained by the technique of star count has been made, particularly at the boundaries of the clouds (Encrenaz et al, 1975). Observations of the $^{12}C^{18}O$ line show that $^{13}C^{16}O$ is not saturated in these parts of the cloud. We have (R. Lucas, 1974) plotted the integrated CO column density asafunction of visual extinction in fig.1. The curve obtained is very similar to the H2/Av relation predicted by (Hollenbach et al, 1973). For Av > 2 mag, CO/Av = 1.6 x 10$^{17}$ cm$^{-2}$/mag, while for 1 < Av < 2 mag the CO column density is much smaller. The agreement with the Copernicus data on ζOph (Av = 1 mag) is excellent.

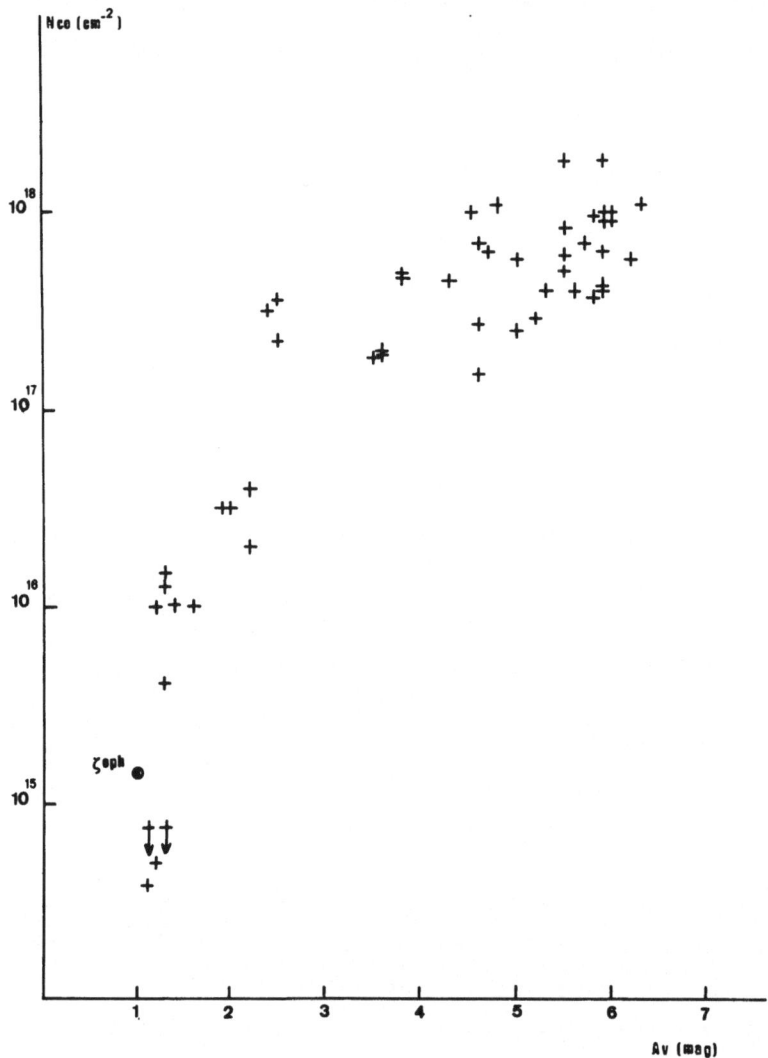

Fig. 1. CO column density versus visual extinction in the
direction of the ρ Ophiuchi cloud. The circle corresponds to the value measured on the star ζ Oph by
the Copernicus satellite.

REFERENCES

- Encrenaz, P.J., Lucas, R., Falgarone, E. to be published,
        1975.
- Hollenbach, D., Werner, M., Salpeter, E.E. 1971, Ap.J.,
        163, 165.
- Lucas, R.1974, Astronomy and Astrophysics, 36, 465.

x The Millimeter Wave Observatory is operated by the Electri-
cal Engineering Research Laboratory, the University of Texas
at Austin, with support from NASA, the NSF and Mc Donald
Observatory.

# 40-350 MICRON EMISSION FROM DUST IN NGC2023

J. P. EMERSON,   I. FURNISS and R. E. JENNINGS

Department of Physics and Astronomy,
University College London, UK

ABSTRACT (only)

A source of far infrared radiation associated with the reflection nebula and molecular cloud NGC2023 has been detected.  Combination of this observation with satellite ultraviolet observations of the exciting star (HD37903, B1.5V) indicates that the energy radiated by dust in the far infrared is balanced by the ultraviolet stellar energy absorbed by the dust.

Comparison of this source with OMC2 and the Kleinmann-Low nebula suggests that the dust temperature is roughly proportional to the gas temperature (which is given by the CO excitation temperature) for ice, silicate or  graphite grains.  If the dust and gas temperatures are equal and the grains are made of ice the gas to dust ratios in all three objects are close to the value in the interstellar medium.

The full text of this paper will appear in the Monthly Notices of the Royal Astronomical Society.

# THE CENTER OF OUR GALAXY

D. Downes

Max-Planck-Institut für Radioastronomie
Bonn, FRG

## ABSTRACT

Interferometer measurements show the galactic center source Sgr A to consist of a non-thermal source, Sgr A East, an H II Region, Sgr A West and an unusual point source. The properties of these sources are given below:

|                          | Sgr A East | Sgr A West | Point Source |
|--------------------------|------------|------------|--------------|
| Flux density at 5 GHz    | 90 f.u.    | 25 f.u.    | 0.6 f.u.     |
| Ang. diam.               | 150"       | 45"        | $0.02'' < \Theta < 0.1''$ |
| Linear diam.             | 6 pc       | 2 pc       | $0.001pc < d < 0.005pc$ |
| Spectral index           | -0.5       | -0.1       | $0.0 \pm 0.1$ |

Sgr A East may be a supernova remnant near the galactic center. The H II Region Sgr A West has the following unique properties:

i)   Its position agrees with the centroid of the 2.2μ radiation, which is believed to come from stars at the galactic center.

ii)  It has a recombination line of halfwidth $\sim$200 km s$^{-1}$, 4 to 5 times larger than that of any spiral arm H II region.

iii) At its center lies the remarkable radio point source, which has a brightness temperature $>10^7$ K, and which may be the actual nucleus of the Galaxy.


The full contents of this talk are to be published in:

    Ekers, R.D., Goss, W.M., Schwarz, U.J., Downes, D., Rogstad, D. 1975, Astron. & Astrophys., in press.

and

    Downes, D. 1974, in "H II Regions and the Galactic Center", ed. A.F.M. Moorwood, ESRO SP-105, Noordwijk, Holland, p. 247.

# THE MOLECULAR CLOUD ASSOCIATED WITH SGR B2

A. H. M. MARTIN

Institute of Astronomy,
Cambridge, England

## ABSTRACT

A greater number of molecular species is seen towards the HII region Sgr B2 than in the direction of any other galactic source. Many
of the more complex molecules have however been observed only with
large beams at a single position; such observations have not as yet
shed much light on physical conditions inside the cloud.

Detailed mapping has been carried out in the 6 cm absorption line
of formaldehyde (Downes et al 1975) and in the 2 mm emission line of
carbon monoxide (Scoville, Solomon and Jefferts 1974). These surveys
reveal a complicated velocity structure with no evidence for uniform
rotation. The strongest component has a velocity near +60 km s$^{-1}$,
close to the velocity of the ionized gas.

The density in the cloud may be estimated from theories of the
molecular excitation. Ammonia has proved particularly useful in this
respect (Morris et al 1973), because transitions from both metastable
and non-metastable levels can be seen; these allow a wide range of
densities to be probed. In the central core densities of more than
$10^6$ cm$^{-3}$ are found, while outside this core the density is around
$10^3$ cm$^{-3}$. Comparison of the ortho and para transitions of ammonia
may provide some information on the past history of the cloud.

An application of Jeans' formula shows the cloud to be gravitationally unstable, and hence a prime site for star formation.  Evidence
that star formation has occurred there in the recent past is provided
by the observation of compact components in the associated HII region
(Martin and Downes 1972).

REFERENCES

Downes, D., Wilson, T. L., Bieging, J. and Martin, A. H. M., 1975,
    in preparation.
Martin, A. H. M. and Downes, D., 1972, Ap. Letters, 11, 219.
Morris, M., Zuckerman, B., Palmer, P. and Turner, B. E., 1973, Ap. J.,
    186, 501.
Scoville, N. Z., Solomon, P. M. and Jefferts, K. B., 1974, Ap. J.(Let-
    ters), 187, L63.

# HIGH RESOLUTION MAPS OF THE GALACTIC CENTER AT 2.2 AND 10 μ

E.E. BECKLIN and G. NEUGEBAUER

Hale Observatories,
California Institute of Technology,
Carnegie Institution of Washington

## ABSTRACT

Photometric maps of the central 1' of the Galactic center are presented at 2.2 and 10μ with a spatial resolution of ∿ 2."5. Most of the
2.2-μ radiation within the central 2 pc comes from discrete sources with
absolute 2.2-μ magnitudes brighter than -8. At 10μ, nine discrete
sources plus an extended ridge of emission are seen. Five of the discrete sources are located within the extended ridge and have properties
similar to the planetary nebula NGC 7027.

(The contents of this paper have been submitted to Astrophysical Journal
Letters.)

# H II REGIONS IN THE NUCLEI OF EXTERNAL GALAXIES

M.-H. ULRICH

Department of Astronomy
University of Texas at Austin
Austin, Texas, U.S.A.

## ABSTRACT

In this article we review the observations of the ionized gas in nuclei of galaxies and the interpretation of the observed data in terms of the physical conditions of the gas. The ambiguities and uncertainties both in the observations and interpretation are outlined.

Comparison is made between optical line intensities in the spectra of normal H II regions and planetary nebulae on one hand and the spectra of nuclei of galaxies on the other hand. The causes of the differences between the spectra of these various types of ionized gas regions are briefly discussed.

# 1. INTRODUCTION

In this article, the word "H II regions" will be used to designate the regions of ionized gas in central parts of galaxies regardless of the ionization and excitation mechanism of the gas: hot stars, shock waves or nonthermal radiation.

In several well-studied nuclei of normal galaxies, it appears most likely that the gas is kept ionized by stellar radiation. The mass of gas present in these nuclei and the number of stars necessary to keep it ionized directly depend on the stellar evolution and the rate of star formation in the central regions. In other nuclei, like Seyfert-type nuclei where the ionization of the gas is probably caused by a nonthermal source, the analysis of the emission line spectrum of the gas gives the intensity and the energy distribution of the nonthermal source, as well as the mass of gas present in these peculiar nuclei. This shows that the study of H II regions in nuclei of galaxies provide important information on the evolution of the stellar population in normal nuclei and on the properties of the nonthermal source in active nuclei.

However the observations of nuclei of galaxies are difficult, mostly for two reasons: (i) lack of linear resolution and (ii) weakness of the fluxes received on earth. For example, no recombination lines have been detected from external galaxies; the only molecular lines observed are those of OH and $H_2CO$ seen in absorption in front of the central continuum sources in NGC 253 and NGC 4945 (Weliachew 1971, Gardner and Whiteoak 1974, Whiteoak and Gardner 1975). Also in most galaxies the rapid changes of the densities of stars and gas with distance from center in the small central regions 50 pc in radius, cannot be observed directly because of insufficient resolution of the ground based observations (1 arc sec corresponds to 50 pc at 10 Mpc). Other difficulties are encountered when one attempts to interpret the observed data in terms of physical parameters of the gas in the central regions of galaxies. These difficulties are summarized in Section II where we review the data which can be observed, the physical parameters

derived from the data and some of the basic uncertainties in
interpreting the observations. The mechanisms of ionization of
the gas in nuclei of galaxies are discussed in Section III.

The H II regions in the nuclei of external galaxies show a
large variety of morphologies and a wide range of emission line
intensities. In the elliptical galaxies where ionized gas is
present the gas is always concentrated in the very central region
and the emission line intensities increase smoothly toward the
center. In that respect the amorphous central regions of Sa and
Sb galaxies are similar to elliptical galaxies, and are devoid of
O and B supergiants; for example no such stars are seen in the
central bulge of M31. In contrast, in Sc galaxies the central
parts are rich in discrete H II regions similar to those found in
spiral arms in that they are ionized by O and B supergiants. In
M33 the nucleus itself appears as a stellar cluster about 1.5 arc
sec in diameter with an integrated spectral type close to A7 and
emitting very weakly in Hα and [N II]λ6584 (Benvenuti, D'Odorico
and Peimbert 1973).

In this article, we will review the properties of H II
regions in the central parts of elliptical, Sa and Sb galaxies
only. The innermost H II regions in Sc galaxies have been extensively studied by Searle (1971) and Smith (1975) in their works
on the composition gradients across the disks of spiral galaxies
and these works are reviewed in this Symposium by Dr. E.M. Burbidge.

Our knowledge of H II regions in nuclei of galaxies primarily
comes from optical observations in the range 3500 - 10000 Å. The
other available data come from infrared observations. About forty
galaxies have been successfully observed by broad band photometry
between 1 μ and 20 μ (Rieke and Low, 1972). Most of them show a
steep energy distribution longward of 5 μ consistent with the
infrared emission being due to re-radiation by dust; in a few
Seyfert galaxies the infrared flux is variable and is attributed
to non-thermal radiation. In some cases the energy distribution

flattens shortward of 5 μ and this is readily explained by the
contribution from the numerous late-type stars in the central
regions of the galaxies (more details on infrared observations of
galaxies can be found in Ulrich 1974). Only two galaxies, NGC 253
and NGC 1068, have been detected between 20 μ and 350 μ. In both
galaxies the energy distribution peaks around 100 μ but the
maximum is very flat for NGC 1068 (Harper 1975 this Symposium)
and sharply peaked for NGC 253 (Harper et al. 1973). Both galaxies
also show spectral features attributable to dust between 5 μ and
15 μ. Rieke and Low (1975) who observed the spectral features in
NGC 253 tentatively attribute them to silicates while in NGC 1068
the absorption features are still unidentified and resemble those
observed in the carbon-rich RV Tauri star AC Her (Jameson, Longmore, McLinn and Woolf, 1974).

## 2. OBSERVABLE DATA AND INTERPRETATION

In this section, we review the spectrographic and photometric
observations which can be made on nuclei of galaxies, and outline
the uncertainties encountered in interpreting the observed data
in terms of the physical conditions of the gas and of the stellar
population.

Seyfert-type nuclei will be mentioned briefly and only for
comparison with other nuclei; Seyfert nuclei are bright and the
numerous emission lines in their spectra, up to 45 in NGC 1068
or NGC 4151, which can be measured allow a sophisticated analysis
of the physical conditions of the gas in these nuclei. In contrast,
in normal or mildly active nuclei which constitute the subject of
this article, less than a dozen of emission lines, some of them
very weak against the background of the stellar population, can
be measured. Several lines particularly useful for determining
the physical conditions of ionized gas are undetected in nuclei of
galaxies, for example the best thermometer of H II regions which
is the line intensity ratio [O III]$\lambda$4363/$\lambda$5007 cannot be used in
the case of nuclei of galaxies because [O III]$\lambda$4363 is usually
very weak or absent and therefore only an upper limit to the
electron temperature $T_e$ can be calculated.

The [S II] lines near 10320 Å,which are used with the lines
at 4072 Å to determine the reddening in planetary nebulae or
Seyfert nuclei,are undetected in normal nuclei. The reddening is
therefore derived by comparing the observed Balmer decrement with
a case B Balmer decrement assuming an electron temperature in the
range 6000 - 20000°K. However the measure of the intensities of
the hydrogen lines in emission in nuclei of galaxies is not
straightforward since correction has to be made for the hydrogen
lines in absorption of the stellar population. The electron
density is derived from the intensity ratio of the doublet
[O II]λ3726,3729 or [S II]λ6717,6731. The absolute intensity of
a line from a galaxy nucleus,measured through an aperture of, say,
10 arc sec,is much smaller than the line intensity emitted by a
homogeneous cloud of gas having a density equal to the value
derived from the doublet ratio and occupying the volume seen
through the 10 arc sec photometer aperture. There are, therefore,
extreme density fluctuations, the gas having the electron density
measured from the doublet ratio being clumped in cloudlets or
filaments. In nuclei of galaxies, the filling factor is of the
order of $10^{-4}$ - $10^{-5}$. Note that in some analyses the determination of the electron density by the doublet intensity ratio (which
requires a good wavelength resolution) is made from spectrographic
observations of the very central region about 2 - 3 arc sec in
diameter while the total line intensity is measured with a photometer, with a diaphragm aperture of about 10 arc sec, and
combining the results of the two types of observations is another
cause of uncertainty in the interpretation of the data since the
density of the filaments at the very center is not necessarily
the same as in the other filaments further away from the nucleus.

One of the most interesting information derived from the
emission line spectra in nuclei of spiral and elliptical galaxies
is the overabundance of nitrogen by a factor 6 (with an uncertainty by a factor 2). This overabundance of N provides the
most likely explanation for the rapid change of the line intensity
ratio Hα/[NII]λ6584 from the value ~ 3 in the spiral arms of Sb
or Sa galaxies to a value ≤ 1 in the nuclei (Peimbert 1968,

Alloin 1973). In the nucleus of the Seyfert galaxy NGC 1068,
Shields and Oke (1975) also found an overabundance of N by a
factor 6, similar to the overabundance observed in normal nuclei.

Combining $T_e$, $N_e$ and the absolute line intensities one can
calculate the total mass of the ionized gas and also the number of
ionizing photons (if the ionization is radiative). The total mass
of the gas, the abundances of the elements in the gas, and the
numbers and types of stars necessary to keep the gas ionized
provide clues to the understanding of stellar evolution and rate
of star formation in nuclei of galaxies and hopefully may eventually help make some progress toward a better knowledge of the
origin of the violent events observed in some galaxies.

Finally the observations of the emission line profiles or
the variation of the wavelength shift with distance from center
provide information on the velocity field in the central region.
From this, in principle, one could calculate the kinetic energy,
and estimate whether the nucleus is loosing mass or not. Although
the wavelength shifts are measured accurately, the three dimension
velocity field cannot be reconstructed unambiguously, and, moreover,
the mass involved in these outward motions is very poorly known.
Outward motions of the ionized gas have been detected in a few
normal nuclei, most notably in M31 (Rubin and Ford 1971), M81
(Goad 1974) and NGC 253 (Demoulin and Burbidge 1968). The values
estimated for the rate of mass loss from the nuclei of M31 and
M81 are very small: $10^{-2}M_\odot$/year leaving the inner 50 pc in M81
(Goad 1974) and $10^{-2}M_\odot$/year leaving the region 400 pc in radius
in M31 (Rubin and Ford 1971).

2.1 Results relative to the best-studied nuclei of galaxies

Tables 1a and 1b give the masses of ionized gas and the line
intensities in the best-studied normal nuclei, which are those of
the nearby spiral galaxies: M31, M51, M81. Data relative to the
nucleus of the Seyfert galaxy NGC 1068 are included for comparison,
together with the line intensities in the normal H II region

Ori III and the planetary nebula NGC 7027.

Table 1a    Masses of ionized gas

| Galaxy | Mass of ionized hydrogen $(M_\odot)$ | Diameter of observed region (arc sec) (pc) | | Reference | Distance* (Mpc) |
|---|---|---|---|---|---|
| M31 | $10^5$ | 240 | 800 | 1 | 0.69 |
| M51 | $3.4 \times 10^3 - 7.9 \times 10^5$ | 20 | 340 | 2 | 3.2 |
| M81 | $1.7 \times 10^3 - 9.8 \times 10^4$ | 5 | 85 | 2 | 3.2 |
| NGC 1068 | $10^5$ | 7 | 770 | 3 | 22 |

Table 1b   Line Intensities

| Object | [O II] (3727) | Hβ | [O III] (5007) | [O I] (6300) | Hα | [N II] (6584) | Reference |
|---|---|---|---|---|---|---|---|
| M51 | 2.8 | 1 | 3.2 | 0.49 | 2.9 | 7.2 | 2 |
| M81 | 1.8 | 1 | 1.2 | 0.62 | 2.6 | 3.7 | 2 |
| Ori III | 2.2 | 1 | 1.35 | 0.013 | 2.8 | 0.9 | 2 |
| NGC 7027 | 0.35 | 1 | 15 | 0.2 | 2.9 | 0.9 | 4 |
| NGC 1068 | 1.38 | 1 | 15** | 0.36 | 3.9 | 5.25** | 3 |

References to Tables 1a and 1b.

1 - Rubin and Ford, 1971
2 - Peimbert, 1968
3 - Shields and Oke, 1975
4 - Oke and Sargent, 1968
*  Distances used in the quoted references
**  Combined intensities of respectively λλ5007,4959 and λλ6584,6548.

The mass of ionized gas in the central regions of the four galaxies listed in Table 1a does not exceed $10^5$ $M_\odot$. This is small compared to the mass of stars in the same volume. For example, in M31, only in the region 8 pc in radius, the total mass is estimated to be 2 x $10^8$ $M_\odot$ (Morton and Thuan, 1973). Since [O I]$\lambda$6300 is seen in the spectra of these nuclei, some neutral gas is present but the amount of neutral gas is unknown. In M51 and M81, the electron density calculated from the doublet [O II]$\lambda$3726-3729 is in the range 500-1000 $cm^{-3}$. In NGC 1068, the gas exists in at least two phases with densities 800 and 2 x $10^5$ $cm^{-3}$.

Table 1b lists the emission line intensities which illustrate best the differences which exist between the normal H II regions and planetary nebulae on one hand and nuclei of galaxies - normal nuclei or Seyfert nuclei - on the other hand. These differences are discussed in the next section.

### 3. DISCUSSION OF THE MAIN DIFFERENCES BETWEEN THE PHYSICAL CONDITIONS OF THE GAS IN NUCLEI OF GALAXIES AND IN NORMAL H II REGIONS OR PLANETARY NEBULAE

3.1 Normal nuclei

Examination of Table 1b shows that the gas in the nuclei of M51 and M81 differ from the H II region in the solar neighborhood Ori III, by (i) a large value of the ratio [N II]$\lambda$6584/H$\alpha$ and (ii) a relatively intense line [O I]$\lambda$6300.

The large ratio [N II]/H$\alpha$ could be produced by an electron temperature $T_e$ = 17000°K much higher than in normal H II regions. However oxygen is such an effective cooling agent that it would have to be underabundant by a factor of at least 20, if such a high temperature were to be maintained. Alternatively, the temperature of the ionized gas in the nuclei could be around 8000°K but then the intensity of the [N II] line would be due to an over-abundance of N by a factor ~ 5 over the solar value. This second alternative is generally accepted as the most plausible explanation

for the high value of [N II]$\lambda 6584$/H$\alpha$ in nuclei of galaxies. As for
the relatively high intensity of [O I]$\lambda 6300$, it has been shown by
Williams (1973) that in dense filaments of photoionized nebulae the
[O I] emission is enhanced by the oxygen-hydrogen charge exchange.
Aside from [N II]$\lambda 6584$ and [O I]$\lambda 6300$ the other line intensities
are similar to those in normal H II regions and thus the spectra
of these nuclei are compatible with the gas being ionized by hot
stars. It is not possible however to distinguish between the
effect of numerous horizontal-branch stars and the effect of a
small number of main sequence stars. From Peimbert (1968) the
ultraviolet radiation field necessary to produce the observed H$\alpha$
emission in the central regions of M51 and M81 has the following
characteristics (regions of 7 arc sec in diameter):
-Number of ionizing photons per sec in the range 0-912 $\overset{\circ}{A}$:

$\quad$ 6 x $10^{50}$ in M51, 1.5 x $10^{50}$ in M81

-Corresponding flux F at maximum of the black body energy distribution for two different temperatures (for M81 only):

$\quad$ 25000°K $\quad$ F = 1.4 x $10^{37}$ erg sec$^{-1}$ A$^{-1}$

$\quad$ 40000°K $\quad$ F = 7.1 x $10^{36}$ erg sec$^{-1}$ A$^{-1}$

If the temperature of the ionizing stars is 40000°K then the
ionizing field can be provided by

$\quad\quad$ 50 $\quad$ stars with R = 10 R$_\odot$

or $\quad$ 10000 $\quad$ stars with R = 0.7 R$_\odot$

## 3.2 Emission lines and Radio Sources

$\quad$ Comparison of the optical and radio properties of nuclei of
galaxies shows that there is a correlation between the presence of
a small radio source in the nucleus and the presence of emission
lines in the optical spectrum of a galaxy. This correlation is
illustrated in Table 2 which gives the results of the investigation
of a complete sample of 54 elliptical galaxies identified with
relatively weak radio sources. (Colla, Fanti, Fanti, Gioia, Lari,
Lequeux, Lucas and Ulrich 1975). Among these galaxies, those
which have a central compact radio source have a larger probability
to have emission lines in their optical spectra than the galaxies
which have no central radio source.

Table 2  Correlation between the presence of compact
central radio sources and the presence of
emission lines in the optical spectra of
<u>elliptical</u> galaxies

| Radio Structure | Total Number of Galaxies | Number with emission lines in optical spectrum | Percentage |
|---|---|---|---|
| Central Radio Source | 33 | 16 | 48% |
| No central Radio Source | 21 | 3 | 15% |

The emission lines in the galaxies of this sample which have central
radio sources are rather weak preventing a detailed analysis of the
physical conditions of the gas and therefore it is not known
whether the gas is ionized by non-thermal radiation or by hot
stars. Perhaps gas is present in most nuclei but it is detected
only in those galaxies which have central radio sources and therefore enough non-thermal radiation to ionize it.

3.3  Recent progrss in the analysis of the spectrum of NGC 1068

Table 1b shows that the spectrum of the Seyfert galaxy
NGC 1068 resembles the spectrum of the planetary nebula NGC 7027
rather than the spectrum of a normal H II region. The presence in
the spectrum of NGC 1068 of lines of highly ionized elements like
Ne V as well as strong lines of O III proves that the ionizing
radiation has a much flatter spectrum than that of a black body,
and it is generally admitted that the gas in Seyfert nuclei is
ionized by non-thermal radiation with a power-law energy
distribution. The spectrum of NGC 1068 differs from that of
NGC 7027 by (i) a larger ratio [N II]$\lambda$6584/H$\alpha$ which is shown to be
caused by an overabundance of N like in normal nuclei and (ii) a
larger intensity of [O II]$\lambda$3727 relative to [O III]$\lambda$5007 in
NGC 1068. I would like to briefly describe some recent work on

this latter point.

If all the gas in the nucleus of NGC 1068 were exposed to the ionizing radiation rich in hard UV photons there would be very little $O^+$ and therefore very weak [O II]$\lambda$3727 contrary to what is observed. Therefore some of the gas is shielded from the ionizing central source by optically thick globules or filaments.

Van Blerkom and Arny (1972) have studied the physical conditions of the gas in the shadow of optically thick globules in planetary nebulae. They found that this shaded gas can be ionized by the diffuse nebular radiation and if He II L$\alpha$ is absent from the radiation field then $O^+$ is abundant in this gas and emits substantial [O II]$\lambda$3727. When this is applied to NGC 1068 it is found that there is still not enough [O II]$\lambda$3727 to match the observed intensity (Shields and Oke 1975). For this reason Shields and Oke have proposed that additional [O II] is produced by filaments initially rich in $O^{++}$ moving into the shadows of optically thick clouds: thus $O^{++}$ recombines into $O^+$ and emits additional [O II]$\lambda$3727; subsequently it recombines into $O°$ and emits [O I]$\lambda$6300, then eventually the cloud goes back into the direct ionizing field and is ionized again. Therefore for each cloud this is a time dependent model but the lines from the nucleus as a whole come from the combined contributions of the clouds in various degrees of ionization. Shields and Oke (1975) have calculated that with such a model applied to NGC 1068 the line intensity ratio [O III]$\lambda$5007/[O II]$\lambda$3727 can be matched to the observed ratio.

## REFERENCES

Alloin, D. 1973, Astr. and Ap. 27, 433.
Benvenuti, P. D'Odorico, S. and Peimbert, M. 1973, Astr. and Ap. 28, 447.
Colla, G., Fanti, C., Fanti, R., Gioia, I., Lequeux, J., Lucas, R. and Ulrich, M-H. 1975, Astr. and Ap. in press.
Demoulin, M-H. and Burbidge, E.M. 1970, Ap.J., 159, 799.
Gardner, F.F. and Whiteoak, J.B. 1974, Nature, 247, 526.
Goad, J.W. 1974, Ap.J., 192, 311.

Harper, D.A. and Low, F.J. 1973, Ap.J., 182, L89.
Jameson, R., Longmore, A., McLinn, J., and Woolf, N. 1974, Ap.J.
190, 353
Morton, D.C. and Thuan, Trinh, X. 1973, Ap.J., 180, 705.
Oke, J.B. and Sargent, W.L.W. 1968, Ap.J., 151, 807.
Peimbert, M. 1968, Ap.J., 154, 33.
Rieke, G.H. and Low, F.J. 1972, Ap.J., 176, L95.
_____ 1975, Ap.J., 196, in press.
Rubin, V.C. and Ford, W.K. 1971, Ap.J., 170, 25.
Searle, L. 1971, Ap.J., 168, 327.
Shields, G.A. and Oke, J.B. 1975, Ap.J., in press.
Smith, H.E. 1975, Ap.J., in press.
Ulrich, M-H. 1974, I.A.U. Symposium No. 58, The Formation and
Dynamics of Galaxies, p. 279, ed. J.R. Shakeshaft.
Van Blerkom, D. and Arny, T. 1972, M.N.R.A.S. 156, 91.
Weliachew, L. 1971, Ap.J., 167, L47.
Whiteoak, J.B. and Gardner, F.F. 1975, Ap.J., 195, L81.
Williams, R. 1973, M.N.R.A.S., 164, 111.

# THE DISTRIBUTION OF HII REGIONS IN THE LARGE MAGELLANIC CLOUD

TH. SCHMIDT-KALER

Astronomisches Institut der Universität
Bochum, FRG

## ABSTRACT

The H$\alpha$ regions of the Large Magellanic Cloud with diameters $D \gtrsim 25" \simeq 6$ pc delineate a clear spiral pattern corresponding to a ScIII-IVp galaxy. The supergiant HII complex 30 Doradus appears to be the mildly active nucleus of the LMC.

The Large Magellanic Cloud (LMC) is usually classified as an irregular galaxy (Irr I). However, the distribution of the centres of the optical HII-regions with diameters $D \gtrsim 25" \cong 6$ pc and of the blue supergiants O-B8 Ia-Iab (fig. 1) shows a clear spiral pattern (Schmidt-Kaler and Isserstedt 1975). The ridge-lines of the radio continuum intensities at 1410 MHz (Mathewson and Healey 1964) and 2650 MHz (Broten 1972) coincide with the principal spiral filaments. Two complex arms dominate the distribution which corresponds to a ScIII-IVp galaxy. The spiral features start from 30 Doradus; they are asymmetric and completely unrelated to the LMC Bar which has so far been considered as the centre of that galaxy. The structure in the surroundings of 30 Dor is especially well visible in the distribution and form of the dark clouds (fig. 2). The gas-to-dust-ratio in the LMC has been determined from UBV photometry of about 700 supergiants avoiding selection effects (Isserstedt and Schmidt-Kaler 1975). The resulting value $N(HI)/E_{B-V} = 6 \cdot 10^{21} cm^{-2} mag^{-1}$ is, contrary to earlier work, in excellent agreement with values derived in our Galaxy.

There is evidence to consider the supergiant HII complex 30 Dor as the mildly active nucleus of the LMC (Feitzinger and Schmidt-Kaler 1975). Except for the stellar density (which is only moderately increased) the situation near the centre of 30 Dor is very similar to that near the galactic centre. In a sequence of activity 30 Dor is placed:

M31 → LMC → Galaxy → M101 → NGC 4258 → M82 → N-galaxies → Seyfert galaxies → radiogalaxies → quasars.

## REFERENCES

Broten, N.W. 1972, Austral. J. Phys. 25, 599.
Feitzinger, J., Schmidt-Kaler, Th. 1975, Astr. and Ap. submitted.
Isserstedt, J., Schmidt-Kaler, Th. 1975, Astr. and Ap. in press.

Mathewson, D.S., Healey, J.R. 1964, I.A.U. Symp. No. 20: The Galaxy and the Magellanic Clouds, ed. F.J. Kerr and A.W. Rodgers (Canberra: Australian Academy of Science), p. 245.

Schmidt-Kaler, Th., Isserstedt, J. 1975, Astr. and Ap. submitted.

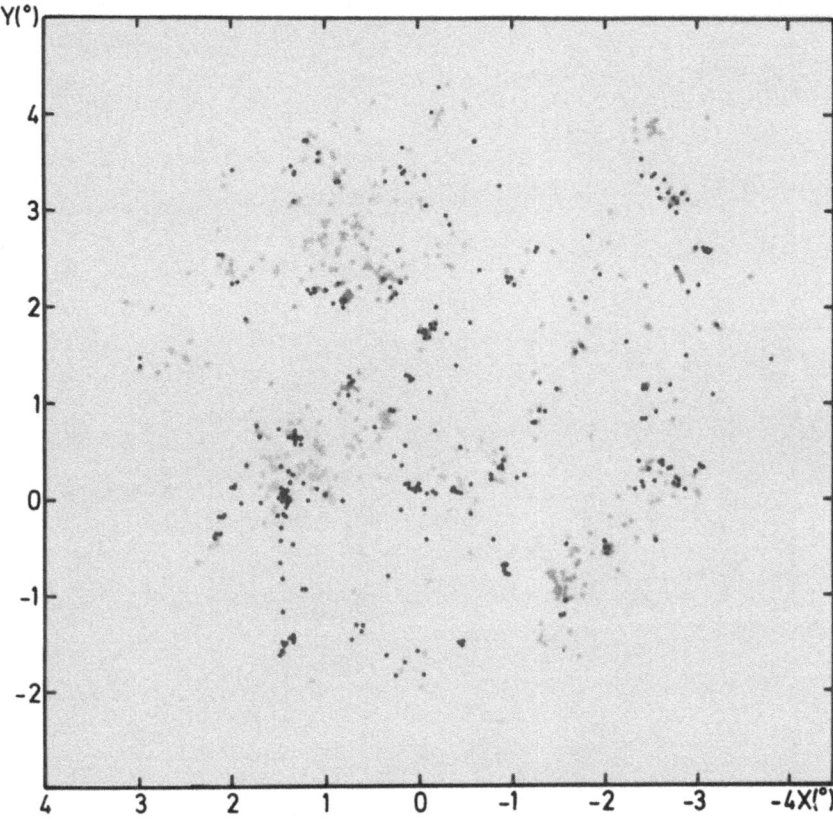

Fig. 1 The spiral structure of the LMC from Hα regions
(with diameters $D \gtreqless 25''$, black dots) and supergiants with $(U-B)_o \lesseqgtr -0.6$ (grey dots).

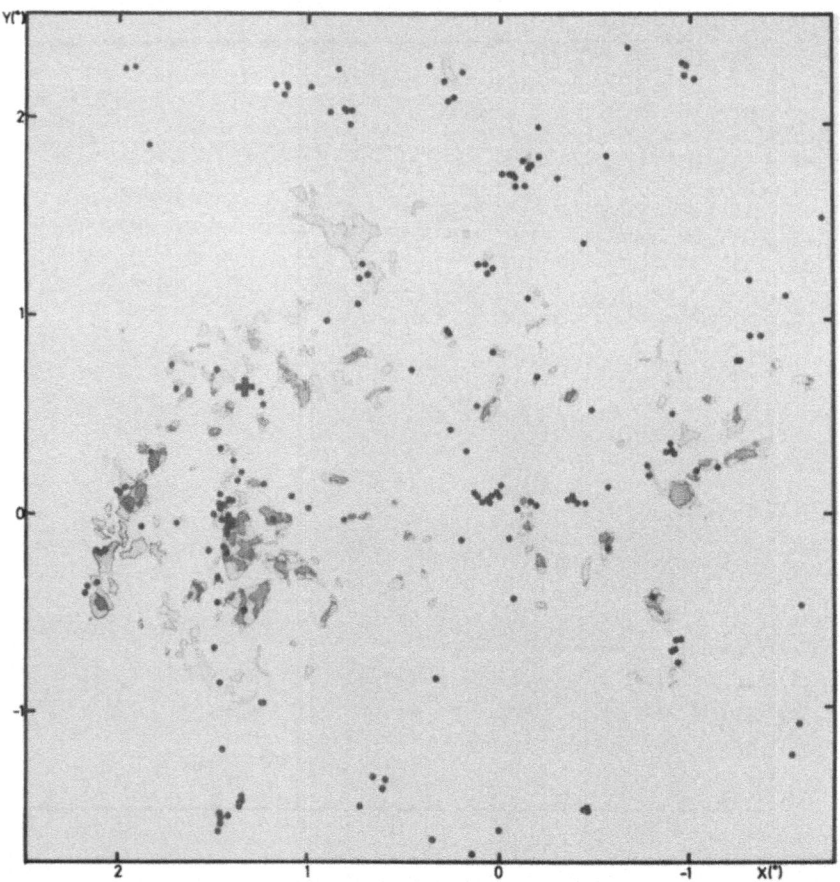

Fig. 2 The dark clouds in the LMC after Hodge 1972 ( \\\ )
and van den Bergh 1974 ( /// ). HII-regions with
D ≳ 25" (black dots) and 30 Doradus (cross). The
magnetic field is in general parallel to the filaments.

# Lecture Notes in Physics

# SPRINGER TRACTS IN MODERN PHYSICS

Ergebnisse der exakten Naturwissenschaften

Editor: G. Höhler

Associate Editor:
E.A.Niekisch

Editorial Board:
S. Flügge, J. Hamilton,
F. Hund, H. Lehmann,
G. Leibfried, W. Paul

Springer-Verlag
Berlin
Heidelberg
New York

## Volume 66

30 figures. III, 173 pages. 1973
Cloth DM 78,—; US $31.90
ISBN 3-540-06189-4

# Quantum Statistics

**in Optics and Solid-State Physics**

**R. Graham:** Statistical Theory of Instabilities in Stationary Nonequilibrium Systems with Applications to Lasers and Nonlinear Optics.
**F. Haake:** Statistical Treatment of Open Systems by Generalized Master Equations.

## Volume 67

III, 69 pages. 1973
Cloth DM 38,—; US $15.50
ISBN 3-540-06216-5

**S. Ferrara, R. Gatto, A. F. Grillo:**

# Conformal Algebra in Space-Time

**and Operator Product Expansion**

Introduction to the Conformal Group in Space-Time. Broken Conformal Symmetry. Restrictions from Conformal Covariance on Equal-Time Commutators. Manifestly Conformal Covariant Structure of Space-Time. Conformal Invariant Vacuum Expectation Values. Operator Products and Conformal Invariance on the Light-Cone. Consequences of Exact Conformal Symmetry on Operator Product Expansions. Conclusions and Outlook.

## Volume 68

77 figures. 48 tables. III, 205 pages. 1973
Cloth DM 88,—; US $35.90
ISBN 3-540-06341-2

# Solid-State Physics

**D. Schmid:** Nuclear Magnetic Double Resonance — Principles and Applications in Solid-State Physics.
**D. Bäuerle:** Vibrational Spectra of Electron and Hydrogen Centers in Ionic Crystals.
**J. Behringer:** Factor Group Analysis Revisited and Unified.

## Volume 69

13 figures. III, 121 pages. 1973
Cloth DM 78,—; US $31.90
ISBN 3-540-06376-5

# Astrophysics

**G. Börner:** On the Properties of Matter in Neutron Stars.
**J. Stewart, M. Walker:** Black Holes: the Outside Story.

Prices are subject to change without notice
■ Prospectus with Classified Index of Authors and Titles
Volumes 36—74 on request

## Volume 70

II, 135 pages. 1974
Cloth DM 77,—; US $31.50
ISBN 3-540-06630-6

# Quantum Optics

**G. S. Agarwal:** Quantum Statistical Theories of Spontaneous Emission and their Relation to Other Approaches.

## Volume 71

116 figures. III, 245 pages. 1974
Cloth DM 98,—; US $40.00
ISBN 3-540-06641-1

# Nuclear Physics

**H. Überall:** Study of Nuclear Structure by Muon Capture.
**P. Singer:** Emission of Particles Following Muon Capture in Intermediate and Heavy Nuclei.
**J. S. Levinger:** The Two and Three Body Problem.

## Volume 72

32 figures. II, 145 pages. 1974
Cloth DM 78,—; US $31.90
ISBN 3-540-06742-6

**D. Langbein:**

# Theory of Van der Waals Attraction

Introduction. Pair Interactions. Multiplet Interactions. Macroscopic Particles. Retardation. Retarded Dispersion Energy. Schrödinger Formalism. Electrons and Photons.

## Volume 73

110 figures. VI, 303 pages. 1975
Cloth DM 97,—; US $39.60
ISBN 3-540-06943-7

# Excitons at High Density

Editors: H. Haken, S. Nikitine
Biexcitons. Electron-Hole Droplets. Biexcitons and Droplets. Special Optical Properties of Excitons at High Density. Laser Action of Excitons. Excitonic Polaritons at Higher Densities.

## Volume 74

75 figures. III, 153 pages. 1974
Cloth DM 78,—; US $31.90
ISBN 3-540-06946-1

# Solid-State Physics

**G. Bauer:** Determination of Electron Temperatures and of Hot Electron Distribution Functions in Semiconductors.
**G. Borstel, H. J. Falge, A. Otto:** Surface and Bulk Phonon-Polaritons Observed by Attenuated Total Reflection.

Selected Issues from

# Lecture Notes in Mathematics